Principles of Scientific Research

Principles of Scientific Research

Editor
Donald E. Franceschetti, PhD

SALEM PRESS

A Division of EBSCO Information Services, Inc.
Ipswich, Massachusetts

GREY HOUSE PUBLISHING

Cover image: The microscope, by Trifonov_Evgeniy

Copyright ©2017, by Salem Press, A Division of EBSCO Information Services, Inc., and Grey House Publishing, Inc.

All rights reserved. No part of this work may be used or reproduced in any manner whatsoever or transmitted in any form or by any means, electronic or mechanical, including photocopy, recording, or any information storage and retrieval system, without written permission from the copyright owner. For permissions requests, contact proprietarypublishing@ebsco.com.

∞ The paper used in these volumes conforms to the American National Standard for Permanence of Paper for Printed Library Materials, Z39.48 1992 (R2009).

Publisher's Cataloging-In-Publication Data
(Prepared by The Donohue Group, Inc.)

Names: Franceschetti, Donald R., 1947- editor.
Title: Principles of scientific research / editor, Donald R. Franceschetti, PhD.
Description: [First edition]. | Ipswich, Massachusetts : Salem Press, a
   division of EBSCO Information Services, Inc. ; [Amenia, New York] :
   Grey House Publishing, [2017] | Series: Principles of | Includes
   bibliographical references and index.
Identifiers: ISBN 978-1-68217-609-2 (hardcover)
Subjects: LCSH: Science--Research.
Classification: LCC Q126.9 .P75 2017 | DDC 507.2--dc23

FIRST PRINTING
PRINTED IN THE UNITED STATES OF AMERICA

Contents

Publisher's Note vii
Editor's Introduction ix

Abductive reasoning 1
ANCOVA 3
ANOVA 6

BACI 9
Brown, Robert 11
Burbidge, E. Margaret 14

Cavendish, Henry 17
Case Study Research 20
Causal Networks 22
Chi-Square Goodness of Fit Test 24
Chi-Square Test of Independence 27
Cluster Sampling 30
Comparative Research 32
Completely Randomized Designs 35
Copernicus, Nicolaus 37
Correlation Modeling 40
Correlational Research 42
Correlations and Causations 44
Crossover Repeated Measures Design 47
Cross-Sectional Sampling 49
Curie, Marie 51

Darwin, Charles 55
Deductive Reasoning 58
Descriptive Research 60
Descriptive Statistics 62
Distributions 64

Einstein, Albert 69
Experimental Research 72
Explanatory Research 74
Exploratory Research 76
External Validity 78

Faraday, Michael 81
Feynman, Richard 84
Field Experiment 87
Fractional Factorial Designs 89
Full Factorial Design 92

Geller, Margaret 95

Hawking, Stephen 99
Higgs, Peter 104
Histograms 107
Holistic Study (Ethnography) 109
Hooke, Robert 112
Hypothesis Testing 115
Hypothesis-Based Study 117

Independent Rariable Manipulation 119
Inductive Reasoning 121
Internal Validity (Causality) 123
Interpretive Methods 125

Kepler, Johannes 127
Kruskal-Wallis Test 130

Laboratory Experiment 133
Latin Square Designs 135
Leakey, Mary 138
Linear and Nonlinear Relationships 140
Longitudinal Sampling 143

Mach, Ernst 147
Mann-Whitney *U* Test 150
Meitner, Lise 153
Mendel, Gregor 156
Mill's Methods of Causal Reasoning 159
Multiple Case Study 161
Multistage Sampling 164

Nested Analysis of Variance 167
Nested Designs 170
Nonequivalent Dependent Variables Design 172
Non-probabilistic Sampling 175
Nonresponse Error 177

Objectivity 181
One-Tailed and Two-Tailed *t*-Tests 183
Oppenheimer, J. Robert 185

Peirce, Charles Sanders 189
Plackett-Burman Design 193
Positivist Methods 195
Posttest-Only Design 197
Pretest-Posttest Experimental Research Design ... 199
Probabilistic Sampling 201

v

Proofs	204
Prospective Cohort Design	206
Quadrat Sampling	209
Qualitative Research	210
Quantitative Research	213
Random Sampling Error	217
Randomization	219
Randomized Complete Block Designs	221
Regression Discontinuity Design	224
Regression Modeling and Analysis	226
Regression Point Displacement	229
Replication, Manipulation, and Randomization	232
Retrospective Cohort Design	234
Sample Frame Error	239
Sampling Design vs. Experimental Design	241
Sampling Design: Randomness and Interspersion	244
Sampling Framework	246

Sampling vs. Census	248
Significance Levels	251
Solomon Four-Group Design	253
Split-Plot Type Designs	255
Stratified Random Sampling / Randomized Block Design	258
Student's t-Test	261
Switching Replications Design	263
Systematic Sampling	266
Time-Series Designs	269
Type I and Type II Errors	271
Turing, Alan	273
Wu, Chien-Shiung	277
Glossary	290
Bibliography	293
Index	301

Publisher's Note

Salem Press is pleased to add *Principles of Scientific Research* as the eighth title in the *Principles of series* that includes *Chemistry, Physics, Astronomy, Computer Science, Physical Science, Biology, and Scientific Research.* This new resource introduces students and researchers to the fundamentals of scientific research using easy-to-understand language, giving readers a solid start and deeper understanding and appreciation of this complex subject.

- The 105 entries include entries that explain basic principles of scientific research, ranging from Abductive Reasoning to Type I and Type II Errors, as well as biographies of key figures in scientific research that include a description of their significant contributions to the field, ranging from Robert Brown to Chien-Shiung Wu. All of the entries, and are arranged in an A to Z order, making it easy to find the topic of interest.

Entries related to basic principles and concepts include the following:
- Fields of study to illustrate the connections between the scientific research and the various branches science research theory and design to experimental design and statistical analysis;
- An abstract that provides brief, concrete summary of the topic and how the entry is organized;
- Principal terms that are fundamental to the discussion and to understanding the concepts presented;
- Basic principles that clarify the essentials of the topic
- Text that gives an explanation of the principles and its importance to scientific research, including theory and practice, benefits and drawbacks, and practical applications;
- Formulas and equations related to the principle;
- Illustrations that clarify difficult concepts via models, diagrams, and charts of such key topics as longitudinal sampling, nested designs, and probabilistic sampling;
- Further reading lists that relate to the entry.

Entries related to important figures in scientific research include the following:
- A brief overview of the individual and their contributions;
- Key dates and biographical data;
- Primary field(s) and specialties;
- Sidebars explaining their significant advances, inventions, or discoveries;
- Text that provides information about the scientist's Early Life, Life's Work, and Impact;
- Further reading lists that relate to the entry.

This reference work begins with a comprehensive introduction to the field, written by editor Donald E. Franceschetti, professor emeritus at the University of Memphis.

The book includes helpful appendixes as another valuable resource, including the following:
- Time Line of Inventions and Scientific Advancements
- Glossary;
- Bibliography; and
- Subject Index.

Salem Press and Grey House Publishing extend their appreciation to all involved in the development and production of this work. The entries have been written by experts in the field. Their names and affiliations follow the Editor's Introduction.

Principles of Scientific Research, as well as all Salem Press reference books, is available in print and as an e-book. Please visit www.salempress.com for more information.

Editor's Introduction

From the *New Atlantis* to What Lies Ahead

Many people in modern society would describe their main activity in life as research. The Random House Collegiate Dictionary defines research as "a diligent and systematic inquiry into a subject in order to discover or revise facts, theories," etc. In this sense research is an activity that is characteristic of modern western societies. It has existed for a long time. More recently, emphasis has been placed on research-based methods, in science, in teaching, in legislation, in economic planning and so on. Exactly what is meant by the modern emphasis on research?

Research and the Scientific Method

The world changed in the seventeenth century. The ancient Greeks and Romans had refined deductive logic to a precise science. In deductive logic one sets out to draw all the conclusions inherent in a set of premises. By far the most impressive success was found in the geometry of Euclid (325 BCE- ?) and some of the philosophical tracts of Aristotle (384 BCE - 322 BCE). Euclid mainly collected results known to the Babylonians and other inhabitants of the Ancient Neat East. His great contributions lay in systematizing results of others so that they followed from a simple set of axioms and postulates. Aristotle was a collector of facts; he a great was also a great organizer, who writings on a great variety of subjects have been preserved for us. His teaching on deductive logic has come down to us as the *Organon*, and had a great influence on philosophy.

The western world put philosophy at the service of religion. The rediscovery of Aristotle's writings presented the now monotheistic world with a number of challenges. Jewish, Christian, and Moslem scholars all adapted the thought of the pagan Aristotle to their own religious systems.

As the Dark Ages ended Aristotle was considered the greatest of philosophers and his ideas could not be brought into question.

The Greeks considered the celestial realm to be perfect but the terrestrial realm was far from perfect. The Greeks had applied mathematics to the motions of celestial objects. The orbits of the planets were perfect circles, because circles were perfect. When it became apparent that the planetary orbits were not perfect circles, a complicated system of epicycles was introduced. The need for a more robust methodology began to be felt. The polish astronomer Copernicus (1473-1543) introduced the heliocentric solar system and doubts began to grow that the earth upon which men walked was the center of the universe.

The hypothetico-deductive method is a variant of deductive logic used particularly by some physicists. A good example is found in the *Philosophiæ Naturalis Principia Mathematica* of Isaac Newton (1642-1727). Newton's famous dictum: *Hypothesis non fingo* or "I do not make hypotheses," turned physics away from physical speculation and toward a precise mathematical description of what occurs leaving the "why" an open question.

Earnest Rutherford (1871-1937) has quipped that science was either Physics or stamp collecting. The stamp collecting side is represented by Carol Linnaeus (1707-1778), who was the first biologist to classify the members of the animal and plant kingdoms. Grouped with Linnaeus must be the early Thomas Edison (1837-1931) the American inventor who found a workable filament material by a process of trial and error. The trial and error method does work sometimes, but does not provide insight as to what might be the most promising theory. A small amount of physical theory, if available, could have greatly simplified Edison's search.

Rutherford's comment became the basis of a classification scheme proposed in a highly influential book entitled *Pasteur's Quadrant: Basic Science and Technological Innovation*, written by Donald E Stokes and published in 1997. Stoke suggested that the prevalent classification of research as pure or applied was misleading and that scientific activity could be better plotted in two dimensions. Along one axis based on whether considerations of use motivated it, and a perpendicular axis by whether it was driven by a quest for fundamental knowledge. In this scheme traditional pure basic research occupied one quadrant. The work of scientists such as Louis Pasteur, which was almost always driven by the need to treat one disease or another, fell into a separate quadrant while the work of Thomas Edison occupied the non-theory driven quadrant.

Sir Francis Bacon

The greatest propagandist for the inductive method was Sir Francis Bacon (1561-1626), contemporary of Galileo (1564-1642) and Newton (1642-1727). Unlike Galileo but like Newton, he was a Protestant, a prolific writer, and a man who aspired to a position of influence. Like Newton he attached far more importance to his social status and his role in government than to his fundamental contributions to science or philosophy. Unlike Newton he was born to the nobility but the early death of his father greatly reduced his financial prospects. He entered the service of Queen Elizabeth I, but the Queen did not favor him. He fared much better under her son, King James I, rising through various administrative posts, becoming Lord Chancellor of England and being raised to the peerage. In the process he made a number of influential enemies. Bacon was forced to resign as Lord Chancellor, but allowed to retain his titles as a peer of the realm: Baron Verulam and Viscount St. Albans. He died shortly thereafter having caught a chill during an early experiment with frozen food.

Bacon's value as a scientist is open to question as he seems unaware the scientific achievements of his time including William Gilbert's modeling of terrestrial magnetism or with Sir William Harvey's discovery of the circulation of blood. Nonetheless his written books—*The Advancement and Proficience of Learning Divine and Humane* (1605), *Novum Organum Scientarium* (1620) and *New Atlantis* (1627) gave voice to the methods of inductive logic, so that it is generally referred to as Baconian induction even though Bacon's methods have been replaced by more modern ones. *New Atlantis* is an incomplete utopian novel that describes a mythical island where a great many citizens are involved in scientific research. This reflects the growing involvement in professional societies and scientific journals. A century later, the mode of existence on the island was parodied by Jonathan Swift in the third book of *Gulliver's Travels*. While some research papers are written in a manner that invites parody, the image of a culture based on scientific research appealed to a great many people. A great many historians cite Bacon's work as a turning point in scientific research.

Immanuel Kant, possibly the greatest philosopher of modern times, includes a motto from Bacon's preface to *Instauratio Magna* in his work, *Critique of Pure Reason*. More recently, Loren Eiseley, popular writer and anthropologist, described Bacon in *Francis Bacon and the Modern Dilemma* (1962) as "the man who saw through time."

Bacon's time saw the emergence of the first scientific societies in Europe; The Accademia del Cimento, to which Galileo belonged, and the Royal Society of London. These learned societies sponsored the first scientific journals and provided national forums where the best scientists could meet to discuss and debate their work. They also sponsored scientific research which until that time, had been the prerogative of royalty.

As the Industrial Revolution (1760–1840) progressed, it became the case that ordinary citizens were capable of amassing fortunes. Governments found this accumulation of wealth to be irresistible and so began to impose inheritance taxes so that wealth could not pass down untaxed from one generation to the next. Wealthy individuals learned that nonprofit foundations were one way to preserve their interests when they could not pass their money directly to their heirs. Industrialists often found their interests best-served by scientific research, leading to the establishment of such foundations as the Rockefeller Foundation, the Ford Foundation, and more recently the Bill and Malinda Gates Foundation.

The way in which science and wealth interacted to some extent reflected national style. In England, the best universities, originally established to serve the nobility, witnessed the growth of new institutions meant to serve those students who were disinclined to devote years to the study of Latin and Greek. In Germany, organic chemistry would flourish. The popular notion of the scientist toiling away, night after night on his latest invention to serve no higher purpose then his own pure curiosity is a trope that actually derives from nineteenth-century German romanticism rather than out of the traditions of the universities of the Middle Ages.

Science and Invention: Edison's Invention Factory, Bell Telephone Labs

The founding fathers of the American nation made provision in Article 8 of the United States Constitution for the new National Government to issue letters patent which granted inventors monopoly rights to their inventions for a limited number of years.

In doing so they were following the example of the British Statute on Monopolies, adopted in England in 1623. Over the centuries this law became very important in the advancement of technology as barriers to information exchange were lowered.

Thomas Edison was a largely self-taught inventor who accumulated over 1000 patents in his lifetime. In 1876, he established a research laboratory which became known as Edison's invention factory.

In the same year, 1876, the Scottish-educated engineer, Alexander Graham Bell invented the telephone. Bell Telephone Labs was established in 1925. Its technical staff would be responsible for numerous patents, also.

As time went on industrial research organizations acquired a secondary mission: filing patents claims in broad areas of interest to the parent companies, and as a result a novel technology often involves the use of patents owned by the parent organization. Before the technology can be brought to market, developers would have to acquire a license to use the patent. Failure to secure the appropriate license can stall or halt development. Today, particularly in the medical realm, a new product may list several hundred items of prior art, requiring the services of specialist in intellectual property law, before it hits the market. In the United States, and most industrialized countries, intellectual property law has become a major area of legal practice, with a separate bar exam, additional educational requirements, and increasingly subject to international treaties as well as federal law.

European Educational Institutions and American Pragmatism

During the Middle Ages, European universities generally took pride in their traditions, many having outlasted several national governments. Universities were administered by their faculties, or by the church, and often enjoyed a measure of extraterritoriality. Bound together by the church and the Latin language they provided a refuge for scholars during the later Middle Ages and into the Renaissance and early Modern Era.

The university system had to adapt to the emergence of the sciences and fields such as engineering. The system of academic degrees: bachelor's, master's, and doctoral was basically retained but the number of subjects in which one could earn a degree was greatly expanded. The once universal use of Latin was discontinued.

The discovery of the New World and the protestant reformation brought many new styles to higher education.

In 1861 Yale University became the first American institution to award the Doctor of Philosophy degree. Until then Americans seeking an advanced degree had to travel to Europe to complete their education.

The main source of training for American scientists remained the British and German universities, until the Second World War.

Science Builds the Bomb

Perhaps no period in the recent history of science has left as great a mark on the organization of science as the Manhattan Project. The story has the elements of high-level science as well as high drama: A group of European scientists, transplanted to the United States as political refugees, realize at the beginnings of the Second World War that it was possible to build an extremely powerful weapon based on discoveries made less than ten years earlier. At the same time, powerful and ruthless enemies had already begun their evil work.

President Franklin Roosevelt makes the decision to build a bomb at the imploring of Albert Einstein, a European refugee fleeing the drumbeat of anti-Semitism spreading throughout Europe as well as (ironically, as it turns out) a dedicated pacifist. The project is confronted with both practical and ethical conundrums–The bomb may not work. If it does work, it might result in the deaths of a hundred thousand civilians. The bomb might work too well and eliminate all human life on earth. A European-trained leader is selected for the project. The team leader, J. Robert Oppenheimer, has a politically questionable past. But in the opinion of General Groves, a key Army officer, Oppenheimer is one of the few men who could make the project work. Is it ethical to exempt so many men at a time of military crisis? Is it ethical to plan on the deaths of so many thousands to bring the war to a conclusion?

The American-led effort was successful and the bombs constructed but the decision to use the bombs fell to Harry Truman, who assumed office on Roosevelt's death and who as vice president had no knowledge of the bomb's existence

Historians and philosophers argue about the bombs' use to this day. Did it save more lives than it destroyed? Did it shorten the course of the war? Did it make the world a safer or more dangerous place? The threat of nuclear war has become an accepted fact the post-WWII world, although the United States' bombing of Japan is the only nuclear act of war to date.

Following the War, the United States, geographically isolated from Europe and Asia, found itself engaged in a debated concerning whether or not it should assume a leadership role or return to isolationism. The Allied nations, left so decimated by WWII, faced an immense rebuilding task. They worried that if the United States abandoned its leadership role, other nations might fall sway to the allure of communism.

Vannevar Bush and the Endless Frontier

The question of what to do with the research facilities built in the United States during the Second World War and dedicated to developing nuclear weapons came up even before the war ended. Franklin Roosevelt asked Vannevar Bush, head of wartime research and development, for suggestions. His findings were reported in the document Science: The Endless Frontier, which described the rewards of investing in research in glowing terms. The book is a bit painful to read now, its optimism reminiscent of the *New Atlantis*. It recommended the establishment of a National Science Foundation to fund basic research in the sciences. It further recommended that the National Science Foundation and other governmental agencies charged with the conduct of research should use the available university faculties to conduct the research and train the next generation of researchers, with the government bearing the full cost of the research. A great many universities embraced the ideal of the research university. Unfortunately, there was the hidden assumption that the exponential growth of the scientific establishment could continue indefinitely. It did from the early years under President Eisenhower until America achieved President Kennedy's dream of landing men on the Earth's moon in 1969.

By the mid-1970s the situation had changed markedly. University science and engineering departments and faculties had grown, while there were fewer American citizens with interest in science and engineering.

Developments since the Second World War

To understand how research has developed since the Second World War we will consider the advances made in communications and computer science. One could argue that the current state of affairs comes about as the result of four critical discoveries. The first was thermionic emission from a heated metal, an observation made by Thomas Edison himself in 1883. Edison could think of no practical application of the effect, so he dutifully recorded his observations in a notebook and went on to other things. Fortunately for electrical science, he was visited by the British Sir John Ambrose Fleming in 1884. In 1889, Fleming began a collaboration with Italian inventor Guglielmo Marconi. In the process, he obtained a patent for vacuum tube rectifier. A few years later the American Lee De Forest added a third element to make a vacuum tube triode. Advances in vacuum tube design and applications certainly fell in Pasteur's Quadrant. Entire new fields of enquiry including many aspects of nuclear and particle physics were opened by the ability to achieve a high vacuum.

The second critical invention has been the digital computer and has its origins in the purely deductive mathematical logic of Aristotle. The mathematical question is subtle: Is there an effective procedure to calculate a number of interest? By the nineteenth century, mathematicians had begun to wonder if all the theorems of mathematics could be derived from a single set of postulates. To answer this question Alan Turing had to refine the notion of an effective procedure. He reduced each effective procedure to a set of instructions for a so-called Turing machine, a machine which read numbers or symbols from a tape, compared the symbol just read, with a symbol stored in a one slot memory, and based on the symbol would write the result and move the tape forward or backward. Turing showed that it was possible to formulate a universal Turing machine that could read in a description of any Turing machine and then emulate it. The surprising results of Turing's analysis were that it was generally impossible to feed the description of a Turing machine into another to determine that the first machine would come to a halt. The important practical result was the design, at least in concept, of a programmable digital computer. The first computers were actually humans who could do calculations by hand with speed and accuracy. During the Second World War these were augmented by vacuum tube devices which, if large, frequently failed due to filament burn out. Electronic computers would make a great leap in power and reliability with the invention, in 1947, of the transistor.

The third major development was the transistor, a substitute for the vacuum tube triode, which was far more reliable and could be miniaturized. Once the basic theory was understood the device quickly moved from the quest for fundamental understanding to application driven research.

The final stage in the evolution of computer and communication technology was that of miniaturization. Perhaps no one of Edison's time could have envisioned that using photolithography on slices of ultra- refined silicon one could manufacture integrated circuits with thousands of transistors per square centimeter and do so cheaply enough that a calculator or cell telephone could become a throwaway item. The giants of high-tech manufacturing yesterday; Oppenheimer and the polymath John von Neuman, who did much of the engineering on the first computers, would have to move over to accommodate teenage college drop-outs, like Bill Gates and Steve Jobs, who not incidentally listed J. Robert Oppenheimer as one of his heroes.

THE CHANGING FRONTIER?

The digital computer, with its birth in pure theory and its development in wartime necessity, is now ubiquitous. Computers can analyze data at very high speeds. Computers can solve some problems, actually a relatively small fraction, but can always be applied to simulate a complex system. Computing still represents an open frontier.

There are other obvious frontiers to be explored. Research areas once cloaked in military secrecy are now open to private enterprise. The space race has given way to cooperative missions. The aspirations of the physics community in the United States, to build a superconducting super collider were scrapped, and the United States must join with European nations to conduct particle physics research. For the moment, the ultimate consideration in how quickly research gets done will be economic. The players may change but scientific discovery remains an endless frontier.

FROM INDUCTION TO ABDUCTION

One more trend is worthy of mention. In *The Rise of Statistical Thinking, 1820-1900*, Theodore M. Porter shows how the idea of uncertainty has come to permeate the sciences, both in the laboratory and in the social realm. With the development of the computer, there has been a realization that the majority of physical problems have no exact solution, and further, that for most purposes a highly accurate solution may not be worth the required effort to obtain it. Thus, the search for accurate solutions has given way to a search for solutions which are, in the words of artificial intelligence pioneer, and Nobel Laureate Herbert Simon, are good enough. Coming to terms with this reality is one of the major challenges facing researchers in all fields.

—*Donald R. Franceschetti*

A

Abductive Reasoning

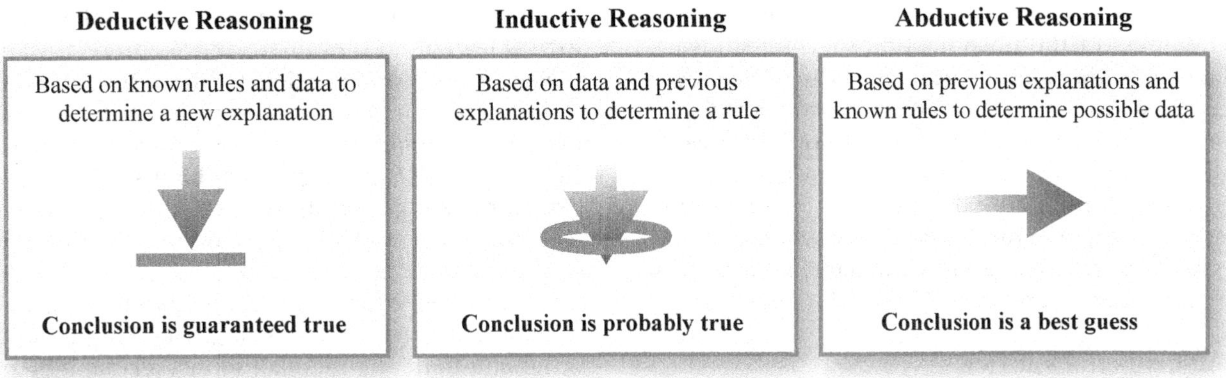

Deductive Reasoning	Inductive Reasoning	Abductive Reasoning
Based on known rules and data to determine a new explanation  Conclusion is guaranteed true	Based on data and previous explanations to determine a rule  Conclusion is probably true	Based on previous explanations and known rules to determine possible data  Conclusion is a best guess

FIELDS OF STUDY

Hypothesis Testing; Research Theory

ABSTRACT

Abductive reasoning is the form of logical inference that is used when available information is incomplete. A hypothesis is developed to explain some observed phenomenon as best as possible. Though less certain than other forms of reasoning, it is valuable and frequently used in both science and everyday life.

PRINCIPAL TERMS

- **abductive logic programming:** an extension of logic programming that permits some predicates to have undefined values rather than definite ones.
- **deductive reasoning:** a type of reasoning in which general rules are applied to a specific situation in order to draw a narrower conclusion; if the premises on which the conclusion is based are true, the conclusion is also deemed to be true.
- **inductive reasoning:** a type of reasoning in which knowledge of specific situations is used to draw a broader and more general conclusion; if the premises on which the conclusion is based are true, the conclusion is deemed to be highly probable but not certain.

ABDUCTION, DEDUCTION, AND INDUCTION

Abductive reasoning, or abduction, is a logical process of inference in which incomplete data is used to draw the simplest and most likely possible conclusion. It is an important concept in fields ranging from law to computer science and is often used by people in everyday life. Yet it is often poorly understood, especially compared with the better-known deductive reasoning and inductive reasoning. While these types of logical inference rely on a systematic building up of premises that lead to a certain or probable conclusion, abduction involves educated guessing based on the best available observations or facts.

In the context of a scientific experiment, deductive and inductive reasoning may be used to analyze facts or phenomena generated by the experiment. Abductive reasoning comes before this, when the researcher forms the hypothesis that the experiment is designed to test. It is therefore a crucial part of the research design process. Abduction bridges the gap between what is already known about the universe, such as natural laws and other forms of information that can be relied on or assumed to be true, and new information that a researcher wishes to obtain. In other words, deductive and inductive reasoning are best used to analyze the results of an experiment, while abductive reasoning is best used to devise a possible explanation for the phenomenon being studied and determine how best to gather data to test that explanation.

1

ABDUCTION IN RESEARCH

Abductive reasoning is a crucial part of research design that all researchers should be familiar with. However, while abductive reasoning always makes do with incomplete data, it is important to recognize that this does not condone ill-informed research or total guesswork. Errors in research are common when researchers do not have enough information about the context of the phenomenon they are studying. This limits the options available when forming a hypothesis about the most likely explanation for an observed event. Researchers must thoroughly investigate all of the potential factors that may be at work in a given scenario before designing an experiment to test an explanation for that scenario. When used correctly, abductive reasoning is a valuable tool for both scientific inquiry and everyday decision making.

ABDUCTION IN PRACTICE

A classic example of abductive reasoning is in a criminal trial. The jury must come to a conclusion based on presented evidence, even if they do not have access to all of the relevant information. Another common example is medical diagnosis. The complexity of health factors on both general and individual levels means that full information regarding a patient is often unavailable. A doctor typically asks questions of the patient to gather as much data as possible, but the patient's complaints may not align exactly with examples in medical literature, or the patient may be totally unresponsive. In such cases the doctor must make an educated guess about the patient's condition based on symptoms and any other known factors.

For example, a patient with a skin irritation might be asked whether they have been exposed to any chemicals or bitten by an insect or snake. The answers, along with the doctor's knowledge, will help narrow down the likely cause of the rash. However, even if the patient says they were bitten by a snake, they may not know what kind of snake it was. If only rattlesnakes live in the area and the patient's symptoms are consistent with a rattlesnake bite, the doctor can safely guess that the patient was bitten by a rattlesnake and provide the appropriate treatment. It is still logically possible that a different kind of snake is to blame, or even that the bite was from a nonvenomous snake and the rash is completely unrelated. A rattlesnake bite is simply the most likely explanation under the circumstances.

ARTIFICIAL INTELLIGENCE

Abductive reasoning has become especially relevant in the field of computer science, particularly with regard to abductive logic programming (ALP). Research into developing artificial intelligence (AI) has shown that guessing or estimating is one of the most difficult tasks to program a computer to perform. For a computer to make such a logical leap, it must evaluate and assign relative importance to multiple rules that have been defined to help guide it. This process is meant to approximate how humans use their awareness of the rules that govern the world around them—water is wet, gravity pulls objects toward the earth, and so on—to conceive of possible explanations for observed phenomena.

The challenge for AI programmers is that there are so many rules that human beings take for granted, it is difficult to define them all so that a computer can refer to them. Even after this is done, programmers must then devise a mechanism that the computer can use to decide which rules are more likely to account for a specific phenomenon. Still, when implemented successfully, ALP allows intelligent systems to better determine and correct faults, plan and make decisions, and revise themselves.

—*Scott Zimmer, JD*

FURTHER READING

Dunne, Danielle D., and Deborah Dougherty. "Abductive Reasoning: How Innovators Navigate in the Labyrinth of Complex Product Innovation." *Organization Studies*, vol. 37, no. 2, 2016, pp. 131–59.

Holyoak, Keith J., and Robert G. Morrison, editors. *The Oxford Handbook of Thinking and Reasoning*. Oxford UP, 2012.

Mirza, Noeman A., et al. "A Concept Analysis of Abductive Reasoning." *Journal of Advanced Nursing*, vol. 70, no. 9, 2014, pp. 1980–94. *Academic Search Complete*, search.ebscohost.com/login.aspx?direct=true&db=a9h&AN=97431963&site=ehost-live. Accessed 9 Mar. 2017.

Tavory, Iddo, and Stefan Timmermans. *Abductive Analysis: Theorizing Qualitative Research*. U of Chicago P, 2014.

Walton, Douglas. *Abductive Reasoning*. 2005. U of Alabama P, 2014.

ANCOVA

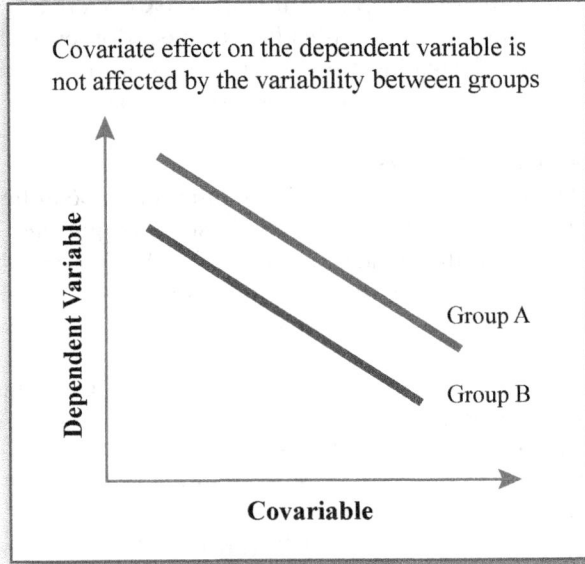

Covariate effect on the dependent variable is not affected by the variability between groups

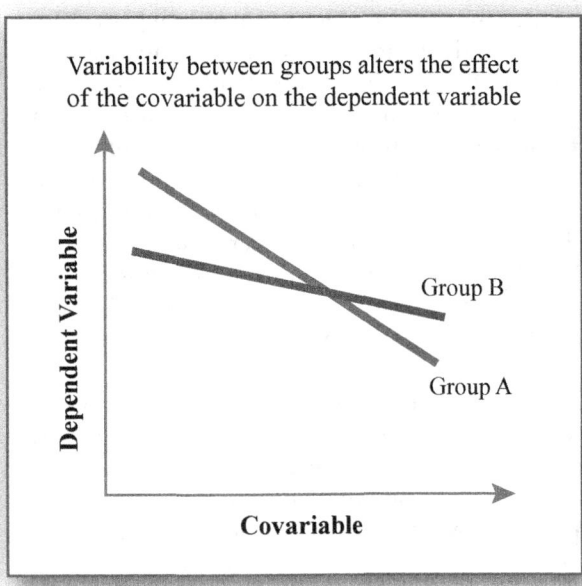

Variability between groups alters the effect of the covariable on the dependent variable

FIELDS OF STUDY

Statistical Analysis; Research Design; Experimental Design

ABSTRACT

Analysis of covariance (ANCOVA) is a statistical method used to adjust group means in order to control for the effects of variables other than the ones being studied. It depends on the use of pretests to establish a baseline that can then be taken into account when analyzing observations.

PRINCIPAL TERMS

- **covariate:** a continuous independent variable other than the one being manipulated that may influence, predict, or explain the effects of the manipulated variable on the dependent variable.
- **interaction:** in statistics, an effect by which the impact of an independent variable on the dependent variable is altered, in a nonadditive way, by the influence of another independent variable or variables.
- **posttest:** an examination that takes place after a treatment or other experimental intervention.
- **pretest:** an examination that takes place before a treatment or other experimental intervention.
- **treatment:** the process or intervention applied or administered to members of an experimental group.

ANCOVA DEFINED AND EXPLAINED

Analysis of covariance (ANCOVA, sometimes ANACOVA or ANOCOVA) is used to analyze and control for the effects of covariates among multiple sample groups in a research study. It combines elements of two other statistical tests, regression and analysis of variance (ANOVA). Regression is used to determine whether and how certain variables are related. ANOVA is used to analyze differences in the means of different sample groups. Accordingly, ANCOVA is used to analyze differences in means of different sample groups after controlling for the effects of any other independent variables that might influence the dependent variable. It is a method of statistical control, rather than experimental design control.

A researcher who plans to use ANCOVA must first identify one or more potential covariates in their study. There are several criteria for selecting effective covariates. First, there must be a linear relationship between the chosen covariate and the dependent variable and that relationship must be the same among all sample groups. Second, the covariate must not be influenced by any other independent variables. If there is any interaction between the covariate and the independent variable being studied, then the covariate cannot be controlled for without significantly altering the results. Third, the covariate must be a continuous, not categorical, variable. In other words, it must be numerically measurable in intervals. If sufficient measurements of multiple potential covariates are taken at the outset, the researcher can choose which best fits the criteria later.

ANCOVA is most often used in pretest-posttest experimental designs. A pretest is conducted before the study to measure the chosen covariate(s). In a clinical trial to determine which dosage of a drug to treat high blood pressure is most effective, for example, subjects might be divided into three sample groups: one receiving 0.5 milligram of the drug, one receiving 1.0 milligram, and one receiving 1.5 milligrams. The independent variable would be the dosage given to each sample group, and the covariate might be each patient's baseline blood pressure. Then, after the treatment has been administered, a posttest is used to measure the dependent variable. In the clinical trial example, the dependent variable would be the patient's post-treatment blood pressure. Given the value of the independent variable, dependent variable, and covariate for each subject, the researcher can conduct ANCOVA to adjust the means for each group to reflect what the results would be if all subjects had the same pretreatment blood pressure. Using the adjusted means, they can then determine whether the null hypothesis should be rejected. In ANCOVA, the null hypothesis states that there is no significant difference among the means of the sample groups after the means have been adjusted to control for the covariate.

Advantages and Disadvantages

ANCOVA, like ANOVA, is governed by certain assumptions. The following assumptions apply to both:

- Residuals (that is, the difference between an observed value in a group and the group mean) in each sample group are normally distributed.
- Variance is homogenous among sample groups.
- Sample groups are fully independent.

Furthermore, ANCOVA also assumes that there is a linear relationship between the covariate(s) and the dependent variable, that the slopes of the lines expressing this relationship are equal among all sample groups, and that the covariate is measured exactly. If any of these assumptions are false, ANCOVA may produce incorrect or misleading results. Moreover, the effects of violating these assumptions are not always clear, which can cause problems during data analysis.

ANCOVA in Practice

The validity of ANCOVA has continued to be upheld in a number of studies. A 2017 study by Jennifer L. Reeves and colleagues used ANCOVA to test the impact of using iPads in prekindergarten (pre-K) classrooms, for example. The scores from the school year's first pre-K assessment were the covariate, and scores from the third assessment were the dependent variable. With that data, the study found that students who used iPads for learning achieved significantly better results in math and phonological awareness than those who did not.

Like ANOVA, ANCOVA can be calculated using sums of squares. With one-way ANCOVA, there are three variation sources: the sum of squares between groups (or "treatment sum of squares"); the sum of squares within groups (or "error sum of squares" or "residual sum of squares"); and the covariance sum of squares. However, because ANCOVA involves adjusting means—and, by extension, adjusting the sums of squares—the calculations involved are far more complex than in ANOVA. As a result, these calculations are most often performed using statistical software. Once the sums of squares are adjusted, they can be used to calculate an F statistic, as in ANOVA, in order to compare the adjusted means.

Significance

Like ANOVA, ANCOVA can be adapted for use in experimental, quasi-experimental, and observational research designs. Both are also used across a wide range of fields, making it important to understand their assumptions, techniques, and use. They are widely available in popular software packages and online applications.

—*Elizabeth Rholetter Purdy, PhD*

Further Reading

Huitema, Bradley E. *The Analysis of Covariance and Alternatives: Statistical Methods for Experiments, Quasi-Experiments, and Single-Case Studies.* 2nd ed., John Wiley & Sons, 2011.

Jennings, Megan A., and Robert A. Cribbie. "Comparing Pre-Post Change across Groups: Guidelines for Choosing between Difference Scores, ANCOVA, and Residual Change Scores." *Journal of Data Science*, vol. 14, no. 2, 2016, pp. 205–29. *Academic Search Complete*, search.ebscohost.com/login.aspx?direct=true&db=a9h&AN=116345929&ehost-live. Accessed 4 Apr. 2017.

Lai, Keke, and Ken Kelley. "Accuracy in Parameter Estimation for ANCOVA and ANOVA Contrasts: Sample Size Planning via Narrow Confidence Intervals." *British Journal of Mathematical and Statistical Psychology*, vol. 65, no. 2, 2012, pp. 350–70. *Academic Search Complete*, search.ebscohost.com/login.aspx?direct=true&db=a9h&AN=74280439&ehost-live. Accessed 4 Apr. 2017.

Randolph, Karen A., and Laura L. Myers. *Basic Statistics in Multivariate Analysis*. Oxford UP, 2013.

Reeves, Jennifer L., et al. "Mobile Learning in Pre-kindergarten: Using Student Feedback to Inform Practice." *Journal of Educational Technology and Society*, vol. 20, no. 1, 2017, pp. 37–44. *Academic Search Complete*, search.ebscohost.com/login.aspx?direct=true&db=a9h&AN=120706100&ehost-live. Accessed 4 Apr. 2017.

Rutherford, Andrew. *ANOVA and ANCOVA: A GLM Approach*. 2nd ed., John Wiley & Sons, 2011.

Trochim, William M. K. "Covariance Designs." *Research Methods Knowledge Base*, 20 Oct. 2006, www.socialresearchmethods.net/kb/expcov.php. Accessed 4 Apr. 2017.

Warner, Rebecca M. *Applied Statistics: From Bivariate through Multivariate Techniques*. 2nd ed., SAGE Publications, 2013.

ANCOVA SAMPLE PROBLEM

Two professors are teaching the same introductory college course in American government. Some students have taken previous classes in American government, but others have not. At the end of the semester, both classes will take the same final exam. A researcher wants to conduct a pretest-posttest study to find out which professor does a better job of teaching the material that will be on the exam. In this study, what would be the independent and dependent variables, and what should the researcher use as a covariate?

Answer:

Each class represents a sample group, so the independent variable is which professor is teaching the class. The dependent variable will be the students' scores on the final exam. Because some students have studied the subject before, the researcher wants to control for prior knowledge. To do this, the researcher will ask both classes to take a pretest at the beginning of the semester that will cover the same range of material as the final exam. The covariate will be the students' scores on this pretest.

ANOVA

FIELDS OF STUDY

Statistical Analysis

ABSTRACT

Analysis of variance (ANOVA) is a method of analyzing the differences between the means of multiple data sets and testing the validity of those differences. It is generally considered the most popular statistical test.

PRINCIPAL TERMS

- **normal distribution:** a probability or frequency distribution in which plotting the values contained in a data set, according to either the probability of their occurring or the frequency with which they occur, results in the appearance of a symmetrical bell-shaped curve, with the majority of values clustered around the middle; also called a bell curve.
- **sum of squares between:** the sum of the squares of the deviation of each group mean from the grand (overall) mean, multiplied by the number of data points in each group; also called the treatment sum of squares.
- **sum of squares within:** the overall sum of the sums of the squares of the deviation of each data point within a group from the mean of that group; also called the error sum of squares or the residual sum of squares.
- **variance:** a measure of how widely data points within a group are dispersed from the group mean, expressed as the average squared distance of each data point from the mean.

ANOVA DEFINED AND EXPLAINED

Analysis of variance (ANOVA) is a common method for analyzing differences among the means of multiple data sets. ANOVA is used to determine if two or more groups are statistically distinct by testing how individual data points vary from group means and from the overall mean. In ANOVA, the null hypothesis is that there are no differences among the means of each group except for those that randomly occur by chance. If the researcher disproves the null hypothesis, then the groups are statistically distinct and that distinction must be explained. ANOVA does not identify any differences that exist among groups; it only shows that they exist.

ANOVA tests may be either one-way or two-way. One-way ANOVA is used to test the impact of one independent variable on two or more groups. Two-way ANOVA tests two or more independent variables. There are two main ANOVA models: fixed effects and random effects. The fixed-effects model is used when the values of the independent variable(s) being tested are deliberately selected by the researchers, either because those are the only possible values or

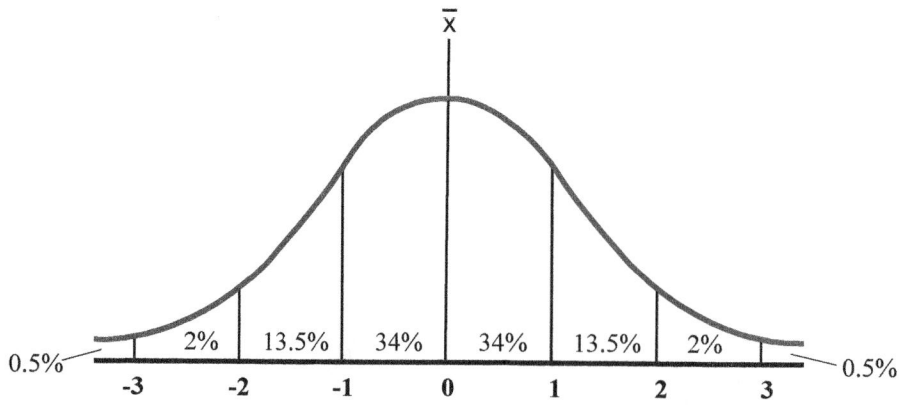

Standard Deviations

they are the only values of interest. For example, a clinical trial that administered different drug dosages to different experimental groups would be analyzed using the fixed-effects model. The random-effects model is used when the values of the independent variable(s) being tested are not chosen by the researchers, typically because subjects were randomly selected from a larger population and the variable(s) being studied are preexisting in that population. For example, a study of the blood-pressure levels of randomly selected individuals would be analyzed using the random-effects model.

Advantages and Disadvantages

Three main assumptions govern the use of ANOVA. The first is that the residuals—that is, the difference between an observed value in a group and the group mean—of each sample group follow a normal distribution. Second, variance (the square of the standard deviation) is homogenous among the sample groups being tested. Third, the sample groups from which the data are gathered are truly independent of one another. While these assumptions must apply in most cases, in a randomized controlled experiment, only the homogeneity of variance is assumed.

ANOVA is often compared to its predecessor, the Student t-test. The t-test was also designed to compare mean scores of groups and shares the assumptions of normal distributions and independence of groups. Since the t-test performs a group-by-group comparison, it is better than ANOVA at identifying details about group differences. However, a single t-test can only test the differences between two groups, and running multiple t tests greatly increases the risk of type I errors. In general, a t-test is preferable when comparing two groups, and ANOVA is the tool of choice for comparisons of three or more groups.

ANOVA in Practice

The validity of ANOVA depends on the findings of the F test, which tests the equality of variances. To conduct the F test, one must determine the sum of squares between groups (SS_B) and the sum of squares within groups (SS_W). SS_B is a measure of the variability between the means of different sample groups and the grand (overall) mean. It is also called the "treatment sum of squares" because this variation is attributable to differences in the treatment or other factor being tested. SS_W is a measure of the variability between data points in each sample group and the mean of that group. It is also called the "error sum of squares" because this variation is attributable to random error. (Adding SS_B and SS_W together gives the total sum of squares, or SS_T, but this is not necessary for ANOVA.)

Next, SS_B and SS_W must each be divided by their respective degrees of freedom. SS_B has $k-1$ degrees of freedom, where k is the total number of sample groups. SS_W has $N-k$ degrees of freedom, where N is the total number of data points from all sample groups. This calculation produces the mean square between groups (MS_B) and the mean square within groups (MS_W), respectively.

Finally, MS_B is divided by MS_W. The resulting value is called the F statistic or F ratio. An F distribution table gives the F critical values based on the degrees of freedom in the numerator ($k-1$) and in the denominator ($N-k$). If the F ratio is greater than the F critical value, the null hypothesis is rejected. Recall that the null hypothesis in ANOVA is that there are no statistically significant differences among the means of each group. An F ratio above the critical value indicates that the differences are, in fact, statistically significant.

Significance

ANOVA is adaptable to a wide range of fields, including medicine, natural science, social science, and business. It may be used in experimental, quasi-experimental, and observational studies. For these reasons, it has become the most popular statistical method of comparing groups by testing the validity of samples taken from overall populations. Assistance in performing ANOVA is also easily accessible, since it is a common function of most statistical software applications. Researchers should keep in mind that ANOVA does not explain any differences that occur but only shows whether differences exist.

—*Elizabeth Rholetter Purdy, PhD*

Further Reading

Cardinal, Rudolf N., and Michael R. F. Aitken. *ANOVA for the Behavioural Sciences Researcher*. Lawrence Erlbaum Associates, 2006.

Madrigal, Lorena. *Statistics for Anthropology*. 2nd ed., Cambridge UP, 2012.

"One-Way ANOVA." *Laerd Statistics*, Lund Research, 2013, statistics.laerd.com/statistical-guides/one-way-anova-statistical-guide.php. Accessed 4 Apr. 2017.

Plansky, M. "Analysis of Variance—One-Way." *Psychological Statistics*, U of Wisconsin–Stevens Point, 1997–2016, www4.uwsp.edu/psych/stat/12/anova-1w.htm. Accessed 4 Apr. 2017.

Randolph, Karen A., and Laura L. Myers. *Basic Statistics in Multivariate Analysis*. Oxford UP, 2013.

Rutherford, Andrew. *ANOVA and ANCOVA: A GLM Approach*. 2nd ed., John Wiley & Sons, 2011.

Tarlow, Kevin R. "Teaching Principles of Inference with ANOVA." *Teaching Statistics*, vol. 38, no. 1, 2016, pp. 16–21. *Academic Search Complete*, search.ebscohost.com/login.aspx?direct=true&db=a9h&AN=112335846&site=ehost-live. Accessed 4 Apr. 2017.

Walker, Jeffrey T., and Sean Maddan. *Statistics in Criminology and Criminal Justice*. Jones and Bartlett Publishers, 2012. Print.

Wall Emerson, Robert. "ANOVA and *t*-Tests." *Journal of Visual Impairment & Blindness*, vol. 111, no. 2, 2017, pp. 193–96. *Academic Search Complete*, search.ebscohost.com/login.aspx?direct=true&db=a9h&AN=121669183&site=ehost-live. Accessed 4 Apr. 2017.

ANOVA SAMPLE PROBLEM

A researcher wants to know if there are any significant differences in the length of periwinkle shells gathered from three different beaches. The researcher has twenty-four shells total: eight from Russia, eight from France, and eight from the eastern United States. The sum of squares between (SS_B) is 13.332, and the sum of squares within (SS_W) is 225.183. Using this information, calculate the *F* ratio, and then use an *F* distribution table to determine if there are any significant differences among the groups. Assume a *P* value of 0.05.

Answer:

Substitute the number of sample groups (*k*) and total number of data points (*N*) into the formulas for the degrees of freedom for SS_B and SS_W:

$$SS_B: k - 1 = 3 - 1 = 2$$

$$SS_W: N - k = 24 - 3 = 21$$

Divide SS_B and SS_W by their respective degrees of freedom to determine the mean square between groups (MS_B) and the mean square within groups (MS_W):

$$MS_B = 13.332 / 2 = 6.666$$

$$MS_W = 225.283 / 21 = 10.723$$

Finally, divide MS_B by MS_W to find the *F* ratio:

$$F = 13.332 / 225.283 = 0.622$$

The *F* ratio is 0.662. According to an *F* distribution table, the *F* critical value for 2 degrees of freedom in the numerator, 21 degrees of freedom in the denominator, and a *P* value of 0.05 is 3.47. The *F* ratio is well below the *F* critical value. Thus, there are no statistically significant differences between the means of the sample groups, and the null hypothesis is accepted.

BACI

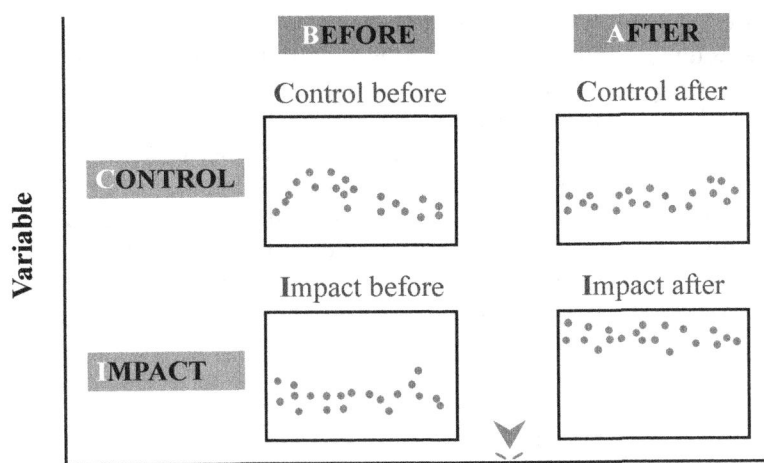

FIELDS OF STUDY

Research Design; Experimental Design; Quasi-experiment

ABSTRACT

The before-after-control-impact (BACI) design is used to analyze environmental impacts over time, aiding researchers in determining if large projects are affecting the environment adversely.

PRINCIPAL TERMS

- **control group:** the subjects in an experiment who do not receive any intervention or who receive a placebo or sham intervention.
- **impact group:** the subjects in an experiment that have received an intervention and are being assessed for any changes caused by that intervention; also called the experimental group.
- **interaction:** in statistics, an effect by which the impact of an independent variable on the dependent variable is altered, in a nonadditive way, by the influence of another independent variable or variables.

UNDERSTANDING BACI DESIGN

The before-after-control-impact (BACI) design is a quantitative, quasi-experimental design. It is often used to study the impact of human activity on the environment over time. In the BACI design, data are collected from an affected site before and after the activity in question. Those measurements are then to those from an unaffected reference site.

The BACI design developed from the before-after design for studying environmental impact. Unlike the BACI design, this design does not compare the affected site to a reference site. Rather, it simply compares the pre- and post-impact data from the affected site to determine what changes have occurred there. Consider the construction of a water-treatment plant that will release heated water into a lake. In such a scenario, researchers might decide to track the effect of water temperature on the fish populations. For their study, the water temperature would be the independent variable. To conduct a before-after study, researchers would study the fish before the plant came online and the water temperature rose. They would measure whichever dependent variables they plan to study. These might include the size, number, and species of fish in the lake. Then, after the heated water was released, researchers measure these dependent variables again. They would use these data to determine what, if any, impacts the higher water temperature may have had on the fish.

The before-after design fails to account for the possibility that any changes in the dependent variables may be unrelated to the independent variable being studied, however. For example, the fish could be affected by disease, weather, predators, or some other, unrelated factor. The BACI design addresses this problem by selecting one or more unaffected sites to serve as controls to be studied along with the impacted area. In the example above, other lakes with similar traits to the lake being impacted would be chosen for this purpose. The fish in the affected lake would form the impact group, while those in the unaffected lakes would from the control group. Data would be collected from all sites before and after the water temperature increased, just as in the before-after design. Researchers could then use these data to determine whether any other factors may be affecting the fish populations. They could also see whether interactions with other independent variables may be exacerbating the effects of the independent variable being studied. Replication through time allows researchers to monitor the response and recovery times.

ADVANTAGES AND DISADVANTAGES OF BACI DESIGN
The BACI design offers many benefits. It is more widely used than the before-after design due to responses that may occur independently because of temporal effects. For example, rain levels can change between the before and after measurements, thus changing the lake's water level. Having an unaffected site for comparison helps researchers determine if such a change is due to the independent variable or to other factors.

The BACI design is ideal for detecting large or permanent changes after an impact. It is good for monitoring for potentially disastrous changes and identifying changes in the mean of the dependent variable(s). However, the BACI design is not suited to identifying small or gradual changes. It is also not ideal for long-term monitoring or for monitoring changes in the variability of the dependent variable(s).

USING BACI DESIGN
The BACI design is useful for environmental research projects where large, lasting effects could occur. For example, researchers could use a BACI design to study the effects of a new highway on elk populations whose migration routes cross the highway. First, they would identify other sites that have similar attributes and are also crossed by migrating elk. Next, dependent variables such as the population size and the number of calves born each year would be selected. These traits would be measured at both the impact and control sites. After the highway was built, data would be collected over time at all sites. Those collected from the impact site would then be compared to data from the control sites. That analysis would help researchers to determine which changes could be attributed to the highway and which could not.

WHY USE BACI DESIGN?
The BACI design is essential for determining environmental impacts over time. Without the ability to identify the environmental impacts of a large project, severe and possibly permanent damage could be done to the environment. The BACI design helps researchers to identify adverse changes to the environment and to determine whether they were caused by the project in question or an unrelated factor.

—*Pamelyn Witteman, PhD*

FURTHER READING
Creswell, John W. *Research Design: Qualitative, Quantitative, and Mixed Methods Approaches.* 4th ed., SAGE Publications, 2014.

Gottelli, Nicholas J., and Aaron M. Ellison. *A Primer of Ecological Statistics.* 2nd ed., Sinauer Associates, 2013.

Orcher, Lawrence T. *Conducting Research: Social and Behavioral Science Methods.* 2nd ed., Taylor & Francis, 2014.

Patten, Mildred L., and Michelle Newhart. *Understanding Research Methods: An Overview of the Essentials.* 10th ed., Routledge, 2017.

Schwartz, Carl James. "Analysis of BACI Experiments." *Course Notes for Beginning and Intermediate Statistics,* Statistics and Actuarial Science, Simon Fraser U, 2015, people.stat.sfu.ca/~cschwarz/Stat-650/Notes/PDFbigbook-JMP/JMP-part013.pdf. Accessed 16 May 2017.

Schwartz, Carl James. *Design and Analysis of BACI Experiments.* 14 Oct. 2012. *Course Hero,* www.coursehero.com/file/14392371/bacipdf/. Accessed 16 May 2017.

Vogt, W. Paul. *Quantitative Research Methods for Professionals.* Pearson, 2007.

Zedler, Joy B. "Ecological Restoration: Guidance from Theory." *San Francisco Estuary & Watershed Science,* vol. 3, no. 2, 2005, escholarship.org/uc/item/707064n0. Accessed 16 May 2017.

> **BACI SAMPLE PROBLEM**
>
> Researchers need to determine the environmental impact of a commercial port being constructed on an estuary. There are five other estuaries on the coast, two of which have existing ports. Identify the groups that might be used for sampling and the variables that might be studied in this situation.
>
> **Answer:**
> In order to use a BACI design to study the research question, impact and control sites must first be determined. The impact site will be the estuary where the new port is being constructed. The three river estuaries that do not have ports will be used as control sites, as these sites represent the closest match to the estuary being studied prior to the building of the port. Groups that might be sampled for the study include the populations of fish, amphibians, birds, insects, and aquatic plants. Dependent variables to be studied could include animal weight, population numbers, plant length, and species distribution.

BROWN, ROBERT

Robert Brown as a young man. [Public domain], via Wikimedia Commons

- **Born:** December 21, 1773; Montrose, Scotland
- **Died:** June 10, 1858; London, England
- **Primary field:** Biology
- **Specialties:** Botany; microbiology; molecular biology

SCOTTISH BOTANIST

Nineteenth-century Scottish botanist Robert Brown collected and classified thousands of previously undocumented plant species during a government-funded scientific and geographical expedition to Australia. He made several important contributions to biology, including the discovery of the phenomenon named for him, Brownian motion.

> **BROWNIAN MOTION: PHYSICAL EVIDENCE OF AN INVISIBLE ATOMIC UNIVERSE**
>
> In 1827, while conducting research on plant reproduction, Robert Brown was using a microscope to examine water-suspended pollen grains from an American wildflower (Clarkia pulchella, called "pinkfairies"). He noticed something peculiar: the tiny particles of pollen seemed to be in constant, random motion.
>
> The continuous movement of miniscule particles—such as dust motes floating through a beam of sunlight—had been observed for centuries, but had never been adequately explained. As early as 1785, Jan Ingenhousz had watched via microscope and commented on specks of coal dust moving across the medium of alcohol, but had drawn no definitive conclusions. The action was variously dismissed as the play of air currents, the effects of evaporation or gravity, or light energy.
>
> Brown tentatively theorized a different causative agency: perhaps the motion was created by the very life forces contained within the pollen. As a scientist, he set out to test his theory. He tried pollen from dead plants, microscopic fragments of fossilized wood, even dust from rocks. It did not matter: organic or inorganic

11

> material, the particles continued to dance to a rhythm of their own.
>
> Unable to develop a satisfactory explanation for his observations, Brown moved on to other research. He never knew before his death how close he had come to the correct interpretation of one of the building blocks of elemental physics. Nonetheless, his studies spurred later scientists to continue the search for the solution to the mind-boggling problem of random particle agitation, which came to be called Brownian movement or Brownian motion after the person best known for first studying the phenomenon in depth.
>
> In 1880, Danish astronomer Thorvald Thiele performed a statistical analysis and advanced a mathematical theory for Brownian motion. Twenty years later, French mathematician Louis Bachelier used Brownian motion as a jumping-off point to propose a process based on probability theory for speculating in stock options.
>
> It took until 1905 before Albert Einstein put forth the theory that has generally been accepted—though many scientists have since worked to refine, clarify, or expand upon Einstein's explanation. In simplified terms, the most logical cause for Brownian motion, Einstein wrote, was the result of fundamental molecular physics: particles continually collide with the atoms contained in the molecules of fluid or gas in which they are suspended. Brownian motion is thus indirect evidence of kinetic energy, which substantiates the existence of molecules comprised of atoms. Furthermore, according to Einstein, in keeping with Avogadro's law governing volume of gases, it is possible to mathematically determine the size, number, and weight of the atoms contained in the particle-suspending sample of fluid or gas.
>
> The following year, Polish physicist Marian Smoluchowski independently reached the same conclusion. In 1909, French physicist Jean Baptiste Perrin (like Einstein, a future Nobel Prize winner) conducted experiments that elegantly verified Einstein's theory of Brownian motion, which in turn confirmed John Dalton's century-old atomic theory.
>
> Since the early twentieth century, the concept of Brownian motion has been applied to various fields of study including mathematics (differential and stochastic processes, probability, and chaos theory), biology and biophysics, astrophysics, and economics.

EARLY LIFE

Robert Brown was born on December 21, 1773, the second son of James Brown, an Episcopalian priest. His mother, Helen Taylor Brown, was the daughter of a Presbyterian minister. As a child, Brown became interested in bryophytes (mosses, hornworts, and liverworts). He received his early education at Montrose Grammar School, where his classmates included future physician and politician Joseph Hume and future historian, economist, and philosopher James Mill.

At the age of fourteen, Brown received a scholarship to study at Marischal College, a preparatory academy at the University of Aberdeen. In 1789, he transferred to the University of Edinburgh to study medicine. His fascination for biology, and botany in particular, was renewed at the university. Brown was influenced by the professor of natural history John Walker, whose students had included such eminent scientists as mathematician John Playfair, botanist Sir James Edward Smith, geologist Sir James Hall, African explorer Mungo Park, and physician Robert Waring Darwin, the father of naturalist Charles Darwin. In 1790, the year before his father died, Brown published his first scientific paper for the Natural History Society. Though he did not earn a degree, he remained at Edinburgh for six years. Brown left medicine to concentrate on natural sciences and developed a strong interest in the new field of biological classification.

In 1795, he enlisted in the Fifeshire Fencibles, a unit in the Scottish defense organization known as the Home Guard. He was commissioned as an ensign and surgeon's mate. The regiment was sent to the North of Ireland, a quiet outpost at the time, which allowed Brown ample opportunity to indulge in a favorite pastime: tramping the woods and hills to collect and study the local flora.

LIFE'S WORK

In 1798, Brown was sent to London on a regimental recruiting mission. While in London, he met Sir Joseph Banks, a wealthy botanist and philanthropist who had sailed aboard the *Endeavour* with British explorer James Cook, working as a naturalist during Cook's first voyage of discovery (1768–71). Banks, a strong advocate for the colonization of Australia (then known as New Holland), was planning a Cook-like expedition to explore the geography and biology of the southern continent. He offered Brown a handsome salary to assume the position of ship's botanist. Brown accepted and left military service in late 1800.

By the middle of 1801, the Australian expeditionary force had been assembled and outfitted. Captain Matthew Flinders commanded the sloop *HMS Investigator*. Brown led the scientific contingent, which also included botanical illustrator Ferdinand Bauer, landscape painter William Westall, gardener Peter Good, mineralogist John Allen, and astronomer John Crosley.

The *Investigator* sailed south from England, stopping for excursions at Madeira and the Cape of Good Hope (where astronomer Crosley, fallen ill, left the ship) before arriving off the coast of Australia. For two solid years, Brown and crew made frequent collecting trips ashore as the ship traversed and mapped the edge of the continent.

By mid-1803, the expedition returned to Port Jackson (present-day Sydney). Many of the crew had come down with scurvy, and the leaky *Investigator* was deemed un-seaworthy. Captain Flinders and a skeleton crew loaded many of Brown's duplicate botanical specimens aboard a small substitute vessel, the *Porpoise*, and headed for England to obtain a new ship and fresh supplies. Brown and illustrator Bauer, meanwhile, remained behind to continue their work until Flinders returned.

Flinders, however, was unlucky. The *Porpoise* was wrecked, Brown's collection of plants was lost, and the surviving crew was marooned for weeks. When Flinders managed to hire a schooner to resume the voyage home, the porous boat barely managed to reach the Indian Ocean island of Mauritius. The French controlled the island and the Napoleonic Wars had broken out, so the British ship was seized and Flinders was imprisoned—he was not released until 1810.

Meanwhile, Brown and Bauer completed their work. They had nearly four thousand specimens—mostly unknown to science—and hundreds of illustrations in hand. They waited in vain for news of Flinders. Finally, with few other available options, they boarded the condemned *Investigator* and limped back to England, arriving in 1805.

Banks was pleased with Brown's work and with the results of the expedition. In 1806, Brown was elected as librarian of the Linnean Society, a post he held until 1822, when he became a fellow of the society. He received a government salary for the five years it took him to describe and catalog the Australian specimens. In 1810—the year he was elected to the Royal Society—he published the initial volume of *Introduction to the Flora of New Holland and Van Diemen's Island*, just one of many botanical publications Brown released during his lifetime.

When Banks died, he bequeathed to Brown, his friend and respected colleague, the use of his home in Soho Square and the care of his massive botanical collection (known as the Banksian Herbarium) for life. A proviso in Banks's will stipulated that after Brown's death the collection would be donated to the British Museum, which during the early 1800s was in the process of a major expansion, following the acquisition of many Egyptian and Greek antiquities and the personal library of English king George III.

Brown, however, used Banks's collection to create a new opportunity for himself. In 1827, he proposed to turn over the entire Herbarium in exchange for a salaried position as keeper of the botanical department at the British Museum. His offer was accepted. Brown, who conducted considerable plant research in his new post, relinquished his Linnean librarianship (though he later served as president of the society for four years), and remained as the museum's botanical keeper until his retirement in 1853. He died in 1858 at age eighty-four, following a severe attack of bronchitis.

IMPACT

Brown conducted pioneering work in taxonomy, ordering, classifying, and naming numerous Australian botanical genera. Though Brown described over 2,200 species—more than 1,700 of which were new—he particularly focused upon multiple species of *Banksia* (a group of spiky, nectar-rich wildflowers), Orchidaceae (the family of orchid plants) and *Dryandra* (shrubs and small trees considered a separate genus until incorporation into the genus *Banksia* in 2007). Brown named some 140 new genera, some of which, following modern techniques of examination, have been reclassified.

From his earliest days of his career, Brown was also a frequent, dedicated user of the microscope in his botanical research. Microscopes enabled Brown to view and differentiate between the reproductive systems of gymnosperms (woody, unisexual plants such as conifers, which bear naked seeds) and angiosperms (bisexual plants, such as flowers, that bear enclosed seeds within an ovary inside fruit). Magnification also allowed Brown to announce in 1813 that he had discovered the cell nucleus (which, as later research would reveal, contains most of a cell's genetic material), and to study the phenomenon later called Brownian motion in his honor. Throughout his life, Brown collected microscopes from both domestic and foreign instrument makers, and made practical suggestions for improvements that helped advance the development of microscopy.

For his body of scientific work, Brown received the highest honor from the Royal Society, the Copley Medal, in 1839.

—*Jack Ewing*

FURTHER READING

Kumar, Manjit. *Quantum: Einstein, Bohr, and the Great Debate about the Nature of Reality.* New York: Norton, 2011. Print. Detailed examination of Brownian Motion and the clashing theories of two twentieth-century scientists about the character and behavior of matter on the molecular and subatomic scale.

Magee, Judith. *Art and Nature: Three Centuries of Natural History Art from Around the World.* Vancouver: Greystone, 2010. Print. Presents an illustrated history of natural science and geographical explorations since 1700, encompassing the results of Brown's Australian expedition and the accompanying botanical drawings of Ferdinand Bauer.

Yoon, Carol Kaesuk. *Naming Nature: The Clash between Instinct and Science.* New York: Norton, 2009. Print. Presents an overview of the history and controversies involved in taxonomy (the science of organizing and classifying living organisms) since Swedish botanist Carl Linnaeus founded the specialty in the eighteenth century.

BURBIDGE, E. MARGARET

BRITISH ASTROPHYSICIST

E. Margaret Burbidge has been keeping an eye on the universe for more than seventy years. An observer by nature, she has made some of the most important astronomical discoveries of the twentieth century, and her work is characterized by a determination to unlock the mysteries of the cosmos.

- **Born:** August 12, 1919; Davenport, England
- **Primary field:** Physics
- **Specialty:** Astrophysics

EARLY LIFE

Eleanor Margaret Peachey was born on August 12, 1919, in Davenport, England. Much like the relationship Margaret and her husband would eventually have, her parents shared a common scientific interest. Her father, Stanley Peachey, was a chemistry teacher at the Manchester School of Technology; her mother, Marjorie Peachey, was a chemistry student at the same school.

When the family moved to London in the early 1920s, Margaret spent much time in her father's new laboratory, looking at the stars through binoculars. When she was twelve years old, her grandfather gave her some astronomy books, which showed her the connection between the stars and numbers, her other fascination. For example, she learned that, after the sun, the nearest star to Earth is 26 trillion miles away.

Peachey enrolled at the University College of London (UCL) in 1936, graduating with a degree in astronomy three years later. She immediately began working on her PhD at UCL, studying stars called B Stars. She studied the stars using a telescope with an attached spectrograph, which identifies the elements present in a star. This research, which earned her a PhD in 1943, would prove to be especially useful several years later in the development of her most well-known theory.

Hoping to continue observing the stars, Peachey applied for a fellowship that would have allowed her to use the advanced telescopes at Mount Wilson Observatory in La Canada, California, but women were not allowed to use the equipment at the time. Jaded by her experience, but no less enthusiastic to pursue her dreams, Peachey returned to UCL to study physics.

That decision turned out to be a good one for Peachey. In 1947, she met Geoffrey Burbidge in a physics class at UCL. The two shared many interests, including tennis, politics, and history. On April 2, 1948, less than a year after they met, Geoffrey and

Margaret married; they remained virtually inseparable until Geoffrey's death in 2010. In fact, because they so often worked on projects together, Margaret and Geoffrey were jokingly referred to as "B squared."

LIFE'S WORK
After relocating to the United States in 1951 to study at the Yerkes Observatory in Williams Bay, Wisconsin, the Burbidges became interested in figuring out the source of the chemical elements that make up everything in our universe. Their friend Fred Hoyle was developing a theory to explain the existence of the chemical elements, and the idea intrigued the Burbidges. They moved back to England in 1953 to help Hoyle and nuclear physicist William Fowler perfect Hoyle's theory.

The Burbidges moved back to the United States in 1955. In 1957, at the age of thirty-eight, and one year after she had given birth to her daughter, Burbidge's most famous work was published in the journal *Reviews of Modern Physics*: "Synthesis of the Elements in Stars," written by the Burbidges, Fowler, and Hoyle, presented a new theory about the composition of matter; the scientists wrote that nuclear reactions in stars create almost all of the chemical elements that make up the universe. Essentially, the theory explained how the universe evolved over billions of years and how it continues to evolve.

On the periodic table of the chemical elements, each element has a distinct atomic number, or weight, equal to the number of protons in the nucleus. According to the theory that the Burbidges and their colleagues developed (which they referred to as the B^2FH theory), there are nuclear reactions occurring in stars all the time, and these reactions cause the nuclei of existing elements present in the stars to fuse. But unlike molecules, which are formed by two or more atoms with discrete nuclei, the fusion that occurs in stars creates new heavier atoms.

When the stars inevitably explode in what is called a supernova, the new elements that were created escape and become part of the universe. That means that every molecule that makes up our universe, every cell in our bodies and every ounce of air we breathe, originated in the stars above us, according to B^2FH.

The theory further identified eight processes through which this synthesis occurs; the processes are called hydrogen burning, helium burning, a-process, the e-process, the s-process, the r-process, the p-processes, and the x-process. Most of the names, claims Burbidge, were thought up on the spur of the moment. R-process, for example, was so named because it took place rapidly. She never expected the names to stick, but they are still being used almost fifty years later.

This theory had astounding implications for the conception of the universe and subsequently earned the Burbidges the Helen B. Warner Prize for Astronomy in 1959. The prize, awarded for contributions to the fields of either observational or theoretical astronomy, was especially appropriate for the pair, since, as Burbidge sometimes said, Geoffrey had the ideas while she did the observing.

Although the Burbidges' theory of nuclear synthesis made them famous (despite a snubbing by the Nobel committee in 1983, which awarded the prize to Fowler only), the research that grew from the theory would prove more controversial. Amid constant opposition and, eventually, indifference, Burbidge would challenge one of the most sacrosanct theories in not only astronomy, but science as a whole: the big bang theory.

Since they were first discovered in 1963, Burbidge has been studying objects called quasars, and her husband was theorizing about their origin before he died. Quasars are star-like objects that are still mysterious, but seem to provide clues as to the nature of the universe. Almost everyone in the scientific community agrees that quasars are extremely far away from Earth. Burbidge says otherwise, and it is on this that her theory refuting the big bang rests.

The theory, which builds on an earlier theory from Hoyle called steady-state theory, says that quasars are closer than we realize but are moving very quickly, which affects their color in the same way that distance would. Burbidge and her late husband are in the minority with their theory of a quasi-steady state universe, which supposes a series of smaller bangs occurring approximately every hundred billion years. However, the theory was dismissed by most of the scientific community.

IMPACT
Burbidge's contributions to the field of astrophysics are extensive. Her work, along with that of her late husband, forced the scientific community to look at the universe and its origin in a completely different way, if only for a moment. In addition, she discovered, of what all things in the universe are made.

She worked around institutional barriers, gender discrimination, and the complexity of her and her husband's parallel scientific careers that threatened to interfere with her research. Burbidge's displeasure with the treatment of women in the field of astronomy led her to create a committee with the American Astronomical Society (AAS) for the status of women in astronomy.

Despite setbacks, Burbidge's career has been long and distinguished. In 1964, she was elected a fellow of the Royal Society of London. She served as president of the AAS from 1976 to 1978. From 1979 to 1988, she was the director of the Center for Astrophysics and Space Sciences at the University of California, San Diego. While there, she helped develop the Hubble Space Telescope.

—*Alex K. Rich*

FURTHER READING

Bartusiak, Marcia. "Margaret Burbidge: Stars, Quasars, Supernovae, Galaxies—If It's Out of This World, She Has Seen It." *Smithsonian* (Nov. 2005). 35. Print. Offers a brief synopsis of Burbidge's career.

Miller, Steve. *The Chemical Cosmos: A Guided Tour.* New York: Springer, 2012. Print. A beginner's guide to the universe and its elements featuring a discussion of Burbidge's contributions.

Panek, Richard. "Two Against the Big Bang." *Discover* 26.11 (Nov. 2005): 48–54. Print. Focuses on the Burbidges' dismissal of the big bang theory, also including remarks on their careers as astrophysicists.

BURBIDGE AND COSMIC EVOLUTION

In 1955, astrophysicists Margaret and Geoffrey Burbidge, William Fowler, and Fred Hoyle, collaborating under their combined initials B2FH, postulated that the elemental composition of everything in the known universe could be traced back to thermonuclear reactions within stars. In 1957, they published their ground-breaking paper "Synthesis of the Elements in Stars" in the journal Reviews of Modern Physics. In it, the team explained that the elements necessary for life—hydrogen, helium, lithium, oxygen, iron, and carbon—are created during the formation and destruction of stars.

For example, a star like the sun converts hydrogen to make helium, which generates heat and light. (More massive stars ignite helium to produce carbon and oxygen as well.) When the star explodes at the end of its life, the elements it formed during its lifetime are thrown into space and used to create new stars. This process, the continuous comingling of cosmic dust and basic elements, provides the basis for the physical universe and everything in it. The profundity of this suggestion was not lost on Burbidge or her coauthors; they began "Synthesis of the Elements in Stars" with a quotation from Shakespeare's King Lear: "It is the stars, the stars above us, govern our conditions."

Their theory, while imperfect, was more sweeping than previous cosmological theories and was appreciated for both its simplicity as well as its sheer enormity. It could now be accurately stated that every atom of a human being, a plant, or a star, excluding hydrogen, was forged inside of another star. It also marked an important and unprecedented exchange between astronomy and nuclear and particle science, laying the foundations of nuclear astrophysics and cosmology.

In 1959, Burbidge and her husband were awarded the Helen B. Warner Prize by the American Astronomical Society. The couple was also awarded the Royal Astronomical Society's gold medal for their contributions to astronomy in 2005. However, only Fowler was awarded the 1983 Nobel Prize for his contribution to the theory. Many have speculated as to why the Burbidges and Hoyle might have been excluded (most pointing to scientists' rejection of the big bang theory), but Burbidge never publicly commented on the matter.

C

CAVENDISH, HENRY

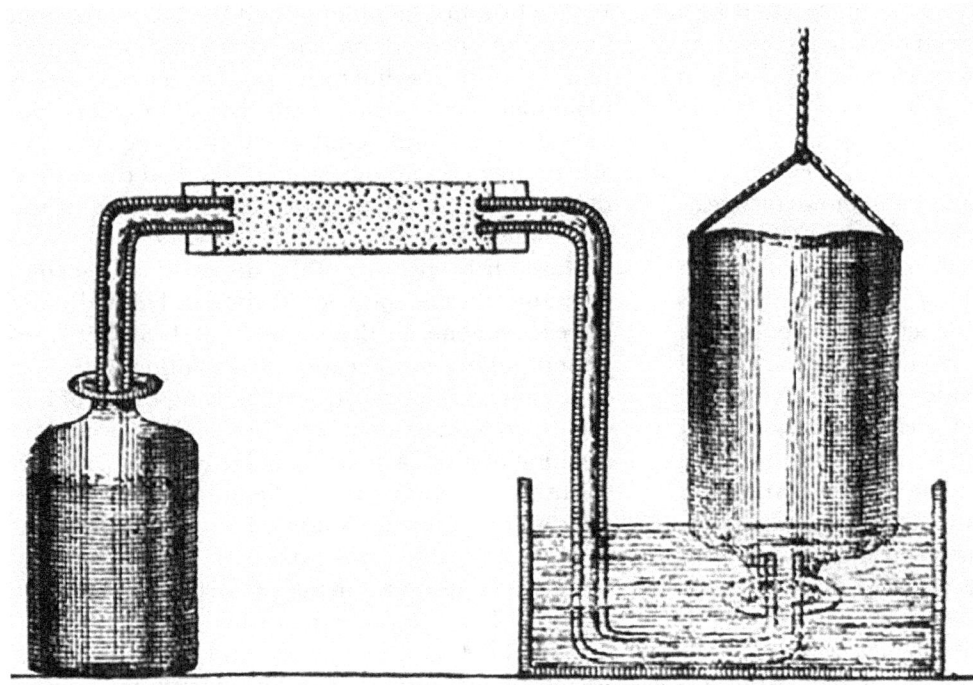

Cavendish's apparatus for making and collecting hydrogen. By Henry Cavendish (Philosophical Transactions (periodical)) [Public domain], via Wikimedia Commons

ENGLISH CHEMIST

Henry Cavendish, a reclusive character, made significant advances in the chemistry of gases and contributed to the study of electrical phenomena. He is noted for discovering hydrogen and for measuring the Earth's density.

- **Born:** October 10, 1731; Nice, France
- **Died:** February 24, 1810; London, England
- **Primary field:** Chemistry
- **Specialty:** Analytical chemistry

EARLY LIFE

Henry Cavendish was born in 1731 into a leading aristocratic British family. His father, Lord Charles Cavendish, was the third son of the duke of Devonshire, while his mother, Lady Anne Grey, was the daughter of the duke of Kent. His mother's death when he was two years old left Cavendish in his father's care. He entered Newcome's Academy in Hackney in 1742 and matriculated at St. Peter's College, Cambridge, in 1749, leaving in 1753 without taking his degree.

After leaving Cambridge, he resided with his father in London. Lord Charles Cavendish was a longtime member of the Royal Society and an avid investigator of meteorological and electrical questions, having received the society's Copley Medal for perfecting a registering thermometer. His father sponsored his election to the Royal Society in 1760; living in an all-male environment permeated with scientific conversation may well have shaped both Cavendish's lifelong fascination with science and his strange social behavior. Throughout his life, Cavendish was reclusive, shunning human society.

Following his father's death in 1783, Cavendish came into a large family inheritance. While very wealthy, he remained frugal in his personal expenditures and oddly indifferent to other uses of money, giving generously to charity but always checking the list of donors and giving precisely the amount of the largest donation. Abhorring any meeting with women—he even left notes for housemaids to avoid personal contact—Cavendish remained an eccentric and elusive person.

Only in the Royal Society, where he attended regularly and participated fully, did he have any public life. Cavendish was responsible for a detailed description and analysis of the society's meteorological instruments in 1776 and served on a committee investigating lightning protection, using his own electrical research to recommend pointed rather than blunt lightning rods. Even at the Royal Society, however, he would flee if approached by a stranger, and he was often seen outside the meeting room, waiting for the moment when he could slip in unnoticed.

LIFE'S WORK

In his lifetime, Henry Cavendish was known primarily for his research in chemistry and electricity, work reflected in a remarkably small number of papers in the *Philosophical Transactions of the Royal Society*. His first public work was in 1776, a series of papers on the chemistry of factitious airs, or gases. Chemistry at the time was dominated by the idea that air was a single element, one of the four Greek elements. British chemists, from Robert Boyle, who first elucidated the gas laws in 1662, through Stephen Hales and Joseph Black in the eighteenth century, had made what they called pneumatic chemistry—manipulating and measuring air in its various states of purity—practically a national specialty.

Chemical research was also carried on around the organizing concept of the phlogiston theory put forward by Georg Ernst Stahl in 1723. Phlogiston was believed to be the element of fire, or its principle, that caused inflammability when present in a body. It was believed to be central to most chemical reactions. Combustion was explained as a body releasing its phlogiston. In this dual context of pneumatic chemistry and phlogiston theory, Cavendish presented his study of factitious airs, or gases contained in bodies. Most important, he isolated and identified "inflammable air"—now called hydrogen. Recognizing the explosive nature of inflammable air, Cavendish went on to identify it as phlogiston itself.

In 1783, Cavendish, having heard of experiments by Joseph Priestley that generated condensation upon exploding inflammable air (hydrogen) and common air (regular air in the Earth's atmosphere), presented a study of an improved eudiometer, a device for testing the quality of air. Cavendish demonstrated that common air was composed of constant proportions of different constituents, rather than being a single elemental substance. This research was followed the next year with two more papers on air, the research for which was principally concerned to find the cause of the absorption of gases in the eudiometer. When he mixed inflammable air with common air, he created an explosion and he noticed condensation on the containing vessel. Where Priestley had mentioned the fact in passing, Cavendish focused on the condensation, noting that "by this experiment it appears that this dew is plain water, and consequently that almost all the inflammable air, and about one-fifth of the common air, are turned into pure water." He had discovered that water was a compound of airs, not one of the four elements.

Cavendish's priority of the discovery of the composition of water was soon disputed. He had done his experiments in the summer of 1781 but postponed publication because the resulting water was contaminated with nitric acid. Solving this problem eventually led him to his last chemical discovery: the isolation of nitric acid in 1785. He had told Priestley of his water results, and a friend had passed the same information to Antoine-Laurent Lavoisier in Paris, who repeated and extended the experiments, while James Watt also made the discovery independently. When all three published their investigations in 1784, Watt, Lavoisier, and Cavendish each laid claim to priority. A brief controversy ensued but was quickly extinguished, with each of the three rather politely deferring to the others. The controversy was rekindled in the mid-nineteenth century and continues in some scholarly circles. Cavendish should probably be conceded the right to claim the discovery of the compound nature of water, although his explanation was given in the context of the phlogiston theory. Lavoisier followed Cavendish chronologically but explained the composition of water in the radically different terms of his new antiphlogiston chemistry, as the union of oxygen and hydrogen.

Lavoisier's antiphlogiston explanation was indicative of a revolution in chemistry that he was leading in Europe. In 1772, when he had weighed the product of calcination (oxidation in the new terminology), there was a weight gain in the calx. He offered the explanation that something was taken up in the process, rather than phlogiston being given off. This "something" he identified as oxygen and thereby created a new chemistry. Cavendish recognized that Lavoisier's oxygen-based chemistry was essentially equivalent to a phlogiston-based chemistry, but he rejected the new ideas to the end of his life. In 1787, Lavoisier introduced his new chemistry in his *Méthode de nomenclature chimique* (Method of chemical nomenclature), and fully elaborated it in 1789 in *Traité élémentaire de chimie* (Elementary treatise on chemistry). Again Cavendish rejected the ideas, stressing that arbitrary names and ideas, as he saw them, could only lead to great mischief. By 1788, Cavendish had published the last of his chemical researches.

The second major area of Cavendish's published research was in electrical phenomena, a near-craze sweeping through scientific and popular circles throughout Europe. In 1771, Cavendish published his first paper on electricity, putting forward a single-fluid theory of electricity, in opposition to the then-popular two-fluid theory. He presented electricity as a single fluid that could be measured according to its compressibility, using the analogy of Boyle's gas law. He provided a quantitative measure of tension as well as quantity, surmising that this fluid followed an inverse square law for repulsion, rather than the first power law displayed by gases. This work was expanded in 1776, with a paper describing his effort to construct a model of the torpedo fish, or electric ray, to analyze whether its electrical effects were similar to electrostatic phenomena, then the center of scientific attention. He demonstrated that they were the same.

In 1783, Cavendish also published a series of papers on heat, focusing on the question of the freezing point of mercury. He was able to expand upon the idea of latent heat presented earlier in the century by Joseph Black. These papers were based on a series of experiments carried out at the request of Cavendish by officers of the Hudson's Bay Company in 1781 and after in northern Canada.

In 1798 Cavendish presented his final paper to the Royal Society, outlining his effort to measure the density of the Earth. Using an experimental torsional balance apparatus devised by John Michel, Cavendish was able to provide an extremely accurate estimate of Earth's density, while also providing the first experimental demonstration of Isaac Newton's gravitational laws.

Impact

Henry Cavendish was widely recognized by his contemporaries as a precise experimenter and a thorough researcher, despite his small output of published works. His international reputation was certified in 1803 by his election as a foreign member of the Institut de France. This reputation was further honored in the founding of the Cavendish Society in 1846 and his family's endowment of the Cavendish Laboratories and the Cavendish Professorship at Cambridge in 1871, where much of the fundamental research in atomic theory was carried out in the early twentieth century.

Unpublished manuscripts discovered after his death have significantly altered the perception of Cavendish's scientific pursuits as limited solely to precise and narrow experimentation. The publication of these papers has deepened the appreciation of his interests and abilities. Russell McCormmach has shown that Cavendish's work was rooted in a coherent and consistent vision of the scientific approach and theories of Newton. Newton had postulated a world made up of interacting particles, attracting and repelling one another according to strict mathematical laws. Cavendish pursued an integrated research program attempting to apply the Newtonian insight to the physical world.

—*William E. Eagan*

Further Reading

Akroyd, Wallace Ruddell. *Three Philosophers: Lavoisier, Priestley, and Cavendish*. Westport: Greenwood, 1970. Print. An account of the chemical revolution in relatively nontechnical terms, focusing mainly on Lavoisier and Priestley, with a short chapter on Cavendish.

Jungnickel, Christa, and Russell McCormmach. *Cavendish: The Experimental Life*. Lewisburg: Bucknell UP, 1999. Print. Examines Cavendish's discoveries within the context of the elite society in which he and his father developed their scientific interests.

McCormmach, Russell. *Speculative Truth: Henry Cavendish, Natural Philosophy, and the Rise of Modern Theoretical Science*. New York: Oxford UP, 2004. Print. Explores the new theories of natural philosophy that emerged in the second half of the eighteenth century, including Cavendish's mechanical theory of heat. Includes an edition of Cavendish's manuscript about the mechanical theory of heat.

Miller, David Philip. *Discovering Water: James Watt, Henry Cavendish, and the Nineteenth Century "Water Controversy."* Burlington: Ashgate, 2004. Print. Describes how Cavendish, Watt, and Lavoisier's independent discoveries that water is a compound of airs and not a combination of elements became an issue of controversy among nineteenth-century scientists.

CAVENDISH'S DISCOVERY OF HYDROGEN

British scientist and pneumatic chemist Henry Cavendish was perhaps best known for his discovery of "inflammable air," or hydrogen, as it came to be known. Cavendish was particularly interested in the relationships between gas properties, chemical reactions, and the composition of matter.

Though seventeenth-century Irish physicist Robert Boyle initially studied and observed hydrogen gas, it was Cavendish who, in 1766, recognized and identified the more complex nature of the gas and the ways in which it behaved. Cavendish first mixed zinc with a poisonous acid, which created a gas he named "inflammable air" (hydrogen), a substance he immediately considered to be a unique chemical element, as it held its own properties. He performed experimental trials in which he combined the hydrogen and oxygen (at that time known as "dephlogisticated air") in a closed space. He noticed that when the two combustible gases merged together, they lost the ability to exist as gases and eventually formed a liquid substance: water. He understood this to mean that when hydrogen and oxygen converge, the result of this convergence is water; or, that water is made up of both hydrogen and oxygen. Once observing this convergence, he next sought to determine the density of hydrogen. He decided to measure the weight of a closed sphere containing the gases. Though his experimental trials later proved to be flawed in terms of mathematical calculations, Cavendish correctly concluded that hydrogen weighs less than oxygen. Not only did Cavendish accurately determine the relationship between hydrogen and oxygen, but also he successfully identified the three main components that make up air. In 1766, Cavendish published a paper explaining his groundbreaking discoveries. He also elucidated his hypothesis that only 1/120 of the Earth's atmosphere contains a substance that is neither oxygen nor nitrogen. This substance was later identified as argon. During the latter half of the eighteenth century, the three substances essential to Cavendish's research were named nitrogen, oxygen, and argon.

In 1787, French chemist Antoine-Laurent Lavoisier completed Cavendish's chemical puzzle by reproducing his original experiments and officially replacing the name of the British scientist's "inflammable air" with "hydrogen." At this time, Lavoisier also formally coined the term "oxygen." Cavendish is widely considered a major pioneer in the field of chemical gases; he has even been nicknamed chemistry's Isaac Newton for his revolutionary contributions to the world of chemistry.

CASE STUDY RESEARCH

FIELDS OF STUDY

Research Design; Research Theory; Experimental Design

ABSTRACT

A case study is a research method that analyzes a particular set of conditions regarding a person or a group of people, an ecosystem, or a geographical location. Case studies ask questions that cannot be answered with statistical data alone but require observation, interviews, and analysis to be understood.

PRINCIPAL TERMS

- **methodology:** a set of practices developed to implement a case study.
- **qualitative data:** information that describes but does not numerically measure the attributes of a phenomenon.
- **quantitative data:** information that expresses attributes of a phenomenon in terms of numerical measurements, such as amount or quantity.

WHAT IS A CASE STUDY?

A case study is the observation, study, and analysis of a real-life situation in order to explore a contemporary phenomenon in depth. Case studies differ from experiments that are set up in a controlled environment and from statistical surveys that draw from a broad dataset. A case study takes place within a natural or normal context. This differs from experimental research, which deliberately controls the environment and manipulates particular elements. Case studies are often used when elements cannot be manipulated. They are also often used to answer questions about how or why a specific phenomenon is occurring. Case studies are also used to study a phenomenon that cannot be repeated, such as a weather event. Case studies can be used to make generalizations to better understand similar phenomena.

Case studies are unique. There is no one method of design or data collection. The first steps to creating a case study are to form the research question and then to develop a methodology. The methodology outlines the research process. The methodology lists the data points that need to be collected and the procedures for data analysis. Case studies can draw from quantitative data in the form of statistical surveys and a review of existing records. However, they often rely heavily on qualitative data, in form of individual interviews, focus groups, or direct observation of the research subjects. A proposition or hypothesis may or may not be necessary to direct the scope of the study.

Researchers must define the specific case to be studied and the scope of the data collection. The case may be an individual person who is representative of a larger group. Or the case may include a group of people, a town or an organization, a program, or an event. A theory of the case (prediction) and rival theories will further inform what data will need to be collected. The method of data collection must also be determined. Data collection may involve interviews, surveys, existing data, or direct observation. If the initial research question does not clearly lead to a specific case, the question may be too broad or vague. It is also important to delimit the time period, subjects, and external data to be used. Researchers need to secure access to the study sites and the consent of the participants.

Why Use a Case Study?

Case study research provides in-depth analysis of a particular individual, event, or group. It is used to explore how or why a specific phenomenon occurs. Case studies are particularly useful when there are many variables and multiple sources of information to be followed over time. When researchers want to understand why a particular phenomenon is occurring, statistics and records may not have enough information to draw a reasonable conclusion. A detailed case study can illustrate the causes of certain behaviors or outcomes in ways statistical analysis cannot. Case studies are not usually used to test hypotheses or compare phenomena in the way that experimental research does.

The most significant disadvantage of a case study is that the conclusions cannot necessarily be generalized to other populations. However, researchers can often determine which factors in a case study apply to other situations. Furthermore, the subjects of case studies are not randomly chosen and the researcher is often in close contact with them. For these reasons, case studies can raise questions of objectivity. Definitions and terms are not always standard between case studies. This can make it difficult to compare variables. Disadvantages also include the vast amount of data that are collected and must be analyzed. Case studies can be time intensive. A case study cannot always determine the cause of an event definitively. However, case studies are useful in identifying potential causes and shared themes.

When Is a Case Study Appropriate?

Case studies are used to study a unique event or situation. The findings of a case study can then be generalized to other areas. For example, a factory that has implemented a new management method and significantly increased production would be a good subject for a case study. A case study could help to determine why is this new method is effective. Researchers might observe the factory floor and the management personnel. They might interview managers and workers. They could use statistical data such as production and employee schedules. All of this data would then be analyzed. This analysis would allow the researchers to draw conclusions about how and why the method was successful.

In this example, it would be difficult to draw meaningful conclusions using only statistical data. Furthermore, if production figures were high but employees were unhappy and looking for new jobs, those findings would not likely be discovered in statistical data. However, this information would likely come to light through observations and interviews. Case studies offer opportunities for researchers to discover and explore additional questions that may not have been apparent at the outset. Case studies are particularly useful in fields in which tightly controlled experiments are unethical or difficult to implement. For this reason, case study research is often used in psychology, anthropology, sociology, and ecology. Although case studies have certain disadvantages compared to experimental research, they remain a valuable research method for illustrating causation.

Maura Valentino, MSLIS

FURTHER READING

Bartlett, Lesley, and Frances Vavrus. *Rethinking Case Study Research: A Comparative Approach*. Routledge, 2017.

Hancock, Dawson R., and Bob Algozzine. *Doing Case Study Research: A Practical Guide for Beginning Researchers*. 2nd ed., Teachers College Press, 2011.

Lapan, Stephen D., et al., editors. *Qualitative Research: An Introduction to Methods and Designs*. John Wiley & Sons, 2012.

Taylor, Steven J., et al. *Introduction to Qualitative Research Methods: A Guidebook and Resource*. 4th ed., John Wiley & Sons, 2016.

Tight, Malcolm. *Understanding Case Study Research: Small-Scale Research with Meaning*. Sage Publications, 2017.

Yin, Robert K. *Applications of Case Study Research*. 3rd ed., Sage Publications, 2012.

Yin, Robert K. *Case Study Research: Design and Methods*. 5th ed., Sage Publications, 2014.

CAUSAL NETWORKS

ABSTRACT

A causal network is a symbolic representation of causal relationships between two entities or nodes on a network. Using mathematics and statistical reasoning, causal networks graphically display the inferences made in order to determine whether or not a causal relationship exists. They are not concerned with how information about observable phenomena is acquired; their focus is on how this information is processed in order to reach a conclusion.

PRINCIPAL TERMS

- **Bayesian network:** a model representing a set of random variables and the conditional dependencies of each one in a directed acyclic graph (DAG).
- **dependency:** a relationship in which the value taken on by one variable is determined by the value of another variable.
- **latent variable:** a variable that is not observed but whose value can be determined by reference to another variable, due to the relationship that exists between the two.
- **parent variable:** a variable that shares some of its characteristics with, and thus forms the model for the design of, another variable known as the child variable.

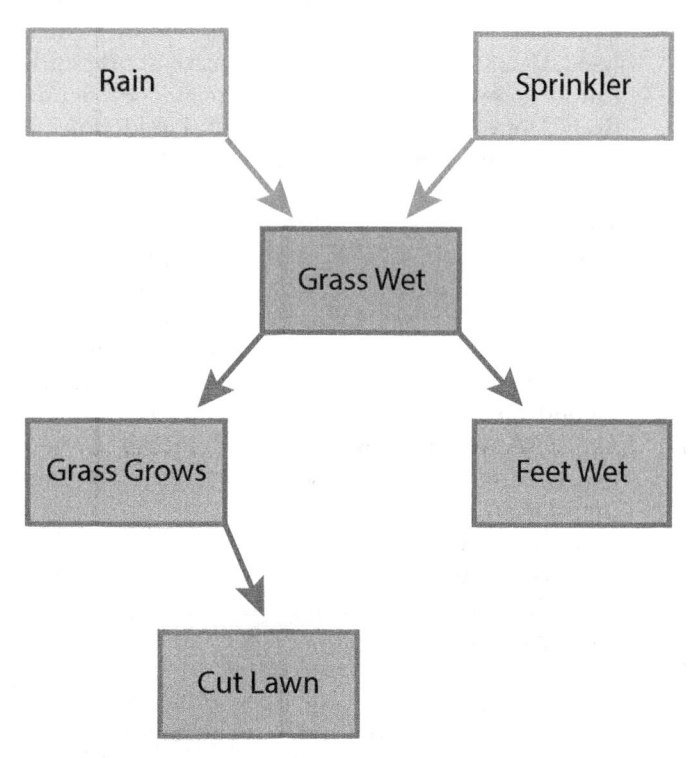

FIELDS OF STUDY

Statistical Analysis; Relational Analysis; Research Theory

Causal Relationships

Causality, or causation, is a basic abstract concept in which one factor (the cause) contributes to another (the effect). Causal networks are graphical models that represent causal relationships between two or more nodes, or points in the network. A causal network is a type of Bayesian network, in which directed acyclic graphs (DAGs)—directed graphs with no loops, or directed cycles—model sets of random variables. Each node is a variable, and as such it can take on different values depending on the circumstances. The nodes are connected by unidirectional arrows that show the direction of the causal relationship. Each arrow begins at a node that causes or contributes to the outcome represented by the node at which the arrow ends.

Causal networks are used to illustrate the functional relationships between different system components at a high level of abstraction. This visualization can provide insights into how systems work that might not otherwise be apparent. Causal networks are essentially depictions of dependencies between variables. A node may have a single dependency, or it may have multiple dependencies. If a node has multiple dependencies, an alteration in its value is the result of change in multiple other nodes. Mathematical functions are used to represent how these multiple causes work in conjunction in order to produce an effect. This can involve the assignment of differing weights to each cause, standing for the relative amount of influence a particular cause has on a specified outcome.

Learning Causality

Much research has gone into furthering the understanding of how human beings acquire information and use it to make inferences about causality. This was long thought to be a form of abstract thought unique to human beings. However, advances in information theory and computer science opened up the possibility of programming computers to be able to perceive causal relationships as well. Causal networks are particularly relevant to those fields, though they can be applied in any system involving causality.

There are at least three stages involved in the process of inferring causality within a particular scenario. First, the observer must learn how the (potential) causal network is structured. This involves asking questions, such as what are the network's nodes, and what is the configuration of its edges (the term for the unidirectional arrows indicating causality). Next, the observer becomes familiar with the parameters of the network. Examples include the values of each node, identification of parent variables, and consideration of any latent variables. The final step, presuming the previous steps were successful, is for the observer to make a determination as to whether or not there exists a causal relationship between the nodes being evaluated.

The Basis of Research

At the most basic level, most experiments are designed to discover whether or not a causal relationship exists between variables. A researcher may set up a causal network to better understand their subject prior to a study. In other cases, a causal network can only be established as understanding of a subject improves.

For example, at one point in history there was no indication that smoking tobacco might be related to developing lung cancer. In fact, smoking was frequently prescribed for relief from the symptoms of various pulmonary ailments, as it was thought that the warmth of the smoke might have a soothing and therapeutic effect on the lungs. Over time, however, evidence of a correlation between smoking and cancer accumulated, and scientists began to investigate a possible causal relationship. Experiments sought to determine if the variable of having lung cancer (which could take on values of "yes" or "no") was to any degree dependent on the variable of having smoked cigarettes (another "yes" or "no" variable). Ultimately, ample signs of such a causal relationship were found. Research also suggested that just the presence of cigarette smoke in one's surroundings might also have a causal relationship with lung cancer, eventually leading to understanding of the dangers of secondhand smoke.

Of course, smoking is not the only cause of lung cancer, and not every single person who smokes or is subject to secondhand smoke develops lung cancer. As with any complex system, many variables are at play. But by establishing a causal network, research on smoking and cancer opened the door to a more complete understanding of the issue.

Applications of Causal Networks

Causal networks underlie much of what may be observed in the natural world. They also provide insight into the mechanisms through which human beings symbolically interpret that world. This insight is particularly relevant as research into artificial intelligence (AI) and other branches of computer science seeks to replicate the mind's ability to understand what it perceives. Along with other Bayesian networks, causal networks are critical to many computing programs and models in numerous fields.

Psychologists, too, are intensely interested in better understanding perceptions of causality. Much of the meaning that people make of their lives is a direct consequence of their perceptions of causality, whether correct or incorrect. People observe the external world and draw conclusions about causal relationships, constructing mental representations of the world based on these conclusions. This process is at the root of consciousness, including what is diagnosed as psychosis. A clearer picture of causal networks and how they are mentally constructed would be beneficial to researchers in virtually every field of inquiry.

Scott Zimmer, JD

FURTHER READING

Cho, Hyunghoon, et al. "Reconstructing Causal Biological Networks through Active Learning." *PLOS ONE*, vol. 11, no. 3, 2016, doi:10.1371/journal.pone.0150611. Accessed 27 Mar. 2017.

Huynh, Van-Nam, Vladik Kreinovich, and Songsak Sriboonchitta. *Casual Inference in Econometrics*. Springer, 2016.

Pearl, Judea. *Causality: Models, Reasoning, and Inference*. 2nd ed., Cambridge UP, 2013.

Xia Liu et al. "Structure-Learning of Causal Bayesian Networks Based on Adjacent Nodes." *International Journal on Artificial Intelligence Tools*, vol. 22, no. 2, 2013. *Academic Search Complete*, search.ebscohost.com/login.aspx?direct=true&db=a9h&AN=87042564&site=ehost-live. Accessed 27 Mar. 2017.

Zeevat, Henk, and Hans-Christian Schmitz. *Bayesian Natural Language Semantics and Pragmatics*. Springer, 2015.

Zeng, Zexian, et al. "Discovering Causal Interactions Using Bayesian Network Scoring and Information Gain." *BMC Bioinformatics*, vol. 17, 2016, doi:10.1186/s12859-016-1084-8. Accessed 27 Mar. 2017.

Zengkai Liu, et al. "An Approach for Developing Diagnostic Bayesian Network Based on Operation Procedures." *Expert Systems with Applications*, vol. 42, no. 4, 2015, pp. 1917–26.

Chi-Square Goodness of Fit Test

Variable	Observed Frequency	Expected Frequency	O – E	$(O - E)^2$	χ^2 $[(O - E)^2]/E$
Category A	21	20	1	1	0.05
Category B	18	20	–2	4	0.2
Category C	14	20	–6	36	0.56
Category D	20	20	0	0	0
Category E	27	20	7	49	2.45

FIELDS OF STUDY

Hypothesis Testing; Statistical Analysis; Variance Analysis

ABSTRACT

The chi-square goodness of fit test is used to study how the distribution of one variable in an observed sample compares to the expected distribution. The test can be used to determine if a desired distribution of values for the categorical variable occurs in the sample without requiring the total population to be studied.

PRINCIPAL TERMS

- **categorical data:** information that can only take on nonnumeric values, such as labels or names.
- **degrees of freedom:** the number of observations in a data set that are free to vary based on certain statistical parameters.
- **expected data:** the values that would be assumed to occur if the null hypothesis were true.
- **goodness of fit:** the degree to which the observed data values are consistent with the expected distribution of the data.
- **independence:** the degree to which two variables in a population are related.
- **observed data:** the values collected from the sample being studied.

Testing Goodness of Fit

The chi-square goodness of fit test is used to study one variable from a population against a hypothesized distribution. To use this test, the variable being studied must represent categorical data. Categorical data consist of nonnumeric values, such as labels or names. For example, the exterior color of an automobile is a categorical variable, as it would hold such values as red, blue, yellow, black, and white. Chi-square tests cannot be used to analyze continuous variables, which can take on any measurable value.

This particular test is used to determine the goodness of fit. Goodness of fit describes the degree to which the distribution of the data in a collected sample (the observed data) is consistent with a probability distribution (the expected data). For example, the test can determine if the numbers of each car color in a sample (observed data) are consistent with the numbers of each color for all cars produced by the factory (expected data).

To conduct the chi-square goodness of fit test, two hypotheses must be stated. The first hypothesis, called the null hypothesis, proposes that there will be no significant difference between the distributions of the observed data and the expected data. The second hypothesis, called the alternative hypothesis, proposes the opposite: that there will be a significant difference between the observed data and the expected data.

Next, the significance level is determined. The significance level represents the probability that the test will conclude the null hypothesis to be false when in fact it is true. The significance level can be any value between 0 and 1. However, a value of 0.01, 0.05, or 0.1 is commonly used.

Then the chi-square (χ^2) test statistic is calculated using the following formula:

$$\chi^2 = \sum \frac{(\text{observed value} - \text{expected value})^2}{\text{expected value}}$$

The expected value for each category of the variable is equal to the product of the total sample size (n) and the expected distribution (percentage) of observations for each category. The summation symbol (Σ) indicates that the calculation must be performed for each observed value of the variable, and the results must be added together.

The next step is to determine the degrees of freedom. This is the number of possible variable values minus one. In the example of the car colors above, where there are five different possible colors (red, blue, yellow, black, and white), the degrees of freedom are four (5 − 1 = 4). Once the chi-square test statistic and the degrees of freedom have been calculated, the P value can be determined using the appropriate chi-square distribution chart. If the P value is less than the predetermined significance level, then there is strong evidence that the null hypothesis should be rejected.

The chi-square test can also be used to test for the independence of variables. This type of chi-square test, called the chi-square test of independence, is used when studying two categorical variables from a single population to see if they are related. For example, this type of test could be used to determine if gender (variable 1) is associated with the color of car purchased (variable 2).

Limitations of the Chi-Square Test

The chi-square test requires that the sample under study be categorical. It cannot compare a continuous variable with another continuous variable or with a categorical variable. In addition, the sample should be selected using simple random sampling. For the test to be statistically meaningful, the observed sample in each category must have a sample size greater than five. It is also important to select a small significance level. This ensures that confidence in the reliability of the test is high.

The Power of One

It is often necessary to compare the distribution of values for a categorical variable in a sample to a specific distribution. For example, a sample of the different products made at in a factory can be checked to ensure that all products made at the factory meet a specific distribution of product types. As it is often difficult or even impossible to check all members of a population for a desired distribution, the chi-square goodness of fit test allows a sample of the total population to be checked to determine if the desired distribution occurs. For this reason, the chi-square goodness of fit test is used in a wide variety of fields.

Maura Valentino, MSLIS

Further Reading

Diez, David M., et al. *OpenIntro Statistics*. 3rd ed., OpenIntro, 2015. *OpenIntro*, www.openintro.org/stat/textbook.php?stat_book=os. Accessed 31 May 2017.

McClave, James T., and Terry T. Sincich. *Statistics*. 13th ed., Pearson, 2017.

Reinhart, Alex. *Statistics Done Wrong: The Woefully Complete Guide*. No Starch Press, 2015.

Salkind, Neil J. *Statistics for People Who (Think They) Hate Statistics*. 6th ed., Sage Publications, 2016.

Triola, Mario F. *Essentials of Statistics*. 5th ed., Pearson, 2015.

Urdan, Timothy C. *Statistics in Plain English*. 4th ed., Routledge, 2017.

CHI-SQUARE GOODNESS OF FIT TEST SAMPLE PROBLEM

A plant is considered heterozygous if it has one dominant gene and one recessive gene for a given trait. If two heterozygous plants are bred, 75 percent of their offspring are expected to display the dominant trait. The other 25 percent are expected to display the recessive trait.

Two heterozygous pea plants that both display the dominant trait (early flowering) have 200 offspring. There are 162 early-flowering offspring (dominant gene) and 38 non-early-flowering offspring (recessive gene). Does the distribution of the dominant and recessive traits match the expected distribution? Use a significance level of 0.05 for the test.

Answer:

First, determine the null and alternative hypotheses:
- Null hypothesis: The percentage of early flowering plants will match the expected distribution of 75 percent.
- Alternate hypothesis: The percentage of early flowering plants will not match the expected distribution.

Determine the expected number of early-flowering plants and non-early-flowering plants based on the expected percentages (75 and 25 percent, respectively). For a sample of 200 plants, the expected number of early flowering plants is

$$200 \times 0.75 = 150$$

and the expected number of non-early flowering plants is

$$200 \times 0.25 = 50$$

Next, calculate the chi-square test statistic (χ^2):

$$\chi^2 = \sum \frac{(\text{observed value} - \text{expected value})^2}{\text{expected value}}$$

$$\chi^2 = \frac{(162-150)^2}{150} + \frac{(38-50)^2}{50}$$

$$\chi^2 = 0.96 + 2.88$$

$$\chi^2 = 3.84$$

Next, determine the degrees of freedom. Because there are two possible values of the variable (early flowering and non-early flowering), there is 1 degree of freedom (2 − 1 = 1). Using a chi-square distribution chart, locate the value closest to the chi-square test statistic (3.84) that corresponds to the degrees of freedom (1), then identify the associated *P* value. In this case, the closest value to the test statistic is 3.841, and the associated *P* value is 0.05. In the chi-square distribution chart, test statistic values increase as *P* values decrease, so a test statistic of slightly less than 3.841, at one degree of freedom, would produce a *P* value of slightly more than 0.05. (A calculator will give a *P* value of 0.0500044.)

Finally, determine whether or not the null hypothesis can be rejected. Because the *P* value (>0.05) is not less than the significance level (0.05), the null hypothesis is not rejected, and is therefore assumed to be true. Thus, the distribution of the sample is not significantly different from the expected distribution.

Chi-Square Test of Independence

	Herb 1	Herb 2	Placebo	Total
Number sick	25	30	35	90
Number not sick	95	100	90	285
Total	120	130	125	375

H_0: Herbs do nothing
H_1: Herbs do something

FIELDS OF STUDY

Hypothesis Testing; Statistical Analysis; Variance Analysis

ABSTRACT

The chi-square test of independence is used to analyze two categorical variables in a single data set to see if there is a significant association between them. The chi-square test of independence provides evidence for or against the null hypothesis, which proposes no relationship exists between the variables.

PRINCIPAL TERMS

- **categorical variable:** an element that can only take on nonnumeric values, such as labels or names.
- **contingency table:** an arrangement of a data set in which one categorical variable is presented in rows and another in columns.
- **degrees of freedom:** the number of observations in a data set that are free to vary based on certain statistical parameters.
- **goodness of fit:** the degree to which the observed data values are consistent with the expected distribution of the data.
- **P value:** the probability of a test result occurring by chance when assuming that the null hypothesis is true.
- **test of independence:** a statistical analysis that is used to determine whether a relationship exists between two variables.

TESTING FOR INDEPENDENCE

The chi-square test of independence is used to see if there is a significant association between two categorical variables. For example, one could determine if gender (male or female) is related to an individual's history of heart attack (yes or no). The variables are related if a change in one variable causes a significant change in the other variable.

To conduct a chi-square test of independence one must first state two hypotheses. The null hypothesis states that the variables are independent and not related. The alternative hypothesis states that the variables are related and are not independent. Then a significance level is selected. The significance level represents the probability that the test will conclude the null hypothesis to be false when in fact it is true. The significance level can be any value between 0 and 1. The most commonly used significance levels are values less than or equal to 0.05.

Next, the test of independence is conducted. First, create the contingency table. For a two-way contingency table, the rows are labeled with the values of one categorical variable, and the columns are labeled with the values of the other. The numbers in the table cells represent how many individuals in the sample fall into the categories. Then determine the value that would be expected for each cell if the null hypothesis were true. The expected value (E) for each cell is equal to the row total multiplied by the column total, then divided by the overall total:

$$E = \frac{\text{row total} \times \text{column total}}{\text{overall total}}$$

Next, calculate the chi-square (χ^2) test statistic. For each cell within the contingency table, use the following formula:

$$\chi^2 = \sum \frac{(\text{observed value} - \text{expected value})^2}{\text{expected value}}$$

The summation symbol (Σ) indicates that the test statistic is the sum of all the values calculated with this equation for each cell in the contingency table. After the test statistic has been calculated, the next step is to calculate the degrees of freedom. For the

27

chi-square test of independence, this is done by multiplying the number of rows (r) minus one by the number of columns (c) minus one:

$$\text{degrees of freedom} = (r-1) \times (c-1)$$

Next, using either a chi-square distribution chart or a *P* value calculator, determine the *P* value that corresponds to the calculated test statistic and degrees of freedom.

Once the *P* value has been determined, it is then compared to the predetermined significance level. If the *P* value is less than or equal to the significance level, then there is strong evidence that the null hypothesis should be rejected. In such cases, the alternative hypothesis can be accepted with a certain degree of confidence.

The chi-square test statistic can also be used to assess the goodness of fit. The chi-square goodness of fit test is used when studying how the distribution of the data in the observed sample compares to the expected probability distribution. This test is applied when there is one categorical variable from a population. For this test, the null hypothesis proposes that the data will be consistent with the expected distribution.

Limitations of the Test of Independence

The chi-square test of independence requires that the variables under study be categorical. The chi-square test cannot compare a continuous variable with another continuous variable or with a categorical variable. In addition, the sample should be selected using simple random sampling. For the test to be statistically meaningful, the observed sample in each category must have greater than five members.

It is also important to select a small significance level. This ensures that confidence in the reliability of the test is high. Last, the test can only assess the existence of an association between variables. It provides no information about causation.

The Power of Comparison

The need to know if a variable has an association with another variable is crucial to many scientific fields. For example, a city planner might want to know if being located under a high-tension power line causes the value of a house to fall beneath the average selling price for all homes. The chi-square test of independence can only indicate a correlation between the two variables. It cannot be used to infer a causal relationship. However, researchers often use the existence of a correlation as an indication that further research to determine causation is warranted.

Maura Valentino, MSLIS

Further Reading

Diez, David M., et al. *OpenIntro Statistics*. 3rd ed., OpenIntro, 2015. *OpenIntro*, www.openintro.org/stat/textbook.php?stat_book=os. Accessed 31 May 2017.

McClave, James T., and Terry T. Sincich. *Statistics*. 13th ed., Pearson, 2017.

Reinhart, Alex. *Statistics Done Wrong: The Woefully Complete Guide*. No Starch Press, 2015.

Salkind, Neil J. *Statistics for People Who (Think They) Hate Statistics*. 6th ed., Sage Publications, 2016.

Triola, Mario F. *Essentials of Statistics*. 5th ed., Pearson, 2014.

Urdan, Timothy C. *Statistics in Plain English*. 4th ed., Routledge, 2017.

CHI-SQUARE TEST OF INDEPENDENCE SAMPLE PROBLEM

Researchers want to test the supplement ABR to determine if it increases the chance that roses will bloom red instead of yellow. Out of 1,000 rose bushes, 566 rose bushes are given supplement ABR and 434 are not. Of the 434 untreated rose bushes, 314 bloomed red, 84 bloomed yellow, and 36 were orange. Of the 566 rose bushes that were treated, 490 bloomed red, 54 bloomed yellow, and 22 were orange.

State the null and alternative hypotheses for this study. Create a contingency table for the data. Then determine whether or not to reject the null hypothesis using a chi-square test of independence with a significance level of 0.05.

Answer:

First, state the null and alternative hypotheses.

- Null hypothesis: Supplement ABR does not significantly affect the color distribution of the roses.

- Alternative hypothesis: Supplement ABR does significantly affect the color distribution of the roses.

Create a contingency table using one variable for the columns and the other for the rows. Find the total for all columns and all rows.

	Red	Yellow	Orange	Row Total
Treated	490	54	22	566
Untreated	314	84	36	434
Column Total	804	138	58	1,000

Next, calculate the expected count (E) for the values in each cell of the contingency table (treated/red, treated/yellow, treated/orange, untreated/red, untreated/yellow, and untreated/orange):

$$E = \frac{\text{row total} \times \text{column total}}{\text{overall total}}$$

$$E_{tr} = \frac{566 \times 804}{1000} = 455.06$$

$$E_{ty} = \frac{566 \times 138}{1000} = 78.11$$

$$E_{to} = \frac{566 \times 58}{1000} = 32.83$$

$$E_{ur} = \frac{434 \times 804}{1000} = 348.94$$

$$E_{uy} = \frac{434 \times 138}{1000} = 59.89$$

$$E_{uo} = \frac{434 \times 58}{1000} = 25.17$$

Next, determine the appropriate degrees of freedom. The treatment has two levels (treated and untreated) and the color has three levels (red, yellow, and orange). Thus, the following formula can be used to calculate the appropriate degrees of freedom:

$$\text{degrees of freedom} = (r-1) \times (c-1)$$

$$df = (2-1) \times (3-1)$$

$$df = 2$$

Next, use a chi-square distribution chart or a P value calculator to determine the appropriate P value for the chi-square test statistic and the degrees of freedom. In this case, the P value is equal to 0.00000014.

Finally, determine whether or not the null hypothesis should be rejected. Since the P value (0.00000014) is less than or equal to the significance level (.05), there is strong evidence that the null hypothesis should be rejected. Therefore, the researchers can assume that supplement ABR does significantly affect the color of the roses.

Then calculate the test statistic (χ^2):

$$\chi^2 = \sum \frac{(\text{observed value} - \text{expected value})^2}{\text{expected value}}$$

$$\chi^2 = \frac{(490-455.06)^2}{455.06} + \frac{(54-78.11)^2}{78.11} + \frac{(22-32.83)^2}{32.83} + \frac{(314-348.94)^2}{348.94} + \frac{(84-59.89)^2}{59.89} + \frac{(36-25.17)^2}{25.17}$$

$$\chi^2 = 31.5542$$

Cluster Sampling

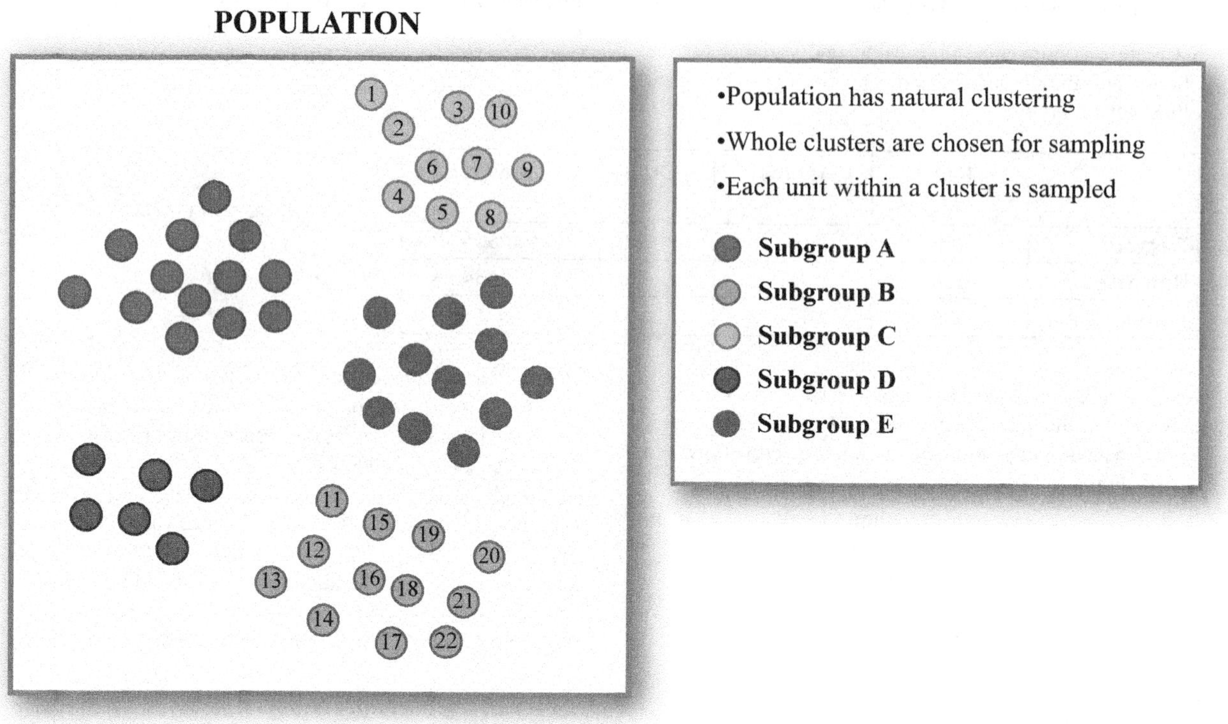

POPULATION

- Population has natural clustering
- Whole clusters are chosen for sampling
- Each unit within a cluster is sampled

● Subgroup A
● Subgroup B
● Subgroup C
● Subgroup D
● Subgroup E

FIELDS OF STUDY

Sampling Design; Sampling Techniques

ABSTRACT

Cluster sampling is a probabilistic sampling technique in which the population of interest is first divided into comprehensive and mutually exclusive subgroups, or clusters, and then a sample of clusters are selected at random for study. This technique is most often used when dealing with a very large or geographically dispersed population or when the population is already concentrated in natural or preexisting clusters. For added randomization, a research study may use two-stage cluster sampling, in which another probabilistic sampling technique is used to select a smaller sample from the randomly chosen clusters.

PRINCIPAL TERMS

- **simple random sampling:** a probabilistic sampling method in which members of the sample group are drawn from the population at random so that each member of the population has an equal chance of being included.
- **stratified random sampling:** a probabilistic sampling method in which the population to be studied is divided into mutually exclusive subgroups or strata, and then random samples are taken from each stratum in numbers proportional to that stratum's share of the larger population.
- **systematic random sampling:** a probabilistic sampling method in which members of the sample group are selected from a sampling frame at regular intervals, beginning with a randomly selected starting point.

OVERVIEW OF CLUSTER SAMPLING

Most research studies use some type of random sampling, also called probabilistic sampling. A probabilistic sampling technique is one that relies on an element of chance. In probabilistic sampling, every member of the target population has a chance of being selected for a sample. The odds of being selected can be calculated for any given individual. The most basic form of probabilistic sampling is simple random sampling. In this method, individuals are simply selected from the target population at random. Every member has an equal chance of being selected.

However, simple random sampling is not always the best sampling method. This may be because of the nature of the study or because of time or cost constraints on the researchers. One alternative probabilistic sampling method is cluster sampling. In cluster sampling, the target population is first divided into subgroups, or clusters, based on preexisting categories within the population. These categories must be mutually exclusive, and they should cover every member of the population. The most common basis for cluster sampling is geographical area. For example, cities might be used as clusters. However, other types of categories may also be used, as long as every member of the population falls into one, and only one, of the clusters.

Once the clusters have been established, the researcher selects a predetermined number of clusters using simple random sampling. The next step depends on whether the researcher plans to use one-stage or two-stage cluster sampling. In one-stage cluster sampling, all members of the selected clusters are included in the sample. In two-stage cluster sampling, a subset of individuals is chosen from each selected cluster using either simple random sampling or systematic random sampling. These randomly chosen subsets form the sample.

Benefits and Drawbacks

Cluster sampling is usually used when other types of random sampling would take too long or cost too much. For example, if the target population is dispersed over a large geographical area, randomly selected individuals might live very far from each other. If the research must be conducted in person, the researcher might not have the time or the money necessary to visit every subject. Geographical cluster sampling would significantly reduce how much the researcher must travel. Another reason to use cluster sampling is if the target population is naturally concentrated into clusters.

The biggest disadvantage of cluster sampling is that it has a higher level of sampling error than other probabilistic sampling methods. One reason for this is that the demographic makeup of the selected clusters may not accurately reflect the diversity of the overall population. In addition, if a certain characteristic happens to be over- or underrepresented in one of the clusters, the researcher may then over- or underestimate its prevalence in the population. However, there are ways to reduce this error. Cluster sampling is most accurate when the clusters are all about the same size and demographically diverse in approximately the same way. Another way to offset sampling error is to increase the sample size. This is often easier and more cost-effective to do with cluster sampling than with other probabilistic sampling methods.

Cluster sampling is sometimes confused with stratified random sampling. In the latter technique, the population to be sampled is divided into groups based on characteristics that the researchers want represented in the study. Individuals are then randomly selected from every group, or stratum, in numbers proportionate to the makeup of the overall population. The main difference is that in cluster sampling, only some of the clusters are sampled, whereas in stratified random sampling, samples are taken from every group. Stratified random sampling works best when the strata are internally homogenous. Cluster sampling works best when the clusters are internally heterogeneous.

Cluster Sampling in Action

An example of a cluster sampling study might be as follows:

A multinational company with multiple office locations worldwide wants to conduct a survey of its employees. However, the company employs tens of thousands of people, and the surveyors do not have time to gather and analyze responses from every single employee. Instead, the surveyors divide the employees into clusters based on work location. They use simple random sampling to select several office locations at random. Depending on how many people work at each location, they may then send the survey to every employee at the selected locations, or they may decide to choose a smaller random sample from each of the selected locations.

The Power of Cluster Sampling

Cluster sampling is an important method for surveying very large or geographically dispersed populations. While it is more subject to sampling error than other probabilistic sampling methods, it also enables researchers to study larger samples with minimal added time or effort. In addition, a number of statistical methods can be used to compensate for sampling error in analysis. If used carefully and with close attention to its limitations, cluster sampling is a powerful tool in the design of research studies.

Trudy Mercadal, PhD

FURTHER READING

Ahmed, Saifuddin. "Methods in Sample Surveys: Cluster Sampling." *Statistical Methods for Sample Surveys*, John Hopkins U, 2009, ocw.jhsph.edu/courses/StatMethodsForSampleSurveys/PDFs/Lecture5.pdf. Accessed 13 Apr. 2017.

Blair, Edward, and Johnny Blair. *Applied Survey Sampling*. Sage Publications, 2015.

Fisher-Giorlando, Marianne. "Sampling in a Suitcase: Multistage Cluster Sampling Made Easy." *Teaching Sociology*, vol. 20, no. 4, 1992, pp. 285–87. *Sociology Source Ultimate*, search.ebscohost.com/login.aspx?direct=true&db=sxi&AN=12826929&site=ehost-live. Accessed 13 Apr. 2017.

Rose, Angela M. C., et al. "A Comparison of Cluster and Systematic Sampling Methods for Measuring Crude Mortality." *Bulletin of the World Health Organization*, vol. 84, no. 4, 2006, pp. 290–96. *Academic Search Complete*, search.ebscohost.com/login.aspx?direct=true&db=a9h&AN=21276056&site=ehost-live. Accessed 13 Apr. 2017.

Valiant, R., Dever, J.A., and Kreuter, F. *Practical Tools for Designing and Weighting Survey Samples*. Springer, 2013.

"What Is Cluster Sampling?" *Stat Trek*, stattrek.com/survey-research/cluster-sampling.aspx. Accessed 13 Apr. 2017.

CLUSTER SAMPLING SAMPLE PROBLEM

A researcher is studying how commute distance influences the number of electric cars purchased in urban areas in the European Union (EU). How might the researcher use cluster sampling to select a sample for this study?

Answer:

First, the researcher compiles a list of all cities in the EU. They may use a minimum population cutoff to decide which cities to include. The population of each city forms a cluster. Next, the researcher uses random sampling to select an appropriate number of cities. The researcher can then conduct their study by determining how many electric cars were purchased in each city and interviewing the owners of the cars.

Alternatively, the researcher could use two-stage cluster sampling. In this method, the researcher might consider each country in the EU to be a cluster. They would select an appropriate number of countries using random sampling, then list all of the cities in the selected countries. Next, the researcher would use random sampling to select an appropriate number of cities from each country.

COMPARATIVE RESEARCH

	Country 1	Country 2	Country 3
Factor 1	⬭	⬭	⬭
Factor 2	▲	▼	▼

FIELDS OF STUDY

Research Design; Hypothesis Testing

ABSTRACT

At its most basic level, comparative research is that which seeks to identify similarities and differences between two or more things. The comparison may be between individual entities, entire populations, or abstract concepts. Comparative research is particularly important in the social sciences. Such studies often involve performing the same experiment in different cultural contexts and examining the results to compare and contrast.

PRINCIPAL TERMS

- **animal models:** animals used in medical research because their anatomy, physiology, or response to disease may provide insight into human health.
- **cross-cultural study:** research that investigates a phenomenon across multiple cultural contexts rather than being confined to one setting.
- **generalization:** the act of applying a characteristic of a sample group to a larger population, or the result of such an action.
- **in vivo testing:** experimental research performed on entire, living organisms rather than on isolated or dead tissue.

OVERVIEW OF COMPARATIVE RESEARCH

Comparisons are used in many aspects of almost every type of research. For example, in a pretest-posttest design, measurements of the experimental group are compared with those of the control group to see if the condition being tested had any effect. This type of comparison is a task performed as part of a broader research design.

Comparative research can also act as a research design of its own. This design is most often used in the social sciences in order to better understand a phenomenon by observing it in different contexts. It is highly useful as a way to combine both quantitative and qualitative methods of research.

Comparative research can serve a number of functions. For example, a researcher might develop a survey to assess attitudes toward aging and then compare individual responses. The researcher might then be able to make a generalization about the attitudes of those surveyed within a certain country and apply it to the country's entire population. Next, a cross-cultural study could reveal similarities and differences among the attitudes of people in different countries, potentially leading to further generalizations or other conclusions. The results can also be compared to assess the effectiveness of the survey itself. People of different cultures might perceive questions differently or encounter other challenges that could affect the research.

APPLES TO APPLES

Perhaps the most important principle of comparative research is that one must compare like entities. A simple example would be trying to compare one group's average height in inches with another's weight in pounds. Not only are the two measurements using different units, but also measuring completely different attributes, so the comparison is not meaningful. An improvement would be to compare group A's average height in inches with group B's average height in centimeters. Here, the attribute is the same for both groups, but the units of measurement must the same to make the groups truly comparable. In this case, a simple conversion of inches to centimeters, or vice versa, makes the comparison possible.

Often, however, comparison problems are more complex and harder to detect. For example, a researcher might try to compare two studies of obesity that were conducted in two countries. Study A finds that, in its country, the rate of obesity is 35 percent, while study B finds that, in its country, the rate is 65 percent. This sizable difference would invite much speculation and further research into its cause. It could even affect health policies. However, looking closer reveals that the two studies defined obesity differently. Study A viewed a person as obese if they were fifty pounds or more overweight, while study B made that determination if a person was twenty-five pounds or more overweight. Therefore, a direct comparison is invalid, and the immediate impression that one country is far less obese must be reevaluated.

This type of variation between studies is common and is a major pitfall that comparative researchers must take care to avoid. For this reason, comparative research is often limited in scope and large comparative studies can be quite complex. Even when data about two different populations appear to align, there may be cultural elements or other factors that are difficult to identify but preclude direct comparison.

Lab Rats

Comparative research is best known as a method of comparing human populations, especially across cultures or states. However, it also has important applications in other fields, such as medical research. Some research, especially on diseases and treatments, require experiments that may be unethical to perform on human beings. Researchers may use animal models during the early stages of this research, so that if adverse consequences occur, no humans are harmed. Doing this allows researchers to move from basic testing of chemical processes into more advanced and informative in vivo testing. If the results are favorable, testing may move on to human subjects. Further comparative research will then determine whether the human subjects experience the benefits seen in the animal models.

Quantitative versus Qualitative

Much comparative research focuses on quantitative rather than qualitative data. This is because it tends to be more straightforward to compare numbers than narrative information obtained from interviews or observation notes. However, qualitative data can be and is used in comparative designs, most notably in comparative survey research. Such cases often require the researcher to process the data further to make it more directly comparable.

Such processing can be accomplished in a variety of ways, one of which is coding. Coding requires the researcher to go through the qualitative data and, based on its characteristics, describe it numerically or using categories. For example, an interview transcript that mentions trauma five times might be classified as "severe," while a transcript that mentions it only once could be classified as "mild." This must be done carefully, lest the researcher introduce their own bias into the study by interpreting the data in a way that makes it more interesting or relevant to the research. Organizing data into generalized categories makes comparison easier and can also help the researcher better identify patterns and trends.

—*Scott Zimmer, JD*

Further Reading

Babones, Salvatore J. "Modeling Error in Quantitative Macro-Comparative Research." *Journal of World-Systems Research*, vol. 15, no. 1, 2015, pp. 86–114.

Engbers, Trent A. "Comparative Research: An Approach to Teaching Research Methods in Political Science and Public Administration." *Teaching Public Administration*, vol. 34, no. 3, 2016, pp. 270–83.

Hantrais, Linda. *International Comparative Research: Theory, Methods, and Practice*. Palgrave Macmillan, 2009.

Kelly, Peter. "Intercultural Comparative Research: Rethinking Insider and Outsider Perspectives." *Oxford Review of Education*, vol. 40, no. 2, 2013, pp. 246–65.

Riosmena, Fernando. "The Potential and Limitations of Cross-Context Comparative Research on Migration." *Annals of the American Academy of Political and Social Science*, vol. 666, no. 1, 2016, pp. 28–45.

Wang, Georgette, and Yi-Hui Christine Huang. "Contextuality, Commensurability, and Comparability in Comparative Research: Learning from Chinese Relationship Research." *Cross-Cultural Research*, vol. 50, no. 2, 2016, pp. 154–77.

Completely Randomized Designs

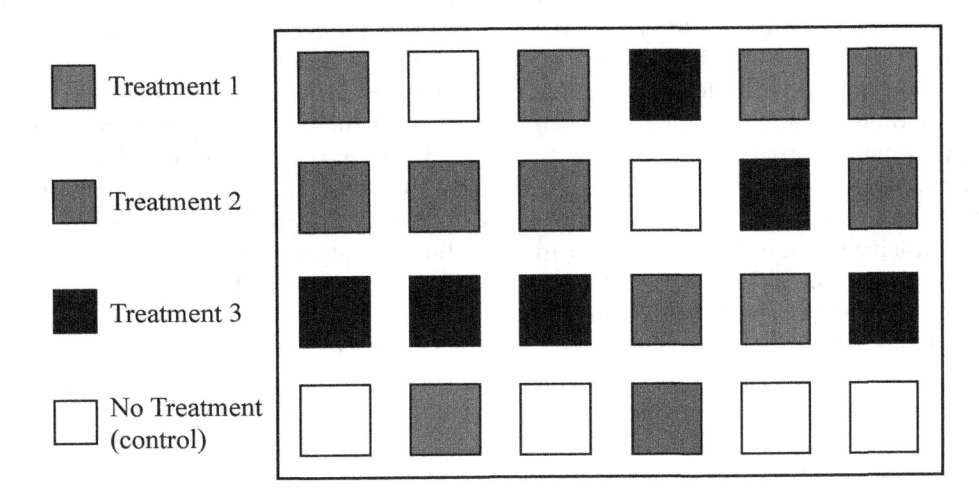

FIELDS OF STUDY

Research Design; Sampling Design

ABSTRACT

Experiments are designed to identify causes and effects. To do this, they must be constructed in a way that ensures that the effects being observed are being caused by the treatment that is the subject of the experiment and not by unrelated factors. Completely randomized designs are one way of minimizing the impact of external factors, including bias, on the outcome of experiments.

PRINCIPAL TERMS

- **control group:** the subjects in an experiment who do not receive any intervention or who receive a placebo or sham intervention.
- **experimental unit:** the smallest unit in an experiment to which a treatment can be applied independently of other units; may be an individual subject or a group of subjects, depending on the experiment.
- **randomization:** the use of methods that are based on chance to assign subjects to the experimental or control groups.
- **treatment:** the process or intervention applied or administered to members of an experimental group.

Goals of Randomization

The goal of most experimental research is to gain information about a small group that can be generalized to a larger group, typically an entire population. Researchers strive to make sure that characteristics that might influence the experiment are distributed among the group the same way they are distributed among the overall population. This helps ensure that the group participating in the experiment accurately represents the larger population.

Randomization is an effective way of making sure that factors external to an experiment are distributed among participants in the same way that they would be in the real world. One common method of randomization is the random assignment of each experimental unit to either the experimental group or the control group. Those in the experimental group receive the treatment while those in the control group receive either no treatment or a placebo. In a completely randomized design, each experimental unit has the same chance of receiving the treatment. Experimental units can be randomly assigned to groups using a random number table or computer program. Researchers can also physically draw numbers.

When to Use Completely Randomized Designs

In a completely randomized design, the researcher works under the assumption that the experimental units involved are fundamentally homogeneous. This means that there should be no known variation between the units. Therefore, this design method is most effective in laboratory experiments rather than field experiments. In a lab setting, the researcher has more control over homogeneity. Otherwise, if a variation is known or environmental factors do exist, a design such as randomized block design might be preferable. This method would ensure that any such factor would be accounted for in the study results.

If used successfully, completely randomized design helps reduce biases that could skew the study's results. It is a simpler and more flexible design in which more than one treatment level can be tested. However, this type of randomization is best suited to smaller experiments. It allows researchers to make a more confident inference about the causal relationship between the treatment and the outcome.

UNINTENDED CONSEQUENCES

In a large population, any inborn characteristic, such as having O negative blood type, occurs randomly at a certain frequency. Thus, if researchers happened to put all of the experimental units with O negative blood in their experimental group, the group would no longer be representative of the population. To avoid this outcome, each person is randomly assigned to belong to either the control group or the experimental group. This could be done by flipping a coin, rolling dice, or any other means of producing a random outcome. If the randomization is effective, the researchers can more confidently expect the same percentage of each group to have O negative blood.

Randomization is important for avoiding unintended and potentially catastrophic consequences. In the example above, it might happen that researchers discover a vaccine that is effective for those with O negative blood but ineffective for all other blood types. Many people could suffer or even die if this was not discovered during the research study and only came to light afterward.

RELIABILITY AND GENERALIZABILITY

Randomization makes it much more likely that the results achieved from a research study will be both reliable and generalizable. Research results are considered reliable if they are accurate representations of what occurred during the course of the research and can be reproduced under similar conditions.

Generalizability is a measure of how accurately the outcome of the experiment reflects what can be expected to occur within the overall population. If 90 percent of those in the experimental group saw an average increase of seven points in their test scores, one would expect similar numbers in the overall population. This would be extremely difficult to achieve if the researchers simply selected participants and assigned them to one group or another based on their whim. It would particularly be an issue if they were drawn from a pool of people being treated for diabetes or hypoglycemia, which are conditions in which people can have very strong negative reactions to ingesting sugars. A far better approach would be to draw participants from a pool that includes a variety of people. Randomization should then also be used to determine whether or not they receive the experimental treatment. Doing this helps avoid intentional, accidental, or subconscious biases of the researchers exerting an influence on the results.

Scott Zimmer, JD

FURTHER READING

Alferes, Valentim R. *Methods of Randomization in Experimental Design*. Sage Publications, 2012.

Chow, Shein-Chung, and Jen-Pei Liu. *Design and Analysis of Clinical Trials: Concepts and Methodologies*. 3rd ed., John Wiley & Sons, 2014.

Christensen, Larry B., et al. *Research Methods, Design, and Analysis*. 12th ed., Pearson, 2014.

Connelly, Lynne M. "Understanding Research: Randomized Controlled Trials." *MEDSURG Nursing*, vol. 25, no. 4, 2016, pp. 281–82. *Academic Search Complete*, search.ebscohost.com/login.aspx?direct=true&db=a9h&AN=117499589&site=ehost-live. Accessed 30 Apr. 2017.

Creswell, John W. *Research Design: Qualitative, Quantitative, and Mixed Methods Approaches*. 4th ed., Sage Publications, 2014.

Maxwell, Scott E. "Completely Randomized Design." *Encyclopedia of Statistics in Behavioral Science*, edited by Brian S. Everitt and David C. Howell, vol. 1, John Wiley & Sons, 2005, pp. 340–41.

Salkind, Neil J., editor. *Encyclopedia of Research Design*. Sage Publications, 2010.

Zhou, Kefei, and Jitendra Ganju. "Inference from Blinded Data in Randomized Clinical Trials." *Clinical Trials*, vol. 10, no. 5, 2013, pp. 744–53. *MEDLINE Complete*, search.ebscohost.com/login.aspx?direct=true&db=mdc&AN=24130201&site=ehost-live. Accessed 30 Apr. 2017.

> **COMPLETELY RANDOMIZED DESIGNS SAMPLE PROBLEM**
>
> Assume that a researcher wishes to study whether the consumption of sugar prior to taking a test influences performance on the test. The research question is, "Does consuming a gram of sugar before taking a test cause one to score higher on the test?" Participants are randomly assigned to either the experimental group—the members of which consume the defined amount of sugar—or the control group, the members of which consume a sugar substitute. What are the dependent and independent variables? What factor is randomized?
>
> **Answer:**
> In this experiment, the independent variable is sugar consumption, and the dependent variable is the score achieved on the test. The factor that is randomized is the assignment of individuals to experimental or control groups. This assignment could be determined by flipping a coin for each subject.

COPERNICUS, NICOLAUS

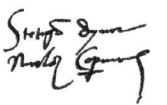

Nicolaus Copernicus' letter in German to Duke Albrecht of Prussia, dated 25 June 1541. Reproduction of the manuscript, as taken from w:Leopold Prowe's (1821-1887) Nicolaus Coppernicus, 1884, Tafel III. [Public domain], via Wikimedia Commons

POLISH ASTRONOMER

Copernicus dismissed the Ptolemaic model of the universe and introduced the theory that the planets, including Earth, revolve around the sun. He defended the rights of the educated to discuss scientific theories, even when those theories contradicted currently accepted beliefs and religious dogma.

- **Born:** February 19, 1473; Thorn, Prussia (now Torun, Poland)
- **Died:** May 24, 1543; Frauenburg, East Prussia (now Frombork, Poland)
- **Also known as:** Mikolaj Kopernik
- **Primary field:** Astronomy
- **Specialty:** Observational astronomy

EARLY LIFE

Nicolaus Copernicus was born Mikolaj Kopernik into a wealthy merchant family in what is now Torun, Poland, in 1473. His father, also named Mikolaj (Nicolaus) Kopernik, was an immigrant from Kraków who married a daughter of a prominent burgher (merchant class) family, Barbara Watzenrode, and like other Thorn merchants, prospered from the exchange of German goods for the wheat, cattle, and other produce of Poland.

Had the elder Kopernik not died in 1483, his sons, Andreas and Nicolaus, would probably have entered into careers in commerce. Their guardianship, however, fell to their maternal uncle, Bishop Lucas Watzenrode of Varmia (Warmia), who was best able to provide for them through a future in church administration. A university education being indispensable to holding church offices, Bishop Lucas sent the boys to study first in Kraków, then in Italy. Nicolaus not only became a master of mathematics and astronomy but also acquired knowledge of medicine, painting, and Greek.

On his return to Prussia in 1503, Nicolaus followed the contemporary practice of Latinizing his name, from Kopernik to Copernicus, and became one of the canons in the Varmia cathedral chapter. As his uncle's physician, assistant, and heir apparent, Copernicus was present during inspection tours, provincial diets (assemblies), and royal audiences. For several years, he managed the diocese efficiently but without enthusiasm. Eventually, Copernicus announced that his interests in astronomy were greater than his ambition to become a bishop.

Life's Work

Copernicus was a humanist whose closest friends and associates were poets and polemicists. His translation of an ancient author Theophilactus Symocatta from Greek into Latin was the first such publication in the Kingdom of Poland, and he dedicated the work to his humanistically trained uncle, Bishop Lucas. Copernicus later used humanist arguments to defend his astronomical theories.

Copernicus was also a bureaucrat whose busy life made it difficult for him to make the observations of the heavens on which his mathematical calculations were based. At one time or another, he was a medical doctor, an astrologer, a cartographer, an administrator of church-owned lands, a diplomat, a garrison commander in wartime, an economic theorist, an adviser to the Prussian diet, and a guardian to numerous nieces and nephews.

In about 1507, Copernicus began work on a theory concerning planetary movements. Contrary to the long-accepted Ptolemaic theory stating that the Earth is the center of the universe, Copernicus believed that the sun was the center of the universe, and that all of the planets, including Earth, revolved around it. His first description of his theory, a pamphlet called the "Commentariolus" (1514), circulated among his friends for many years. Eventually, it came into the hands of Cardinal Nikolaus von Schönberg, who wrote a letter asking Copernicus to publish a fuller account. This letter was ultimately published in *De revolutionibus orbium coelestium* (1543; *On the Revolutions of the Heavenly Spheres*, 1952) as a proof that high officials in the papal curia approved of scholars discussing the existence of a solar system. Copernicus made no answer, however. Instead, he asked his bishop to assign him light duties at a parish center, where he could make his observations and concentrate on mathematical calculations.

For several years, his work was interrupted by war. In 1520, the last grandmaster of the Teutonic Order, Albrecht of Brandenburg, made a final effort to reestablish his religious order as ruler of Prussia. Copernicus led the defense of Allenstein (Olsztyn) and participated in the peace negotiations. In 1525, Albrecht, defeated at every turn, secularized the Teutonic Order in Prussia and became a Protestant vassal of the king of Poland. Albrecht later called on Copernicus's services as physician, and in 1551, he published a volume of Copernicus's astrological observations.

In 1539, a Lutheran mathematician at Wittenberg, Rheticus, made a special journey to Frauenburg to visit Copernicus. Finding him ill and without prospect of publishing the manuscript he had completed at great labor, Rheticus extended his stay to three months so that he could personally copy the manuscript. He then arranged for the publication of *Narratio prima de libris revolutionum* (1540; *The First Account*, 1939) in Danzig and of the mathematical section in Nuremberg in 1542. Unable to supervise the printing of the theoretical section personally, Rheticus gave that task to another scholar, Andreas Osiander of Wittenberg.

Osiander was at a loss as to how to proceed. He saw that Copernicus had not been able to prove his case mathematically. Indeed, it would have been difficult for him to do so without inventing calculus (which was later created by Gottfried Wilhelm Leibniz and Sir Isaac Newton independently of each other for the very purpose of calculating the elliptical orbits of the planets). Consequently, Copernicus had defended his ideas by demonstrating that Ptolemy's was not the only ancient theory describing the universe; indeed, there were ancient philosophers who believed that the sun was the center of the solar system. Moreover, he had argued that free inquiry into science was as necessary as the freedom to write literature or produce fine art. In this respect, Copernicus was presenting his case to Renaissance humanists, especially to the well-educated pope to whom he dedicated his book, as a test of free thought.

Osiander, who perceived that the Catholic world was hostile to all innovations and was equally well aware of the debates raging in the Protestant world over biblical inerrancy, saw that Copernicus was treading on dangerous ground by suggesting an alternate view of the universe than the one presented in the Bible. Consequently, there was a real danger that the theory would be rejected entirely without having been read. To minimize that possibility, he wrote an unauthorized introduction that readers assumed was by Copernicus. This stated that the solar system was merely a hypothesis, a way of seeing the universe that avoided some of the problems of the Ptolemaic system. This led to much confusion and angered Copernicus considerably when he saw the page proofs. Copernicus, however, was too weak and ill to do anything about it. He died shortly after his book was published in 1543.

Impact

Copernicus's theory was not immediately accepted, and not because of the controversies of the Reformation alone, although they made it dangerous for any scientist to suggest that the biblical descriptions of the heavens were incorrect. Copernicus's idealistic belief that God would create only perfectly circular planetary orbits made it impossible for him to prove his assertions mathematically. Nevertheless, Copernicus's theory was the only one to offer astronomers a way out of a Ptolemaic system of interlocking rings, which was becoming impossibly complex. His insights undermined the intellectual pretensions of astrology and set astronomy on a firm foundation of observation and mathematics.

Although Copernicus's defense of the freedom of inquiry was less important in the struggle against religious dogmatism than later demonstrations of the existence of the solar system, Copernicus became a symbol of the isolated and despised scientist who triumphs over all efforts by religious fundamentalists to silence him.

—*William Urban*

Further Reading

Barrett, Peter. *Science and Theology Since Copernicus: The Search for Understanding.* London: Continuum International, 2004. Print. Traces the legacy of Copernicus over four hundred years. Examines the history of the debate between science and Christianity, attempting to fashion a philosophical basis for the simultaneous embrace of scientific method and religious faith in the modern world.

Gingerich, Owen. *The Book Nobody Read: Chasing the Revolutions of Nicolaus Copernicus.* New York: Walker, 2004. Print. An examination of every copy of the original printing of Copernicus's *De revolutionibus* in existence, demonstrating who read the work, what they thought of it, and how exactly Copernicus's ideas spread throughout Europe.

Henry, John. *Moving Heaven and Earth: Copernicus and the Solar System.* Cambridge: Icon, 2001. Print. Argues that Copernicus's discovery had revolutionary effects for the cultural status afforded to theoretical science and mathematics in Western culture.

ON THE REVOLUTIONS: COPERNICAN AND THE CELESTIAL SPHERES

Nicolaus Copernicus is best known for ousting Earth from its position (in the minds of scholars) at the center of the universe. Before Copernicus's theories, it was commonly accepted that Earth was stationary while all the stars and planets, including the sun, revolved around it.

While in Italy studying law, Copernicus was encouraged by a mathematics professor to pursue astronomy. His heliocentric ("sun-centered") concept of the universe probably came about during his education, when he noticed inconsistencies between astronomical observations and the two most popular cosmologies of his day (Aristotelian and Ptolemaic, two versions of the ancient geocentric or Earth-centered model of the universe).

His ideas were first codified in a manuscript called the "Commentariolus" (Little commentary), written sometime before 1514. Though he showed his work to some of his closest friends and colleagues, this brief outline of his heliocentric ideas was never intended for publication. Nevertheless, it represented a statement of intent on Copernicus's part to pursue this line of thinking, and functions as a prelude to his masterwork, De Revolutionibus Orbium Coelestium (1543; On the Revolutions of the Celestial Spheres, 1952). The bulk of the work on his heliocentric theory was probably finished some twenty years after "Commentariolus." Rumors began to circulate about Copernicus's big and potentially controversial theory. But he did not publish his complete, formalized work for another decade.

In 1543, after much urging from his apprentice Rheticus, Copernicus finally published On the Revolutions. Through the six books of On the Revolutions, Copernicus laid out a new cosmology, distinct from Aristotle's and Ptolemy's, in which the motions of the stars, planets, sun, and moon are all explained by placing the sun at the center of the universe. Earth, he argued, rotates both around the sun and on its own axis, which accounts for the apparent motion of the various objects in the heavens.

Copernicus did not live to see the impact of his work; he died the very year it was published. On the Revolutions is often credited with founding modern astronomy, and the fallout from its publication has become known as the Copernican Revolution. Some scholars even credit this book with kicking off the entire scientific revolution of the sixteenth and seventeenth centuries.

CORRELATION MODELING

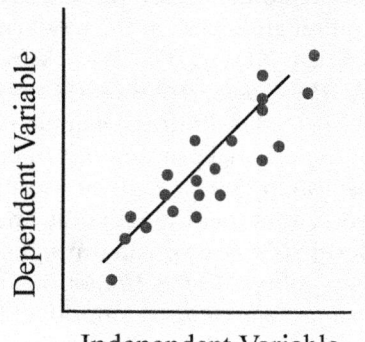

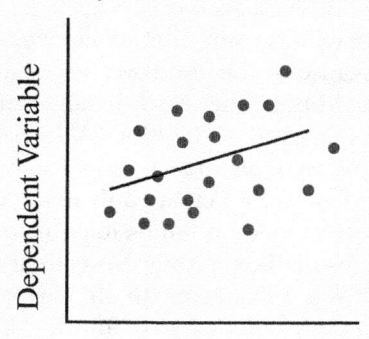

FIELDS OF STUDY

Statistical Analysis; Variance Analysis

ABSTRACT

Correlation modeling is used to analyze whether an increase in one variable is accompanied by a corresponding change in another variable. Correlation modeling is also used to determine whether the change in the second variable is small or large. Correlation modeling cannot be used to determine causation. However, it is useful for identifying trends in the data and relationships between variables.

PRINCIPAL TERMS

- **causation:** a relationship between two events, actions, or other phenomena in which one event is determined to have directly caused or otherwise affected the other.
- **hidden variables:** unidentified factors that have influenced other variables and that can cause correlations between other variables.
- **trend line:** a line drawn approximately through the center of the data values on a two-axis graph to show correlation.
- **variable:** in research or mathematics, an element that may take on different values under different conditions.

WHAT IS CORRELATION MODELING?

Correlation modeling is used to analyze the associations, called correlations, between two variables. Correlation modeling is used to examine whether a change in the value of one variable affects the value of another variable. If it does, a correlation between the two variables is said to exist. A correlation between two variables can be modeled using a two-axis graph. Each axis of the graph represents the value of one of the variables being analyzed. Each data point collected can be located on the graph to create a scatter plot. Once all of the available data points have been graphed, a line with the minimal distance from the data points can be drawn through the center of the data points. This line is called the trend line or line of best fit. It illustrates the general tendency of the data.

The orientation of the trend line demonstrates the correlation between two variables. If the trend line is horizontal, then no correlation exists. If a change in one value is associated with a small change in the other value, then the correlation is considered weak. In this case, the trend line will only have a slight slope. If a change in one value is associated with a large change in the other value then the trend line will be steeper. This indicates a strong correlation.

If the left-most endpoint of the trend line is above the right-most endpoint, then the correlation is negative. Negative correlation occurs when an increase in the value of one variable is associated with a decrease in the value of the other variable. If the left-most endpoint of the trend line is below the right-most endpoint, then the correlation is positive. Positive correlation exists when an increase in the value of one variable is associated with an increase in the value of the other variable.

The relative strength and direction of a correlation can also be quantified mathematically by calculating the correlation coefficient (r). The correlation coefficient can range in value from -1 to 1. A correlation coefficient value close to 1 indicates a strong positive correlation. A correlation coefficient value close to -1 indicates a strong negative correlation. A correlation coefficient value of 0 indicates no correlation.

The coefficient of determination (r^2), pronounced R-squared, can be calculated from the correlation coefficient. The R-squared value is greater than 0 but less than 1. It represents the percentage of the data that are close to the trend line. From this value, researchers can determine how much the variation of one variable is caused by a change in the other variable and how much of the variation is unexplained.

LIMITATIONS OF CORRELATION MODELING

It is important to note that correlation does not necessarily imply causation. Correlation simply implies that there is an association between the values of two variables. It does not mean that the change in the value of one variable is causing the change in the other variable. There may be hidden variables that the researchers have not identified that are affecting the changes in both variables.

For example, a researcher may hypothesize that children who live close to the ocean are more likely to know how to swim than children who live far from the ocean. A survey of a population of children could be conducted to collect information on how far from the ocean each child lives and their ability to swim. The data could then be analyzed using correlation modeling to determine if a correlation exists between the two variables. However, a strong correlation alone would be insufficient to determine whether a child's proximity to the ocean is the cause of the ability to swim. An unidentified factor, such as family wealth, could be influencing both variables. If coastal properties are expensive and only wealthy people live on the coast, then children who live near the ocean may know how to swim because their families can afford lessons. Correlation modeling is useful for determining whether further research into a relationship between two variables is warranted. However, correlation alone cannot be used to draw definitive conclusions about causation.

THE VALUE OF CORRELATION MODELING

A large portion of research studies are concerned with determining the cause of a certain phenomenon. Correlation modeling allows researchers to determine whether a relationship exists between two variables. It helps researchers visualize the relative strength of any association between two variables. While valuable in demonstrating the existence and strength of a correlation, this information cannot be used rule out the influence of other, unknown variables. Thus, correlation modeling alone is insufficient to prove that the change in one variable was caused by the change in the first. However, it can indicate whether further research is worth pursuing.

Maura Valentino

FURTHER READING

Caldwell, Sally. *Statistics Unplugged*. Wadsworth Cengage Learning, 2012.

Diez, David, M. and Christopher D. Barr. *OpenIntro Statistics Third Edition*. OpenIntro Publisher, 2015.

McClave, James, and Terry Sincich. *Statistics*. 12th ed., Pearson, 2016.

Reinhart, Alex. *Statistics Done Wrong: The Woefully Complete Guide.* No Starch Press, 2015.

Salkind, Neil J. *Statistics for People Who Think They Hate Statistics.* Los Angeles, Sage. 2016.

Triola, Mario, F. *Essentials of Statistics.* 5th ed., Pearson, 2014.

Urdan, Timothy, C. *Statistics in Plain English.* 4th ed., Routledge, 2017.

CORRELATION MODELING SAMPLE PROBLEM

Researchers hypothesize that the more time a person spends sitting will cause an increase in the person's weight. The researchers decide to collect data and use correlation modeling to determine whether an association exists between these two variables. The researchers know that they cannot infer causation from correlation. However, they intend to use the relative strength of correlation to determine whether further research into the question is warranted. If the correlation is strong, then additional research will be taken to determine if causation exists.

After all of the data are collected, the researchers determine that the correlation coefficient between the two variables is .78. Based on this information, determine whether the correlation between the variables is strong or weak and positive or negative. Calculate the coefficient of determination from the correlation coefficient. Should the researchers continue their research into the question of causation based on this information?

Answer:

The correlation coefficient can range in value from −1 to 1. A value closer to 1 indicates a strong positive correlation. A value of .78 indicates a strong positive correlation. The researchers calculate the coefficient of determination by finding the square of .78, which is equal to .60. This value indicates that 60 percent of the variation in a person's weight can be explained by time spent sitting. The other 40 percent of the variation in weight is unexplained. Based on these figures, the researchers should conduct further research to determine whether time spent sitting causes weight gain or if the correlation is due to other factors.

CORRELATIONAL RESEARCH

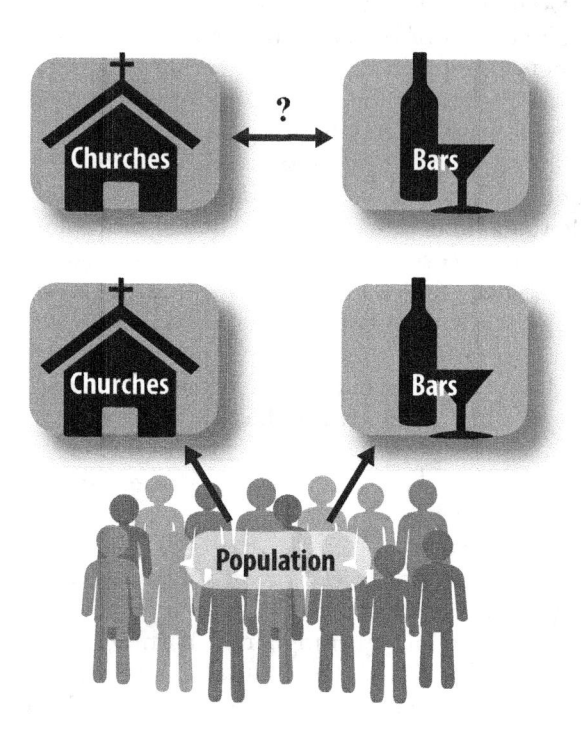

FIELDS OF STUDY

Research Theory; Research Design

ABSTRACT

Correlational research is a type of quantitative study that investigates whether and how two or more variables are related to one another. If a correlation is found, this does not necessarily mean that a change in one variable *causes* a corresponding change in the other variable. This point is made by the oft-repeated phrase "Correlation does not imply causation." When a correlation is found, further investigation is needed to determine whether causation is present or whether an outside force is acting upon both variables.

PRINCIPAL TERMS

- **causation:** a relationship between two events, actions, or other phenomena in which one event is

determined to have directly caused or otherwise affected the other.
- **negative correlation:** a relationship between two variables wherein if one variable increases, the other decreases, and vice versa.
- **positive correlation:** a relationship between two variables wherein if one variable increases, the other also increases, and if one decreases, the other also decreases.

Relationships between Research Variables

Correlational research is a type of quantitative study that investigates whether and how two or more variables are related to one another. Correlational research gathers and analyzes information but does not seek to change variables, as in experiments. The variables being studied must pertain to the same subjects. Thus, if one were researching a possible correlation between the grade achieved on a midterm and students' hair color, one would not collect test scores from one subject group and hair color data from another, as this would be nonsensical. The proper approach would be to select one group and collect both their hair color and their scores. Hypotheses for correlational research take the form of "There is (or is not) a relationship between x variable and y variable."

When a correlation between variables is found, it may be described as either "positive" or "negative." In a positive correlation, if one variable increases, the other increases, and if one decreases, so does the other. In a negative correlation, a decline in one variable is met with an increase in the other and an increase in one is followed by the other's decline. A negative correlation is sometimes referred to as an "inverse relationship."

Correlation may be expressed as a correlation coefficient, which is a number between −1 (perfect negative correlation) and +1 (perfect positive correlation). A value of zero indicates a lack of correlation. Experimental results can also be analyzed for correlation by calculating the correlation coefficient for the variables used.

Correlation versus Causation

When a study finds that two variables share a correlation, the temptation is to assume that one variable is causing the other to change. However, this assumption is often false. One might wonder, then, why correlational studies are conducted at all, if one cannot infer causation based on their results. Despite that limitation, correlational research remains useful for its predictive value and for its ability to rule out causation where no correlation exists. When found, correlation can allow doctors and other researchers to predict changes in a variable's value based on their observation of a different, correlated variable. For example, high levels of triglycerides in a person's blood are correlated with having a higher risk of heart disease. For several decades, there has been debate in the medical community over whether or not there is a causal relationship between these two variables. Regardless of causation, however, doctors can reliably use a test of a person's triglyceride levels to assess whether that person is at risk for heart disease, due to the correlation.

Hasty Conclusion

An example of the problems with correlational research can be seen in the research into uses for hormone therapy (HT). HT is primarily used for treating menopause symptoms, but evidence from various observational studies suggested that it might also be effective at reducing women's risk of heart disease. More specifically, taking HT was associated with a much lower chance of heart attack or stroke in postmenopausal women. Once doctors realized this connection, the assumption was made that more women should be placed on HT, even if they did not require treatment for menopause symptoms, because of its ability to bolster heart health. In other words, the correlation between the variables, incidence of heart disease and receiving HT, was taken as proof of causation (that is, that HT causes heart disease risk to fall).

Later, large experimental studies among otherwise healthy women—namely, the Heart and Estrogen/Progestin Replacement Study (HERS) and the Women's Health Initiative (WHI) trials—showed the opposite result. In fact, the rates of heart attacks, stroke, or other heart-related events rose by 22 percent among women on HT in the WHI trials. Both studies were discontinued early due to these and other harms. These studies had included thousands of healthy subjects, thus producing findings that could be generalized to the overall population. Both concluded that healthy postmenopausal women should not begin long-term HT use for heart health.

It is difficult to overstate the importance for researchers of remaining aware of the pitfalls of correlational research. The stakes can be quite high, so those evaluating a study must ensure that they have accounted for all other factors that could be responsible for an observed effect. Neglecting this duty can, depending upon the nature of the research, jeopardize lives and livelihoods, in addition to damaging the researcher's reputation and credibility. The goal of research is to add to the sum of human knowledge, but failing to scrutinize correlational research adequately can create confusion.

CORRELATIONAL RESEARCH GOES DIGITAL

An exciting, controversial realm for correlational research is the potential of "big data" to inform behavioral and scientific research. "Big data" refers to pieces of information gathered about individuals digitally. While there is great opportunity for expanded correlational research using such data, some researchers caution that big data analyses challenge traditional notions of what constitutes causation and that findings may be only relevant for a brief period, given the changing nature of the Internet.

Scott Zimmer, JD

FURTHER READING

Altman, Naomi, and Martin Krzywinski. "Association, Correlation and Causation." *Nature Methods*, vol. 12, 2015, pp. 899–900.

Bleske-Rechek, April, Katelyn M. Morrison, and Luke D. Heidtke. "Causal Inference from Descriptions of Experimental and Non-Experimental Research: Public Understanding of Correlation-versus-Causation." *Journal of General Psychology*, vol. 142, no. 1, 2015, pp. 48–70.

Cowls, Josh, and Ralph Schroeder. "Causation, Correlation, and Big Data in Social Science Research." *Policy & Internet*, vol. 7, no. 4, 2015, p. 447.

Manson, JoAnn E., et al. "Menopausal Hormone Therapy and Health Outcomes during the Intervention and Extended Poststopping Phases of the Women's Health Initiative Randomized Trials." *JAMA*, vol. 310, no. 13, 2013, pp. 1353–68. *JAMA Network*, jamanetwork.com/journals/jama/fullarticle/1745676. Accessed 23 Feb. 2017.

Miller, Michael, et al. "Triglycerides and Cardiovascular Disease: A Scientific Statement from the American Heart Association." *Circulation*, vol. 123, 2011, pp. 2292–33. *American Heart Association*, circ.ahajournals.org/content/123/20/2292. Accessed 23 Feb. 2017.

Mueller, Jon F., and Heather M. Coon. "Undergraduates' Ability to Recognize Correlational and Causal Language before and after Explicit Instruction." *Teaching of Psychology*, vol. 40, no. 4, 2013, pp. 288–93.

Siegle, Del. "Introduction to Correlation Research." *Educational Research Basics*, Neag School of Education, University of Connecticut, 11 Oct. 2015, researchbasics.education.uconn.edu/correlation. Accessed 23 Feb. 2017.

Wall Emerson, Robert. "Causation and Pearson's Correlation Coefficient." *Journal of Visual Impairment & Blindness*, vol. 36, no. 3, 2015, pp. 242–44.

CORRELATIONS AND CAUSATIONS

FIELDS OF STUDY

Research Theory

ABSTRACT

Correlation and causation are both relationships that may exist between variables in a research study. Correlation is present when a change in one variable is consistently accompanied by a similar change in another variable. Causation exists when a change in one variable is responsible for a change in another variable. These relationships are sometimes confused with one another, creating the possibility of mistakenly assuming that causation exists where, in fact, there is only correlation.

PRINCIPAL TERMS

- **negative correlation:** a relationship between two variables wherein if one variable increases, the other decreases, and vice versa.
- **positive correlation:** a relationship between two variables wherein if one variable increases, the other also increases, and if one decreases, the other also decreases.

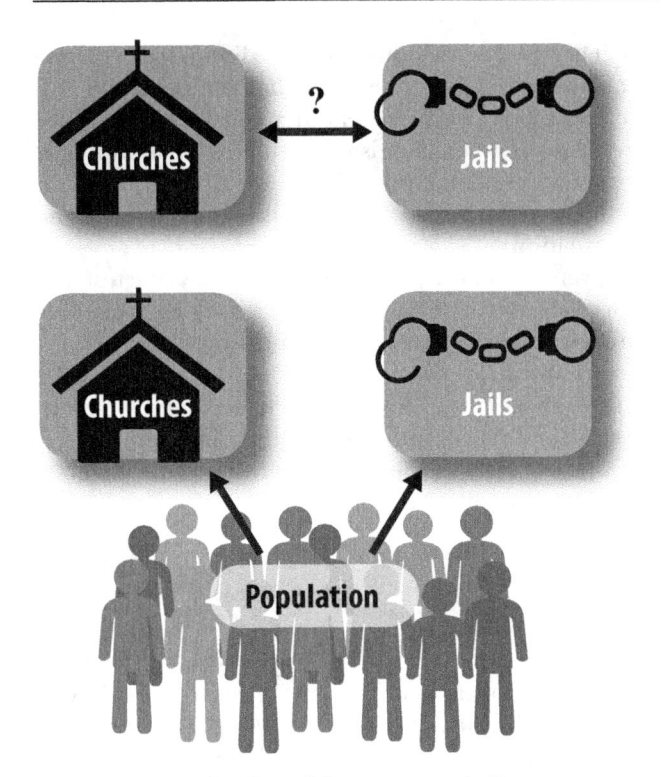

Correlation Is Not (Necessarily) Causation

"Correlation" and "causation" describe relationships that may exist between variables. Correlation exists when a change in one variable is consistently accompanied by a change in the other. Causation exists when a change in one variable causes a change in the other. In statistics, "correlation" refers solely to a linear relationship between variables, and both correlation and causation are specific types of a broader concept of association (any relationship in which two variables are not completely independent). Thus, two variables may exhibit causation without correlation, if the relationship between them is causal but nonlinear. Outside of a purely statistical context, "correlation" and "association" are often used interchangeably, so that establishing causation between two variables necessarily implies correlation (that is, association). Regardless of the definitions used, correlation does not imply causation—but it can be a good indication that further study may identify a causal link, either between the two variables or with another associated variable.

There are many reasons why two variables may exhibit correlation but not causation. One possibility is simple coincidence. Another is that some outside force acted on both variables at once. In this case, although the outside force is the cause of the change in both variables, it might appear to an observer that one of the changes caused the other. It is even possible to correctly identify a causal relationship but fail to understand the direction of the causality, asserting that A caused B, when really B caused A.

Positive and Negative Correlation

Correlation between variables is characterized by direction and strength. The direction of correlation can be positive or negative. Positive correlation occurs when two variables move in the same direction. For example, if a study finds that students who spend more time studying achieve higher test grades on average, then the studying time and the resulting test grade are positively correlated, because both increase in tandem. Negative correlation occurs when an increase in one variable is accompanied by a decrease in the other, or vice versa. For example, a study may find lower rates of illness among populations with higher vaccination rates. Vaccination rates and illness rates are thus negatively correlated, because when one increases, the other decreases.

Strength is a measure of how closely two variables are correlated. If a positive correlation is present but weak, both variables may show a general increasing (or decreasing) trend, but the degree to which an increase in one is accompanied by an increase in the other may vary significantly between data points. A strong correlation increases the likelihood that a causal relationship may exist.

Both direction and strength can be calculated in the form of a correlation coefficient, r. Correlation coefficients range in value from -1 to 1. A value of -1 is a perfect negative correlation, 1 is a perfect positive correlation, and 0 is no correlation. The closer r is to -1 or 1, the stronger the correlation. Note that correlation coefficients only apply to linear relationships. Two variables with a correlation coefficient of 0 may still exhibit a strong association along a curve, for example.

Controlled Experiments

Demonstrating a causal relationship based on correlation can be challenging. The best method for proving causation is generally a controlled experiment. For example, if medical researchers want to determine whether drugs alone, physical therapy alone, or a combination of both is the best treatment for a knee injury, they would set up a controlled experiment in which subjects are randomly assigned to one of the three experimental groups or to the control group. The control group would receive no intervention or only a placebo. However, other factors could influence the effectiveness of each treatment. People who have physically demanding jobs and constantly put weight on their injured knees may respond differently from office workers who sit most of the time. Sex, substance use, weight, and diet could also be factors.

The researchers will work to identify any such factors, whether by reviewing previous literature or by conducting a preliminary trial, and then separate each experimental or control group into smaller subgroups based on which factors might influence the results. If there is a marked difference in response rates between sexes, for example, the researchers might choose to study men and women separately, by conducting two trials or by evaluating the men's responses separately from the women's. Once the sex variable has thus been controlled, if the results still show a strong correlation between one of the treatments and the rate of recovery, then that treatment is more likely be the cause of the recovery. Such a result may require further testing as confirmation, but it is a valuable step in the process of discovery.

A Caution on Correlation versus Causation

While many people presume causation based on correlation, some err in the opposite direction. It is good to be cautious in attributing causation to correlated variables, but one must also be wary of dismissing causation simply because a causal link is not readily apparent. Even if one variable did not directly cause a change in the other, a correlation can provide valuable evidence to help guide research in a new direction.

The proper response to the discovery of a correlation is to first determine that the potential correlation is real. The researcher must next consider every possible mechanism by which causation might be operating. Each possibility must then be tested in order to determine whether it is responsible for the correlation.

Even if all of the possibilities fail to pass the test, the correlation may still be useful evidence. Correlations are especially worthy of attention in the social sciences and in medicine, as they tend to arise during the study of complex phenomena. Assigning significance to correlations only becomes dangerous when researchers fail to approach them cautiously and test them thoroughly.

Scott Zimmer, JD

Further Reading

Altman, Naomi, and Martin Krzywinski. "Association, Correlation and Causation." *Nature Methods*, vol. 12, no. 10, 2015, pp. 899–900. *Academic Search Complete*, search.ebscohost.com/login.aspx?direct=true&db=a9h&AN=110030895&site=ehost-live. Accessed 8 Mar. 2017.

"Causation and Observational Studies." *UF Biostatistics Open Learning Textbook*, U of Florida, bolt.mph.ufl.edu/6050-6052/unit-2/causation-and-observational-studies. Accessed 8 Mar. 2017.

Grinthal, Ted. "Correlation vs. Causation." *American Scientist*, Mar.–Apr. 2015, p. 84. *Academic Search Complete*, search.ebscohost.com/login.aspx?direct=true&db=a9h&AN=101076568&site=ehost-live. Accessed 8 Mar. 2017.

Urdan, Timothy C. *Statistics in Plain English*. 4th ed., Routledge, 2017.

Veličkovič, Vladica M. "What Everyone Should Know about Statistical Correlation." *American Scientist*, Jan.–Feb. 2015, pp. 26–29. *Academic Search Complete*, search.ebscohost.com/login.aspx?direct=true&db=a9h&AN=99967194&site=ehost-live. Accessed 8 Mar. 2017.

Crossover Repeated Measures Design

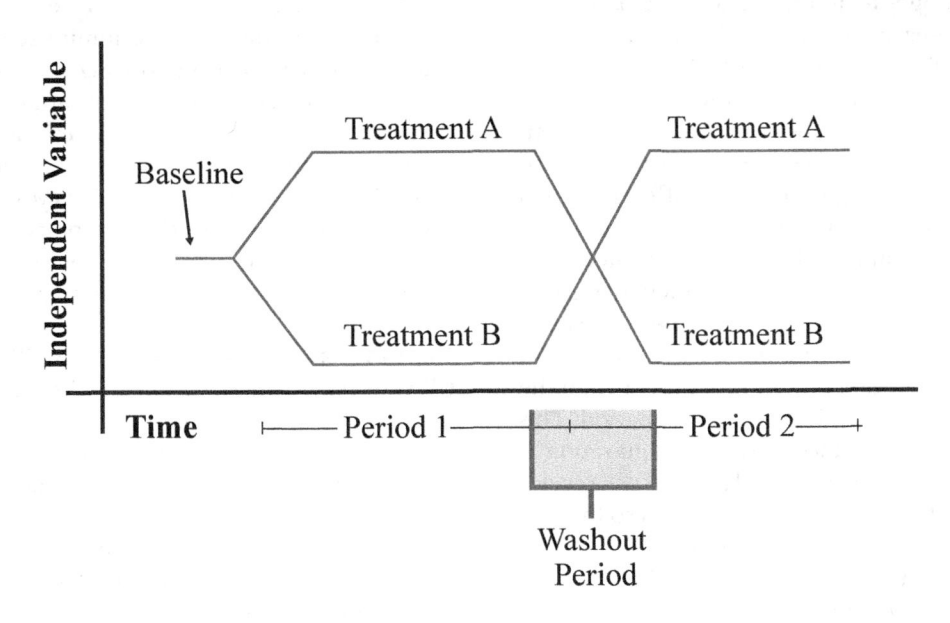

FIELDS OF STUDY

Research Design; Experimental Design

ABSTRACT

The crossover repeated measures design is a controlled design that randomly assigns a group of subjects into two or more groups. Each group will receive all treatments being tested, with a certain amount of time elapsing between administrations. The main goal is to observe differences in the effects of each treatment, both over time for individual subjects and across multiple subjects at once.

PRINCIPAL TERMS

- **carryover effect:** a consequence of a treatment applied in a previous part of a crossover repeated measures experiment that influences the results of a different treatment in a subsequent part of the experiment.
- **experimental unit:** the smallest unit in an experiment to which a treatment can be applied independently of other units; may be an individual subject or a group of subjects, depending on the experiment.
- **repeated measures:** multiple measurements made on the same experimental unit.
- **treatment:** the process or intervention applied or administered to members of an experimental group.
- **washout period:** the specified time period between treatments, during which the previous treatment effect is removed before starting the next treatment.

TESTING DIFFERENT TREATMENT EFFECTS

A crossover repeated measures design is a randomized, controlled design in which the same group of subjects serves as both a control group and a treatment group. Multiple measurements are collected over time. Each subject, or experimental unit, will receive one treatment level during each phase of the experiment, in a random order, until each one has received all treatment levels under study. The number of groups should be equal to the number of treatments being tested (or the number of treatments plus one, if the researcher desires a control group in each phase).

Tests are conducted before, after, and between each phase of the experiment. This enables researchers to assess any changes in the effect of an independent variable on the dependent variable. An independent

47

variable is one that is manipulated by the researcher. A dependent variable is one whose values change in response to changes to an independent variable. For example, if a researcher wishes to learn if a person's weight has an effect on their systolic blood pressure, the weight would be the independent variable and the systolic blood pressure would be the dependent variable. The crossover repeated measures design is often used when seeking to discover differences in a variety of treatment effects.

The most commonly used crossover repeated measures design is the two-period, two-treatment design. In this design, subjects are randomly assigned to receive either treatment 1 in the first period (or phase) and treatment 2 in the second period, or treatment 2 in the first period and treatment 1 in the second period. This design is considered balanced. The randomization is planned so that an equal number of subjects will receive both treatments by the end of the study and so that the order of treatments will not skew results.

The statistical method of analysis for this design is the repeated measures analysis of variance (ANOVA). Statistical assumptions for the ANOVA hold true for the crossover repeated measures design. The independent variable is anticipated to follow a normal distribution.

Advantages and Limitations

The crossover repeated measures design is has certain advantages over independent, parallel experimental designs. In non-crossover designs, each subject belongs to either the control group or an experimental group, and each experimental group will receive only one treatment. Even when a study population seems homogenous, there will always be between-subject variability that can affect the analysis of the results, especially in a non-crossover design. Experiments using the crossover repeated measures design minimize this between-subject variability by having the same subjects act as their own controls for different treatments. Additionally, this type of study is less costly because it requires fewer participants.

However, conducting a crossover design can often be logistically difficult, and it may not yield an accurate assessment of treatment effects. Furthermore, subjects may not be able to participate throughout the duration of the study. Such dropouts can skew the results because the sample size may become too small. If this factor is anticipated in the design phase, it is recommended to use a different experimental design, such as the mixed effects model.

While designing a crossover repeated measures study, it is also important to minimize any carryover effects. This can be done through the incorporation of a washout period between treatment administrations. This period must be long enough so that any effects of the first treatment have subsided before the second treatment is applied. Carryover effects can invalidate or confound the outcome measurements. This also means that the treatments that are tested using this design must be temporary.

Crossover Designs and Clinical Trials

Clinical trials designed to find the best medication to treat symptoms involved with chronic conditions often make use of the crossover repeated measures design. The two-period, two-treatment design has been used in studies to determine the differing effects of formoterol and salbutamol on asthma, for example. It has also been employed to test the effects of different drugs on migraines.

Application of Design

The crossover repeated measures design is extensively used in medicine to help investigators determine the best treatment for a disease or rule out ineffective treatments. It can also be used in other fields, such as education, agriculture, and psychology, where a washout period can reasonably be achieved or where a parallel-group design is not sufficient.

Mandy McBroom-Zahn

Further Reading

Crowder, Martin J., and David J. Hand. *Analysis of Repeated Measures.* Chapman & Hall/CRC, 1990.

Indrayan, Abhaya. *Medical Biostatistics.* 3rd ed., CRC Press, 2013.

Jones, Byron, and Michael G. Kenward. *Design and Analysis of Cross-Over Trials.* 3rd ed., CRC Press, 2014.

Lui, Kung-Jong. *Crossover Designs: Testing, Estimation, and Sample Size.* John Wiley & Sons, 2016.

Raghavarao, Damaraju, and Lakshmi Padgett. *Repeated Measurements and Cross-Over Designs.* John Wiley & Sons, 2014.

Verma, J. P. *Repeated Measures Design for Empirical Researchers.* John Wiley & Sons, 2016.

CROSSOVER REPEATED MEASURES DESIGN SAMPLE PROBLEM

Researchers are seeking to observe the effects of different fluids on exercise capacity. They have twelve subjects. The treatments being tested are (1) 0.47 liters (16 ounces) of Gatorade and (2) 0.47 liters (16 ounces) of water. Create a timeline for a two-period, two-treatment crossover repeated measures experiment. Include treatments, observations, and a washout period.

Answer:
1. First, the subjects will be randomly assigned to one of two groups, with six subjects in each group.
2. Next, the subjects' baseline exercise capacity is tested by having them undergo a forty-five-minute exercise trial. This is the pretest.
3. In the first period, subjects in group 1 will drink the Gatorade, and subjects in group 2 will drink the water. Then both groups will exercise to exhaustion, and each subject's time to exhaustion will be recorded. This is the first treatment phase. Observations are made during this phase.
4. Both groups will wait for one week before beginning the second period of the study, to ensure that they are fully recovered from exertion. This is the washout period.
5. In the second period, subjects in group 2 will drink the Gatorade, and subjects in group 1 will drink the water. Then, once again, the subjects will exercise to exhaustion while the researchers record their time. This is the second treatment phase. As before, observations are made during this phase.
6. The researchers may choose to test the subjects again (posttest) after the experiment is concluded.

CROSS-SECTIONAL SAMPLING

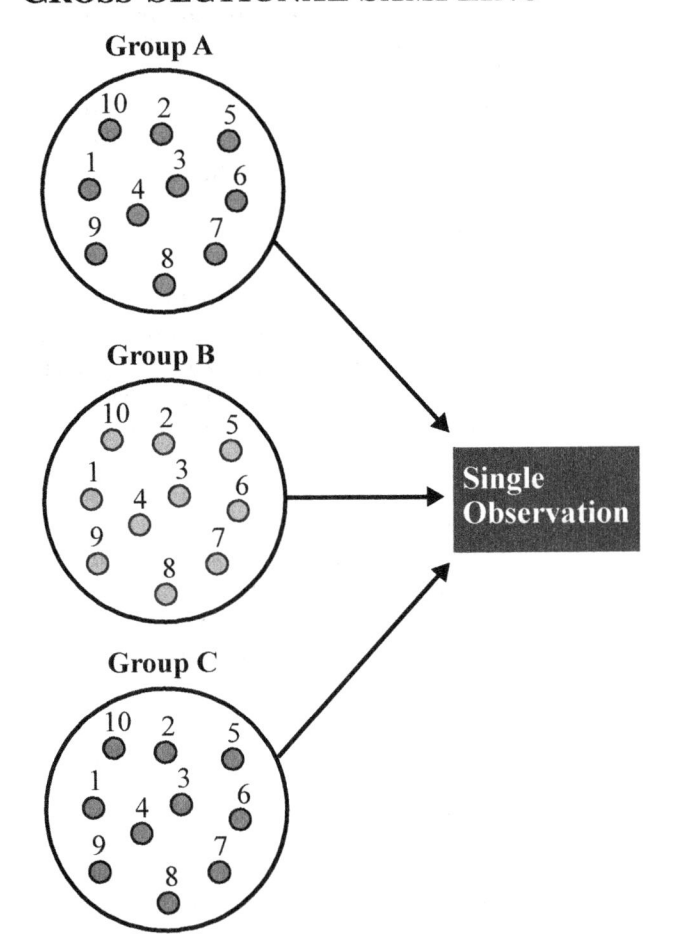

FIELDS OF STUDY

Sampling Design; Sampling Techniques; Experimental Design

ABSTRACT

Cross-sectional sampling is a sampling method that examines a portion of a population at a given moment in time. It is particularly used in descriptive research and to form hypotheses for future research. It is often combined with other sampling methods as part of a research project's sampling design.

PRINCIPAL TERMS

- **case-control study:** a retrospective study that consists of a group of individuals possessing the trait under investigation and at least one control group of individuals who do not possess that trait and are drawn from the rest of the population.
- **dependent variable:** in experimental research, a measurable occurrence, behavior, or element with a value that varies according to the value of another variable.
- **independent variable:** in experimental research, a measurable occurrence, behavior, or element that

is manipulated in an effort to influence changes to other elements.

Cross-Sectional Studies

Cross-sectional sampling takes data from a population either at the same time period or, as appropriate in certain studies, without regard to time differences. The population is usually large but, in some cases, may be small. The length of the data collection period depends on the kind of data involved and the study for which the data are to be used.

Cross-sectional data usually represent a snapshot in time. They are the opposite of time-series data. A time-series study takes data from the same small unit of study at many points in time to observe change. Time-series studies are prospective, or forward-looking.

Cross-sectional studies are also distinct from case-control studies. Case-control studies take a small sample of a specific population as a control group against which to compare a group of individuals with a specific trait being studied. For instance, in a case-control study of cervical cancer, the case is the type of cancer. The control group would include healthy people with similar traits, such as sex and age. These studies are retrospective, looking back at possible causes or risk factors.

Cross-sectional studies are most often descriptive studies. Unlike in experimental research, there is no independent variable manipulated to induce change in a dependent variable. But cross-sectional samples may also be part of a multistage sampling design for experimental research. For instance, cross-sectional and cluster sampling may be used together. Cluster sampling divides a large region into geographic groups, or clusters, before randomly selecting clusters for inclusion in a study.

The Limits and Benefits of Cross-Sections

Outside of purely descriptive research, cross-sectional sampling is usually paired with another research design. This is because a cross-sectional sample provides considerable breadth of information but not evidence of causal relationships. It does not involve an experimental condition that would test a hypothesis about the relationship between variables. It also does not indicate change over time.

Cross-sectional sampling has its advantages, however. Analysis of data from multiple cross-sections can show association between factors. In that way, cross-sectional sampling can lead to hypothesis formation for further research. Such studies also require less time, involve few ethical or moral concerns, and cost less than other types of research. As with all sampling methods, the key is to ensure that it is appropriate to the research being conducted.

Mortality after the Invasion of Iraq

A 2006 study used cross-sectional sampling to determine the scale of Iraqi mortality after the US-led invasion from March 2003 to September 2004. In order to revise the estimate of excess mortality (beyond what would have occurred anyway) of 100,000 deaths during the invasion itself, both cross-sectional and cluster sampling were used. Fifty clusters of forty households each were surveyed in a cross-sectional sample. Researchers collected death information from January 2002 up to 2006 in order to compare pre- and post-invasion mortality rates. Extrapolation from these samples showed that mortality had roughly doubled since the pre-invasion period. While the excess mortality directly attributable to invading forces had declined, comparison of causes of pre- and post-invasion mortality indicated that the invasion caused general instability and violence in the region.

Using Cross-Sectional Data

The Iraq mortality study demonstrates both the use of cross-sectional sampling and how sampling methods are often combined in real-world studies to yield samples that are both accurate and manageable from large populations. Surveying every household in Iraq would not have been plausible. On the other hand, taking a random sample from the entire population without clustering first and treating the individual as the unit, rather than a cluster of households, could have meant missing out on information that household members could provide. In addition to establishing mortality rates, cross-sectional sampling is used in the social sciences, medicine, and epidemiology.

Bill Kte'pi, MA

Further Reading

Burnham, Gilbert, et al. "Mortality after the 2003 Invasion of Iraq: A Cross-Sectional Cluster Sample Survey." *The Lancet*, vol. 368, no. 9545, 2006, pp.

1421–28. *EBSCOhost*, search.ebscohost.com/login. aspx?direct=true&db=edswao&AN=edswao.389899844&site=eds-live. Accessed 20 Mar. 2017.

Cochran, William G. *Sampling Techniques*. 3rd ed., John Wiley & Sons, 1977.

Drukker, David M., editor. *Missing Data Methods: Cross-Sectional Methods and Applications*. Emerald Group Publishing, 2011.

Elhorst, J. Paul. *Spatial Econometrics: From Cross-Sectional Data to Spatial Panels*. Springer-Verlag Berlin Heidelberg, 2014.

Smith, Gary. *Essential Statistics, Regression, and Econometrics*. 2nd ed., Academic Press, 2015.

Strang, Kenneth D., editor. *Palgrave Handbook of Research Design in Business and Management*. Palgrave Macmillan, 2015.

Wooldridge, Jeffrey M. *Econometric Analysis of Cross Section and Panel Data*. 2nd ed., MIT P, 2010.

Curie, Marie

Maire Curie. [Public domain or CC BY 4.0 (http://creativecommons.org/licenses/by/4.0)], via Wikimedia Common

POLISH PHYSICIST

Polish physicist and chemist Marie Curie conducted pioneering research into radioactivity. She and her husband, Pierre Curie, discovered the radioactive elements radium and polonium. Curie was the first woman awarded the Nobel Prize, which she received in both 1903 and 1911.

- **Born:** November 7, 1867; Warsaw, Russia (now Poland)
- **Died:** July 4, 1934; Savoy, France
- **Also known as:** Maria Salomea Sklodowska; Madame Curie
- **Primary field:** Physics; chemistry
- **Specialties:** Nuclear physics; radiochemistry

Early Life

Marie Curie was born Maria Sklodowska on November 7, 1867, in Warsaw, Russia (now Poland). Her father, Wladyslaw Sklodowska, was the director of a secondary school, and her mother, Bronislawa Boguska Sklodowska, was the headmistress of a girls' school.

Curie excelled at her studies. Despite winning awards for her scholastic achievements, Curie was unprepared to attend a Polish university when she graduated from secondary school at the age of fifteen. In Poland, girls were not taught the prerequisite subjects, so she had to leave the country to further her formal education. She decided to attend university in France.

By the time Curie had earned enough money to travel to France, by working as a teacher and a governess, she was twenty-four years old. She traveled to Paris in 1891 and enrolled at the Sorbonne (University of Paris), one of the few European universities that permitted women to attend. Curie adopted the name Marie, the French equivalent of Maria, and mastered the French language so that she could understand her lectures. She took classes on physics and other sciences, always sitting in the front row and taking detailed notes.

After eighteen months at the Sorbonne, Curie took her final exams, scoring top in her class. She received a degree in physics, the first woman at the Sorbonne to do so. An opportunity to earn a second degree presented itself when Curie received a scholarship from Poland. She reenrolled at the Sorbonne, and earned a degree in mathematics in 1894.

While Curie was studying at the Sorbonne, she met Pierre Curie, a professor at Sorbonne's school of physics. The two formed a friendship and began courting. After much deliberation over whether or not to return to Poland, Marie chose to stay in Paris to be with Pierre. They married on July 26, 1895.

LIFE'S WORK

Shortly after her honeymoon, Curie joined her husband at the School of Physics and Chemistry in Paris, where she began research into magnetism. Her presence impressed the faculty that a woman could have the fortitude to perform laboratory experiments and present meticulous results. In September 1897, Curie had her first child, Irène. Preparing her research for publication while raising her newborn daughter, she released her studies on the magnetic properties of steel on April 12, 1898.

Curie next pursued a doctoral degree in physics. She took an interest in X-rays, selecting them as the subject of her doctoral thesis. X-rays had been discovered by German physicist Wilhelm Conrad Röntgen in 1895 when he made a cardboard-encased vacuum tube produce light against a wall with the application of electricity. He believed these X-rays were invisible rays of light able to pass through solid objects. The French physicist Antoine Henri Becquerel then discovered that the element uranium produced X-rays without the aid of electricity. It was uranium that became the focus of Curie's research.

In a storeroom at the School of Physics and Chemistry, Curie conducted her experiments on uranium using an electrometer. She first determined that the strength of X-rays depended on the amount of uranium present in a chemical compound. This led to her hypothesis that X-rays were created by an atom's internal process, rather than the configuration of atoms in a molecule. Next, she determined that thorium was the only other element besides uranium to emit X-rays. She soon discovered that X-rays emitted by a mineral called uraninite, or pitchblende, were four times stronger than they should have been for the amount of uranium and thorium contained in pitchblende. She concluded that a previously undiscovered element was present in her sample.

With her husband assisting her, Curie spent long hours in a small, poorly ventilated workshop grinding down pitchblende slag to find the mysterious element. In the summer of 1898, the Curies were able to isolate the element, which Curie named polonium. In December 1898, the Curies isolated another new element from pitchblende. Using the term radioactivity to describe the rays it emitted, they determined that it was even more radioactive than polonium. They named this element radium.

Curie's discoveries met with skepticism until 1902, when she was able to distill enough pure radium to prove that the element existed. The Curies were unaware of the negative effects of radium on their health, however. By the time Curie had completed her thesis on radioactivity and was awarded her doctoral degree in 1903, becoming the first women in Europe to receive one, both she and Pierre had begun to suffer from radiation sickness.

In December 1903, the Curies were honored with the Nobel Prize in Physics, along with Becquerel, for their work on radioactivity. Curie was the first woman to be presented with a Nobel Prize. In 1904, Pierre was offered a professorship at the Sorbonne. His wife became his laboratory chief and returned to her research on radium.

Curie gave birth to her second daughter, Eve, in December 1904. At the time, she was weakened from years of exposure to radium. She was also juggling motherhood with her work at the Sorbonne and as a teacher at a girls' school. On April 19, 1906, Pierre was killed in a carriage accident. Being the only person qualified to assume Pierre Curie's vacant post at the Sorbonne, Curie took over his classes. She was the first woman to lecture there, filling the lecture halls with the press and curious onlookers as well as students.

Curie continued her research on radium, eventually developing a means to measure its purity and isolate it in metallic form. Her lifetime of work on radium won her the Nobel Prize in Chemistry in 1911, in addition to numerous awards and honorary memberships.

Curie's next goal was to create a school that would be the central location of the newest and most intensive research on radium. The Pasteur Institute for Medical Research funded Curie's Radium Institute in Paris, because radiation had been found to diminish cancerous tumors. The institute's construction was completed in 1914, at the start of World War I. Curie took advantage of her knowledge of X-rays and devoted herself to the war effort by organizing and running mobile X-ray units on the battlefield. The Radium Institute opened at the end of the war in 1918. Curie also founded the Curie Institute in Warsaw in 1932. She worked at the institute in Paris until May 1934, when she became bedridden. Radiation exposure had taken its toll, and she had developed leukemia. Curie died on July 4, 1934, at the age of sixty-six.

IMPACT

Since X-rays were first discovered in 1895, the use of radiation has extended to the medical field as a tool for looking inside the human body and for radiation therapy to treat cancer. Radioactivity has numerous uses in other fields as well. Curie's Radium Institute, renamed the Curie Institute after her death, became one of the world's leading medical and biophysical research centers and also operates as a cancer treatment hospital. The Curie Institute in Warsaw was renamed the Maria Sklodowska-Curie Institute of Oncology after World War II. It has since become the leading cancer research and treatment center in Poland.

Although Curie is known for her discovery of radium, her influence extends beyond the science of radioactivity. Curie advanced the credibility of female scientists in an era when gender equality did not exist. By teaching women to educate themselves in subjects formerly reserved for men, she paved the way for women to enter the physical sciences. Her daughter, Irene, won her own Nobel Prize in 1935 for the discovery of artificial radiation.

The Marie Curie Fellowship was named in Curie's honor. The research grant is awarded by the European Commission to pre- and postdoctoral researchers in various disciplines. The curie, a basic unit of measurement for radioactivity, was named in honor of the Curies as well, although it was later replaced by an International System of Units (SI) unit named for Becquerel.

—*Jamie Aronson*

FURTHER READING

Goldsmith, Barbara. *Obsessive Genius: The Inner World of Marie Curie.* New York: Norton, 2005. Print. Biography of both the public and private figure of Curie, compiled through family interviews and Curie's letters, diaries, and workbooks. Illustrations, notes, bibliography.

Ogilvie, Marilyn Bailey. *Marie Curie: A Biography.* Amherst, NY: Prometheus, 2010. Print. A biography of Curie's life, including a discussion of her discovery of radium and her Nobel Prize awards. Illustrations, bibliography, index.

Segrè, Emilio. *From X-rays to Quarks: Modern Physicist and Their Discoveries.* 1980. Mineola, NY: Dover, 2007. Print. Includes a detailed chapter on the discoveries made by Becquerel and the Curies. Illustrations, bibliography, indexes.

Walter, Alan E., and Hélène Langevin-Joliot. *Radiation and Modern Life: Fulfilling Marie Curie's Dream.* Amherst, NY: Prometheus, 2004. Print. Covers the science behind radiation and the numerous applications of radiation since its discovery. Illustrations, glossary, index.

MARIE CURIE DISCOVERS RADIUM AND ITS USES

Polish physicist and chemist Marie Curie received her doctor of science degree from the Sorbonne in Paris, France, in 1903. This was the same year that she and her husband, physicist Pierre Curie, jointly received the Nobel Prize in Physics, along with Henri Becquerel, for their studies of radioactivity particularly regarding the study of radium. After Pierre's death Marie was made professor of general physics at the Sorbonne, taking his place in the faculty. This was the first time a woman had received such an honor.

In 1911 Curie received the Nobel Prize in Chemistry for the discovery of the elements radium and polonium, as well as her further study of radioactivity. She had first noted the presence of radium and polonium when working on uranium, having observed the intense radioactivity of the mineral pitchblende and realizing that it was more than could be accounted for by the uranium itself. She and Pierre then had to spend months purifying these two radioactive compounds from tons of rocks—sometimes twenty kilograms of raw materials—by purification and crystallization techniques. For example, a ton of pitchblende only yielded one tenth of a gram of radium chloride crystals. It was not until four years later that Curie was able to get the pure radium and isolate it in metallic form. She was then able to examine the separated radium and study it for its properties and applications.

Curie's research on radium led to advances in areas including X-rays and radiology as medical tools, which were utilized during World War I. Marie also espoused the medical uses of radium as a treatment for cancer and tumors. As she continued to work, Curie also refused to patent or make profit from her scientific discoveries, instead allowing her research to be accessible to all so that scientific progress could proceed unhindered.

She went on to establish institutes in her native Poland, such as the Curie Institute in Warsaw, to continue the study of radium. Curie also founded the Radium Institute in Paris, France, in 1932. It later became the Curie Institute and still operates in the twenty-first century. Additionally, one of Marie daughters, Irène, would build upon her research, eventually receiving the 1935 Noble Prize in Chemistry for her work in the synthesis of new radioactive elements, along with her husband Frédéric Joliot.

Marie Curie died in France on July 4, 1934, from aplastic anemia as a result of prolonged exposure to radioactive material. Curie's quiet and dignified demeanor made her highly esteemed among the scientific community and the world. Her adopted homeland of France honored her with museums commemorating her groundbreaking science as well as by burying her body at the Panthéon in Paris.

D

Darwin, Charles

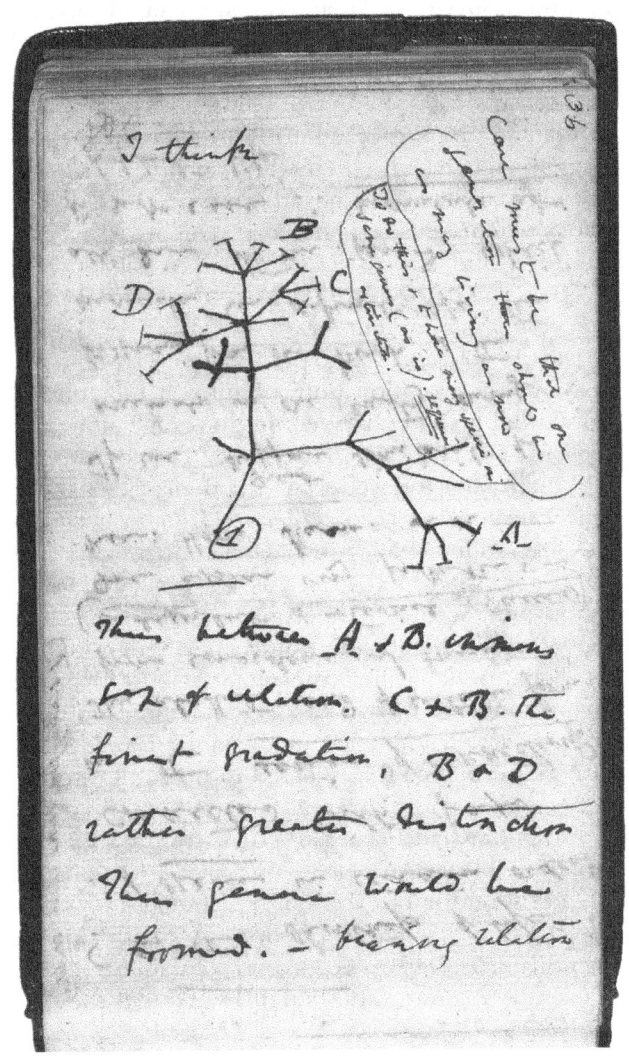

Charles Darwin's 1837 sketch, his first diagram of an evolutionary tree from his First Notebook on Transmutation of Species (1837). Interpretation of handwriting: "I think case must be that one generation should have as many living as now. To do this and to have as many species in same genus (as is) requires extinction. Thus between A + B the immense gap of relation. C + B the finest gradation. B+D rather greater distinction. Thus genera would be formed. Bearing relation" (next page begins) "to ancient types with several extinct forms." Charles Darwin [Public domain], via Wikimedia Commons

BRITISH NATURALIST

Charles Darwin is best remembered for his theory of natural selection, which explains the evolution of animals and humans. His most famous written work, On the Origin of Species (1859), remains a landmark in the field of natural history.

- **Born:** February 12, 1809; Shrewsbury, England
- **Died:** April 9, 1882; London, England
- **Primary field:** Biology
- **Specialties:** Evolutionary biology; geology; zoology; botany

Early Life

Charles Robert Darwin was born in Shrewsbury, a town in Shropshire, England, on February 12, 1809. His father was Robert Darwin, a doctor, and his paternal grandfather was the noted scientist and writer Erasmus Darwin. Darwin's maternal grandfather was Josiah Wedgwood, proprietor of the Wedgwood pottery business. His mother, Susannah, died in 1817.

Darwin was an avid scientist from an early age; among other childhood pursuits, he and his brother enjoyed experimenting in chemistry. Darwin was educated at Shrewsbury School and later at Edinburgh University. Neglecting his medical studies at Edinburgh University, Darwin instead found time to learn taxidermy from a freed slave named John Edmonstone. In addition, he attended lectures on natural sciences, and nurtured his interests in botany, anatomy, zoology, and other natural sciences. Darwin's father became irritated at his son's indifference to medicine, so he sent him instead to study theology at Christ's College, at the University of Cambridge.

By the early 1830s, Darwin had graduated from Cambridge with honors. Despite his fascination with the natural world, he set about preparing for a life in the clergy. Before this, however, he began making plans for an excursion to the islands off the Portuguese coast. In preparation for this voyage, he took a geology class from Reverend Adam Sedgwick, and accompanied Sedgwick on a geological expedition in Wales. This sparked the young scientist's interest in geology.

Upon his return from the geological trip, Darwin found that his former tutor, Reverend John Stevens Henslow, had recommended him to be an assistant to Captain Robert FitzRoy aboard the HMS *Beagle*. Initially, Darwin's father refused to allow his son to join the expedition, but a close relative made a convincing case that the expedition would be in the young man's best interests. Darwin's father relented, and, in September 1831, Darwin traveled to London to begin his life as a shipboard naturalist.

The *Beagle* was scheduled to embark on a two-year voyage to perform hydrographic surveys for the creation of water navigation maps in South America. It set sail in late December 1831, setting in at ports in Brazil, then heading south to Tierra del Fuego and the Falkland Islands on the far southern coast of the continent. Between 1832 and 1836, the *Beagle* sailed the coastline of South America, eventually heading northwest to the Galapagos Islands, off the coast of Ecuador.

Along the way, Darwin observed a variety of life forms and ecologies. He experienced a wealth of adventures worthy of any scientist's dreams: he witnessed volcanic activity, collected fossils and exotic plant specimens, and observed animals and plants that had yet to be catalogued in science texts. Although Darwin's findings aboard the *Beagle* would later prove to be of great significance, he did not know that his observations would lead to a theory of how species evolve by natural selection, or just how revolutionary and controversial that theory would be.

LIFE'S WORK

Darwin returned home from his nearly five-year-long expedition in October 1836. He cataloged his fossils and animal specimens, was admitted into the Royal Geological Society (that same organization gave Captain FitzRoy a gold medal), and began writing about his experiences in South America.

Darwin's early writings about the voyage—often called the "Journal of Researches" or "Journals and Remarks"—were published as part of the multivolume book called *Narrative of the Surveying Voyages of H.M.S.* Adventure *and* Beagle, written by Captain FitzRoy in 1839. Darwin's section appeared under his own name later that year; this same text is commonly known as *The Voyage of the* Beagle.

Darwin married his cousin, Emma Wedgwood, in January 1839. They had ten children together, seven of whom survived. Their combined family wealth ensured that the Darwins and their children would live in extreme comfort for the rest of their lives, giving Darwin the luxury of being able to devote a substantial amount of time to developing his theories on what he observed while aboard the *Beagle*. In the meantime, he enjoyed renown as a geologist, publishing his ideas on how coral reefs are formed in 1841. His book *Geological Observations on South America* was published in 1846 and helped cement his reputation among his contemporaries.

Darwin was reluctant to publish, or even publicize, his views on evolution until he himself fully understood what he had observed on his voyage along the coast of South America. He had good reason to wait, as evolutionary theory was not a popular idea at the time. The book *Vestiges of the Natural History of Creation*, although widely read, provoked a controversy when it was published in 1844; it was denounced as sacrilegious and socially dangerous. But as early as the 1830s, Darwin had begun wondering about evidence for what was then called transmutation (the idea that species mutated into other species), a radical notion that was widely ridiculed at the time. Darwin had gathered evidence that seemed to prove that something, perhaps not transmutation itself but some sort of evolution, had occurred that accounted for the variations in species.

After several more years of research and observation, Darwin decided to publish his own theory. In the mid-1850s, Darwin began to sketch out his theory of evolution; it was published in 1859 under the title *On the Origin of Species by Means of Natural Selection, or the Preservation of Favoured Races in the Struggle for Life*.

The publication of Darwin's book was greeted with enthusiasm by scientists, critics, and the public. But not everyone agreed with Darwin that natural forces, rather than a divine power through the act of providence, were responsible for evolution. Among Darwin's detractors were his former tutor, Adam Sedgwick; the prominent biologist Richard Owen; and Robert FitzRoy, the captain of the *Beagle*.

Although there are many facets to Darwin's theory of evolution by natural selection, at its core is the idea that a species' ability to evolve is based on its natural selection of traits that enable survival, while traits that interfere with survival are gradually weeded out during the process of reproduction. This process of selection has come to be equated with "survival of the fittest," a term not coined by Darwin and which did not appear in his own work until later editions of *On the Origin of Species*.

Darwin based his theory on his own observations that geographic distribution and physical variations of plants and animals could be explained by a natural process rather than a divine act of transformation. He cited as only one example the different types of tortoise in the Galapagos Islands, each of which he believed was descended from the same species; the notion that these changes were supernatural seemed absurd to Darwin. In addition to his observations while aboard the *Beagle*, Darwin continued to conduct experiments in animal and plant breeding, questioning fellow scientists about their own views on the subject.

Impact

Darwin was not the first person to believe in natural selection, and he did not create the idea of natural selection. Preceding (and often influencing) Darwin were centuries of scientists who believed that biological change occurred only as a result of natural means. Closer to his own time were the controversial naturalist Jean-Baptiste Lamarck, who wrote about the idea of the inheritance of traits, and Darwin's own grandfather, Erasmus, who was known as a proponent of transmutation (although the amount of direct influence he had over Darwin's own views is not fully known).

Darwin continued his scientific work for the rest of his life. He followed *On the Origin of Species* with several works that added greatly to his arsenal of evidence and further established his importance in the natural sciences: *The Descent of Man and Selection in Relation to Sex* (1871), which expounded his ideas on human evolution; *The Expression of the Emotions in Man and Animals* (1872); *The Different Forms of Flowers on Plants of the Same Species"* (1875); and many others. Darwin died April 19, 1882.

—*Craig Belanger*

Further Reading

Berra, Tim M. *Charles Darwin: The Concise Story of an Extraordinary Man*. Baltimore: Johns Hopkins UP: 2009. Print. A biography of Darwin that includes information on his family life, his travels on the *Beagle*, and his work and publications related to evolution.

Darwin, Charles. *On the Origin of the Species and the Voyage of the Beagle*. London: Everyman's Lib., 2003. Print. Two of Darwin's most celebrated and influential books on evolution and natural selection. Includes an introductory essay by evolutionary biologist Richard Dawkins.

Desmond, Adrian, and James Moore. *Darwin: The Life of a Tormented Evolutionist*. New York: Norton, 1994. Print. A detailed biography of Darwin that explores his life's work and its historical and social context, as well as his character, theology, and views on evolution.

Ruse, Michael. *Defining Darwin: Essays on the History and Philosophy of Evolutionary Biology*. New York: Prometheus, 2009. Print. Contains essays on subject including Darwin's life, pre-Darwinian theories, evolutionary ethics, evolutionary development, and scientists such as Alfred Russel Wallace, Herbert Spencer, Baptiste de Lamarck, and Gregor Mendel.

DARWIN'S THEORY OF NATURAL SELECTION

Biologist and naturalist Charles Darwin played a major role in the development of the modern biological sciences with the publication of On the Origin of Species by Means of Natural Selection (1859), a comprehensive hypothesis of species formation based on differential survival and reproduction. Darwin lived in an age when scientists were debating whether organisms were fixed in their current forms by divine will or whether they changed over time with new species emerging due to biological processes.

Darwin began formulating the theory of natural selection in 1831 when he served as naturalist aboard the HMS Beagle, an exploratory vessel that visited the islands off the coast of South and Central America. From his observations of living species and fossils collected during this voyage, Darwin began to see patterns that illuminated the history of biological change and development.

The theory of evolution by natural selection holds that species are generated by differential survival. Each organism has a unique compliment of traits as a result of the blending of traits from both of the organism's parents. Darwin theorized that these variations may affect an organism's ability to survive in its environment and may thereby determine the number of offspring a certain individual will pass onto the next generation. As variations are passed from parents to offspring, the differential survival and reproduction of species leads to the accumulation of changes over evolutionary time, eventually leading to the emergence of a new species from a parent species.

Darwin theorized that variations in species result from an interplay between an organism and its environment such that certain environments favor the evolution of certain traits. For this reason, organisms from different geographical areas containing similar environmental features will tend to develop similar behaviors and physical traits. In this way, evolution can create similar species in similar environments even if the species in question have different ancestors. When Darwin initially published his theory of natural selection, the genetic mechanism for the inheritance of traits was not yet known, but the theory of natural selection was later expanded and revised to explain how traits generated at the genetic level were filtered through differential survival and reproduction to appear in subsequent generations.

As Darwin prepared his theory for publication, he received word that fellow biologist Alfred Wallace was working on a similar theory derived from Wallace's studies of the Malaysian Archipelago. Wallace's research prompted Darwin to publish his theory earlier than originally intended, but rather than competing, the two scientists cooperated and published early versions of their papers simultaneously. While both men independently reached similar findings regarding the origin of species, Darwin's presentation was far more detailed and comprehensive and for this reason he is often considered the primary discoverer of the evolutionary mechanism. Darwin's theory of natural selection became a unifying theory in the biological sciences, uniting paleontology, animal behavior, psychology, medicine, botany, and zoology under a common theory that explains the generation and maintenance of variation and diversity within the biosphere.

DEDUCTIVE REASONING

Deductive Reasoning
Based on known rules and data to determine a new explanation
Conclusion is guaranteed true

Inductive Reasoning
Based on data and previous explanations to determine a rule
Conclusion is probably true

Abductive Reasoning
Based on previous explanations and known rules to determine possible data
Conclusion is a best guess

FIELDS OF STUDY

Hypothesis Testing; Research Theory

ABSTRACT

Also known as top-down logic, deductive reasoning is the arrival at a logical conclusion through the use of premises, each of which is a statement that may be true or false. If all of the premises are true, then the conclusion must be true also. A typical example of deductive reasoning is a set of two statements and a conclusion. The first statement defines the requirements of a class, the second statement describes an element that may or may not belong to the class, and the conclusion expresses whether or not the element is included in the class.

PRINCIPAL TERMS

- **abductive reasoning:** a type of reasoning in which incomplete knowledge of a specific situation is used to draw a conclusion about the simplest and most likely explanation for that situation; if the premises on which the conclusion is based are true, the conclusion is deemed to be possible or probable but not certain.
- **inductive reasoning:** a type of reasoning in which knowledge of specific situations is used to draw a broader and more general conclusion; if the premises on which the conclusion is based are true, the conclusion is deemed to be highly probable but not certain.
- **mathematical proof:** a logical and valid argument that demonstrates the truth of a mathematical statement.

Reasoning with Certainty

Deductive reasoning, or top-down logic, is one of the primary logical methods for making inferences and conclusions. A simple way to describe the process of deductive reasoning is to say that it proceeds from the general to the specific. Deductive reasoning begins with the presentation of one or more broad rules or statements that are assumed to truthfully describe some aspect of how the universe works. Then these rules or statements, known in this context as premises, are applied to a particular case or example. Finally, a conclusion is made based on the application of the premises to the case. Unlike inductive reasoning or abductive reasoning, deductive reasoning leaves no uncertainty about the conclusion that is reached. If the premises are true, the conclusion must be true.

Deductive reasoning is used in every type of research, regardless of discipline. It is an important part of the process of testing hypotheses and drawing conclusions from experiments and data analysis. One important application of deductive reasoning is in the formulation of mathematical proofs, which are central to mathematics. Deductive reasoning is also familiar to many because of the prominent, if often exaggerated, role it plays in detective fiction. Investigators gather evidence, compare it with known laws of science and other accepted truths, and draw conclusions about a case.

Testing Theories

Deductive reasoning is the best tool with which to test a theory that is already known, because it moves from the general (the theory) to the specific (the case at hand). It is not useful for attempting to generate a theory based on observations, because doing so involves moving from the specific (the observations) to the general (a hypothesis that explains the observations, but which may be incorrect). To generate theories, the more appropriate logical tools are inductive and abductive reasoning.

Deductive reasoning is largely concerned with causality. It tends to begin with reference to a hypothesis that can be tested after data has been collected. Deductive reasoning generally involves the use of quantitative data instead of qualitative data, although there are exceptions to this. Deductive reasoning is also distinctive because it often builds more and more complex arguments by accumulating simple assertions. This deceptively simple approach allows one to create descriptions of complex phenomena by combining together strings of statements in the form of, "If A then B, if B then C, but if B not D," and so on.

Deductive Reasoning in the Real World

Examples of deductive reasoning can be quite simple. The classic case of deductive reasoning was presented by the Greek philosopher Aristotle regarding his fellow philosopher Socrates: men are mortal, and Socrates is a man, so Socrates must be mortal. A similar example could be seen in everyday life. If an employee misses the bus, they will be late for work. If they are late for work, they will miss a meeting. Therefore, if they miss the bus, they will miss the meeting. Such simple deductive combinations of two premises are called syllogisms.

One of the major limitations of deductive reasoning is that every conclusion is dependent on the truth of the premises. Any flaws or errors in the premises will therefore lead to a flawed conclusion. While this may not be a problem for simple examples, it can be for long and complex arguments. Many inferences, whether in formal scientific research or everyday life, use inductive reasoning to develop premises. These premises are then used to make deductive conclusions. In many cases the premises can be assumed to be true, meaning the conclusion must also be true. However, if a premise is later proved untrue, the conclusion can also be untrue.

LOGICAL FALLACIES

A major advantage to understanding deductive reasoning is that it enhances one's ability to spot logical fallacies. These are errors in thinking that cause one to reach a conclusion that is not supported by one or more of its premises. Typically a fallacy is the result of a faulty premise—that is, one that is either inaccurate or more nuanced than it appears. For example, an obvious fallacy is present in the following deductive chain: all fish can swim; bears can swim; therefore, bears are fish. The fallacy at work here is the assumption that fish are the only creatures that can swim. In order to avoid being taken in by the fallacy, one must realize that it is possible for a creature to be able to swim even if it is not a fish. In other words, all fish are swimmers, but not all swimmers are fish. It can be quite challenging to unpack the assumptions that a logical fallacy is built on, as these are usually ideas that one takes for granted and never thinks of questioning.

Scott Zimmer, JD

FURTHER READING

Evans, Jonathan St. B. T., and David E. Over. "Reasoning to and from Belief: Deduction and Induction Are Still Distinct." *Thinking & Reasoning*, vol. 19, no. 3–4, 2013, pp. 267–83. *Academic Search Complete*, search.ebscohost.com/login.aspx?direct=true&db=a9h&AN=92006224&site=ehost-live. Accessed 13 Mar. 2017.

Foresman, Galen A., et al. *The Critical Thinking Toolkit*. Wiley-Blackwell, 2017.

Güss, C. Dominik. "Deductive Reasoning." *Encyclopedia of Human Development*, edited by Neil J. Salkind, vol. 1, Sage Publications, 2006, pp. 347–49.

Lande, Nelson P. *Classical Logic and Its Rabbit-Holes: A First Course*. Hackett Publishing, 2013.

Schechter, Joshua. "Deductive Reasoning." *Encyclopedia of the Mind*, edited by Harold Pashler, vol. 1, Sage Publications, 2013, pp. 226–30.

Weston, Anthony. *A Rulebook for Arguments*. 4th ed., Hackett Publishing, 2009.

DESCRIPTIVE RESEARCH

FIELDS OF STUDY

Research Theory; Research Design

ABSTRACT

Descriptive research is most remarkable for what it does not do. It does not answer questions about how or why various phenomena occur or which event causes a result. Instead, descriptive research simply describes phenomena and populations by identifying their defining characteristics. The hallmark of descriptive research is that it defines rather than explains.

PRINCIPAL TERMS

- **correlation:** an association between two variables under study.
- **quantitative data:** information that expresses attributes of a phenomenon in terms of numerical measurements, such as amount or quantity.
- **sample size:** the number of subjects included in a survey or experiment.

OVERVIEW

Unlike other types of research, descriptive research defines rather than explains. Some descriptive research describes a concrete phenomenon, such as a chemical reaction. Other descriptive research projects describe a relationship, whether a correlation or a causal relationship, between two or more phenomena or abstract concepts.

Descriptive research is paired with quantitative research, qualitative research, or both. Quantitative research uses numerical data to describe what is happening objectively. Qualitative research uses text to describe findings and may be more subjective. The use of both quantitative and qualitative research elements is known as "mixed methods research." Averages, frequency, standard deviation, and other statistics are often reported in descriptive research. Because it is difficult for human beings to extract meaning from raw numbers, many descriptive research projects present their findings through charts, graphs, and other visual aids.

Features common to other research such as hypotheses and control and experimental subject groups are not needed in a descriptive study. At least one variable or constant is needed for a descriptive research study, however. A theory may be formulated at the end of the study, based on the findings. Some descriptive research studies are designed upon a descriptive theory and use their findings to test the basic concept behind that theory.

STRENGTHS AND LIMITATIONS
The purpose of descriptive research is to capture the characteristics of the target population as they normally conduct themselves in their natural environment. Elements that might distort the authenticity of the data is generally to be avoided. Data collection in descriptive research attempts to be as unobtrusive as possible, because when subjects are reminded that they are being interviewed or studied, inevitably some of them will adjust their answers to fit with what they think researchers would like to hear. This is known as "response bias." Some descriptive research disguises itself as something other than data collection in an effort at obtaining the most honest results possible.

Many different instruments, including surveys, focus groups, interviews, and ethnography, can be used to collect data for descriptive research. One challenge with the familiar survey is to collect large enough numbers of responses for later researchers to be able to make reliable generalizations from the data. Because response rates are often low, survey distributors may offer an incentive to those who are willing to spend their time sharing information. Another challenge with such instruments is that questions must be carefully crafted in order to elicit appropriate responses.

Observations are another useful approach for data collection in descriptive research. Observations require researchers or their assistants to monitor the population of interest directly and to note what they see, hear, and otherwise perceive. This type of data collection can be very effective, especially when there is a likelihood that subjects will tend to underreport or overreport a behavior being studied—in a study of how often drivers exceed the speed limit, for example. Observations must be recorded with caution and after adequate training, however, lest the researchers recording the observations inadvertently introduce their own biases into the data.

US CENSUS
The census data periodically collected by the federal government in the United States is a prime example of descriptive research. Each year, the US Census Bureau collects information from a random sampling of households across the nation, not as part of an experiment, but in an effort to produce a statistical snapshot of the country at a particular moment in time. The information that is collected by the Census Bureau's American Community Survey is extremely diverse. It includes basic information about household structure, such as the number of adults living in the home, the number of children, and so forth. There is also a large amount of information about lifestyle, such as the number of hours worked per week, dwelling characteristics, and form of transportation used for commuting. Demographic characteristics are covered as well, including individual household members' race or ethnicity, age, sex, citizenship, occupation, disability, veteran status, and level of education.

A more limited set of questions is put to every household in the nation every ten years. All of this information is combined into vast data sets that can later be used by researchers to attempt to arrive at a deeper understanding of the population of the United States. These researchers can cross-reference different pieces of data to answer questions like "How many people over the age of sixty are employed full-time" or "What percentage of college graduates are female?" Moreover, the large scale of the census is such that generalizations made from the data tend to be more reliable than they would be with a study relying on a smaller sample size.

APPLICATIONS
Descriptive research can act as the catalyst for experimental research that poses questions about the phenomena that were revealed and then attempt to arrive at answers to those questions. It might be used to raise public awareness or to change policy about social issues that may be in need of attention, such as the rate of dropout in low-income high schools. Or, it might lay the groundwork for further medical or scientific study. For example, knowing how many people have a particular disease or disorder might then lead to research into how and why they acquired it.

Scott Zimmer, JD

FURTHER READING

De Vaus, David. "The Context of Design?" *Research Design in Social Research.* Sage, 2001.

Fawcett, Jacqueline, and Joan Garity. *Evaluating Research for Evidence-Based Nursing Practice.* F. A. Davis, 2009.

Greener, Ian. *Designing Social Research: A Guide for the Bewildered.* Sage, 2011.

Hernández, Ebelia. "What Is 'Good' Research? Revealing the Paradigmatic Tensions in Quantitative Criticalist Work." *New Directions for Institutional Research,* vol. 2014, no. 163, 2015, pp. 93–101.

Nassaji, Hossein. "Qualitative and Descriptive Research: Data Type versus Data Analysis." *Language Teaching Research,* vol. 19, no. 2, 2015, pp. 129–32.

Willis, Danny G., et al. "Distinguishing Features and Similarities between Descriptive Phenomenological and Qualitative Description Research." *Western Journal of Nursing Research,* vol. 38, no. 9, 2016, pp. 1185–1204.

Descriptive Statistics

FIELDS OF STUDY

Statistical Analysis; Research Design

ABSTRACT

Descriptive statistics are a collection of measurements that provide an overview of a data set. Descriptive statistics provide a statistician with an overall sense of the contents of a distribution. While some detail is lost if one only looks at descriptive statistics without studying the rest of the data, descriptive statistics are a useful way of quickly summarizing the contents of the data set.

PRINCIPAL TERMS

- **central tendency:** the most commonly occurring value in a distribution, such as the mean, median, or mode.
- **dispersion:** the extent to which the values in a distribution are spread out from one another.
- **distribution:** a mathematical function that defines all possible values of a variable according to either the probability of their occurring or the frequency with which they occur.
- **quantitative data:** information that expresses attributes of a phenomenon in terms of numerical measurements, such as amount or quantity.
- **standard deviation:** a measurement of the degree of dispersion present in a distribution; greater dispersion results in a larger standard deviation.

OVERVIEW

Descriptive statistics are used to analyze, organize, and summarize data. There a several different ways to describe a set of quantitative data. For example, say a researcher has surveyed one hundred people about their ages. There are a number of statistics that can give the researcher a general understanding of the kinds of different responses to the survey without looking at each value individually. One useful step is to determine the difference between the largest value (oldest age) and the smallest value (youngest age). The extent to which the values in a distribution are spread out from one another is known as dispersion. Some data sets have values that are widely scattered along the distributions. Other data sets have values that are more tightly clustered.

It can also be helpful to know the distribution's central tendency. The central tendency of a distribution describes how the data cluster together in the distribution and which values are most common. There are several different ways to evaluate the central tendency of a distribution. The mode describes the value that appears most frequently. The mean calculates the average of all values in the distribution. The median is the value that marks the central point of the distribution, with an equal number of values greater than it and less than it.

When researchers have access to all of these descriptive statistics, it is easier to make general conclusions about the distribution. For example, if one group of one hundred study subjects has an average age of seventy-eight and another has an average age of twenty-six, it is immediately clear that the first group has a larger number of elderly people.

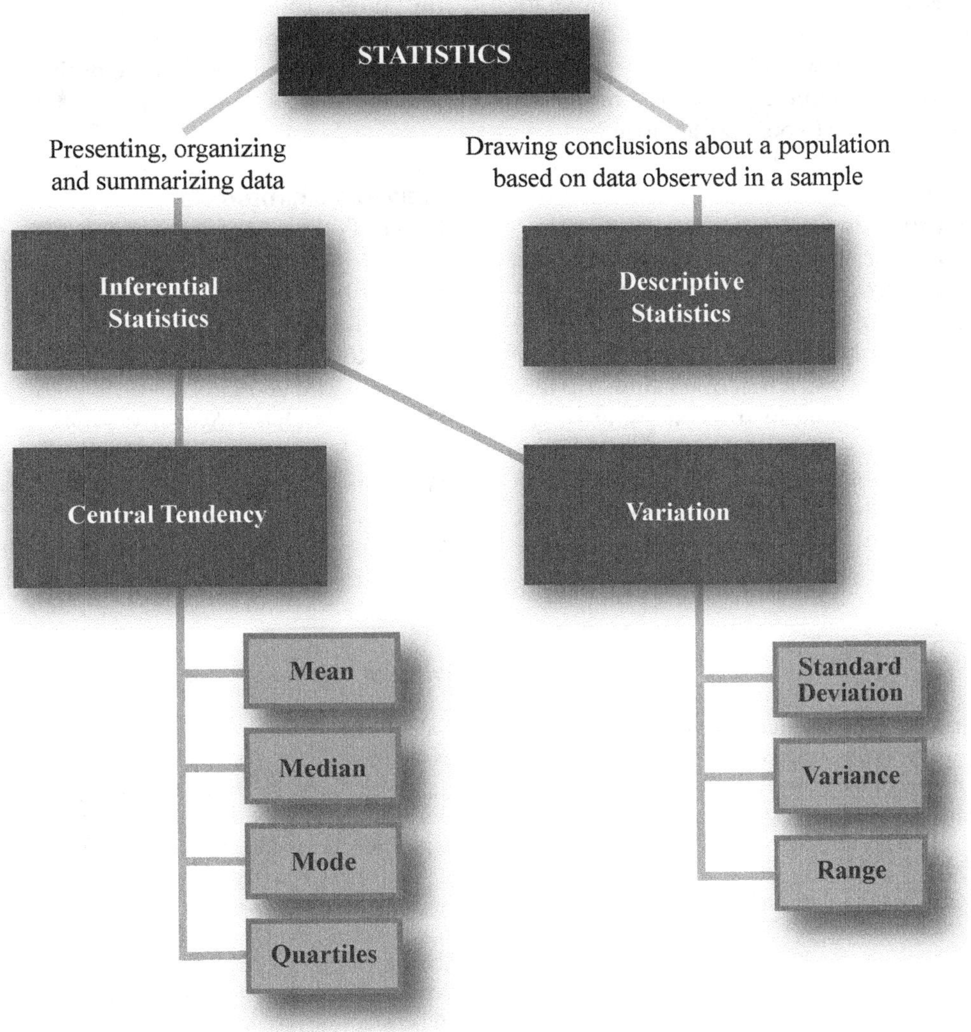
STATISTICS
Presenting, organizing and summarizing data
Drawing conclusions about a population based on data observed in a sample
Inferential Statistics
Descriptive Statistics
Central Tendency
Variation
Mean
Median
Mode
Quartiles
Standard Deviation
Variance
Range

Mean, Median, and Mode

Central tendency sounds complicated, but it actually describes a fairly simple idea. Central tendency describes the most typical values in a data set. There are several ways of measuring central tendency. One approach is to calculate the average of all the values in the distribution. In statistics this is known as calculating the mean. The mean is calculated by adding together all of the values in the distribution and then dividing the sum by the total number of values. In the example above, the researcher would find the sum of the respondents' ages and then divide by one hundred respondents to find the mean. The mean provides a simple method for comparing distributions.

Another type of central tendency measurement is the median. The median is the value that is closest to the middle of all of the other values. It has the same (or nearly the same) number of values greater than it and less than it. So, if a distribution contained eleven unique values, the median would be the value for which there are five values larger than it and five values smaller than it. In most distributions there are some duplicate values, so the median does not always have exactly the same number of values greater and less than itself.

A third measurement of central tendency is known as the mode. The mode is the value that appears most frequently in the distribution. In the age example, if five people are sixty years old, three people are forty-five, and all other responses are unique, then the mode for the distribution would be sixty—the value that occurs most often. No single measure of central tendency tells the whole story of a distribution. Thus, it is common for mean, median, and mode to be reported together.

DISPERSION

Descriptive statistics also include measures of dispersion. Dispersion describes how the values of a data set are spread out from one another. There are several different measures of dispersion, including variance, range, and standard deviation. To calculate the standard deviation, one must first determine the variance. Variance is calculated in three steps. First, calculate the mean of the entire data set. Second, subtract the mean from each number in the data set and square the result. Third, calculate the average of the square differences. Once the variance is known, the standard deviation can by determined by calculating the square root of the variance. The standard deviation is useful for determining which values in a data set are normal and which are outliers.

Scott Zimmer, JD

FURTHER READING

Bruce, Peter C. *Introductory Statistics and Analytics: A Resampling Perspective.* John Wiley & Sons, 2015.

Holcomb, Zealure C. *Fundamentals of Descriptive Statistics.* Taylor and Francis, 1998.

Kirk, Roger E. *Statistics: An Introduction.* 5th ed., Thomson, 2008.

Sullivan, Michael. *Statistics: Informed Decisions Using Data.* 5th ed., Pearson, 2017.

Weiss, Neil A. *Introductory Statistics.* 10th ed., Pearson, 2017.

Wheelan, Charles. *Naked Statistics: Stripping the Dread from the Data.* W. W. Norton, 2014.

SAMPLE PROBLEM

Suppose that researchers are studying family structure and conduct a survey asking respondents how many children they have. Although the survey is sent out to almost one hundred people, only fifteen responses are received. The responses are collected into a distribution and are 0, 1, 1, 1, 1, 2, 2, 2, 3, 3, 4, 5, 5, 7, 8.

Calculate the mean, median, and mode for this data set.

Answer:

To calculate the mean, all of the values in the distribution are added together and divided by the total number of values. Here, the sum of all the values equals 45. This total is then divided by 15 for an answer of 3.

The mode is the value that appears most frequently in the distribution. Here, that value is 1 because the answer "1" was given four times, more than any other response.

The median is the response that has equal or nearly equal numbers of responses that are greater than it and less than it. Here, the median has a value of 2 because when the responses are arranged in numerical order, the eighth response—the one in the middle, having seven values on either side of it—is 2.

DISTRIBUTIONS

FIELDS OF STUDY

Statistical Analysis; Variance Analysis

ABSTRACT

A distribution is a mathematical function that defines all possible values of a variable according to either the probability of their occurring or the frequency with which they occur in a data set. Different types of variables may take on different values. For example, some variables can have values that are not whole numbers, while others must be whole numbers or values such as "red" or "positive." Distributions can be used to show the degree of variation within a data set for the specified variable. Distributions can help predict the likelihood of the variable having a given value.

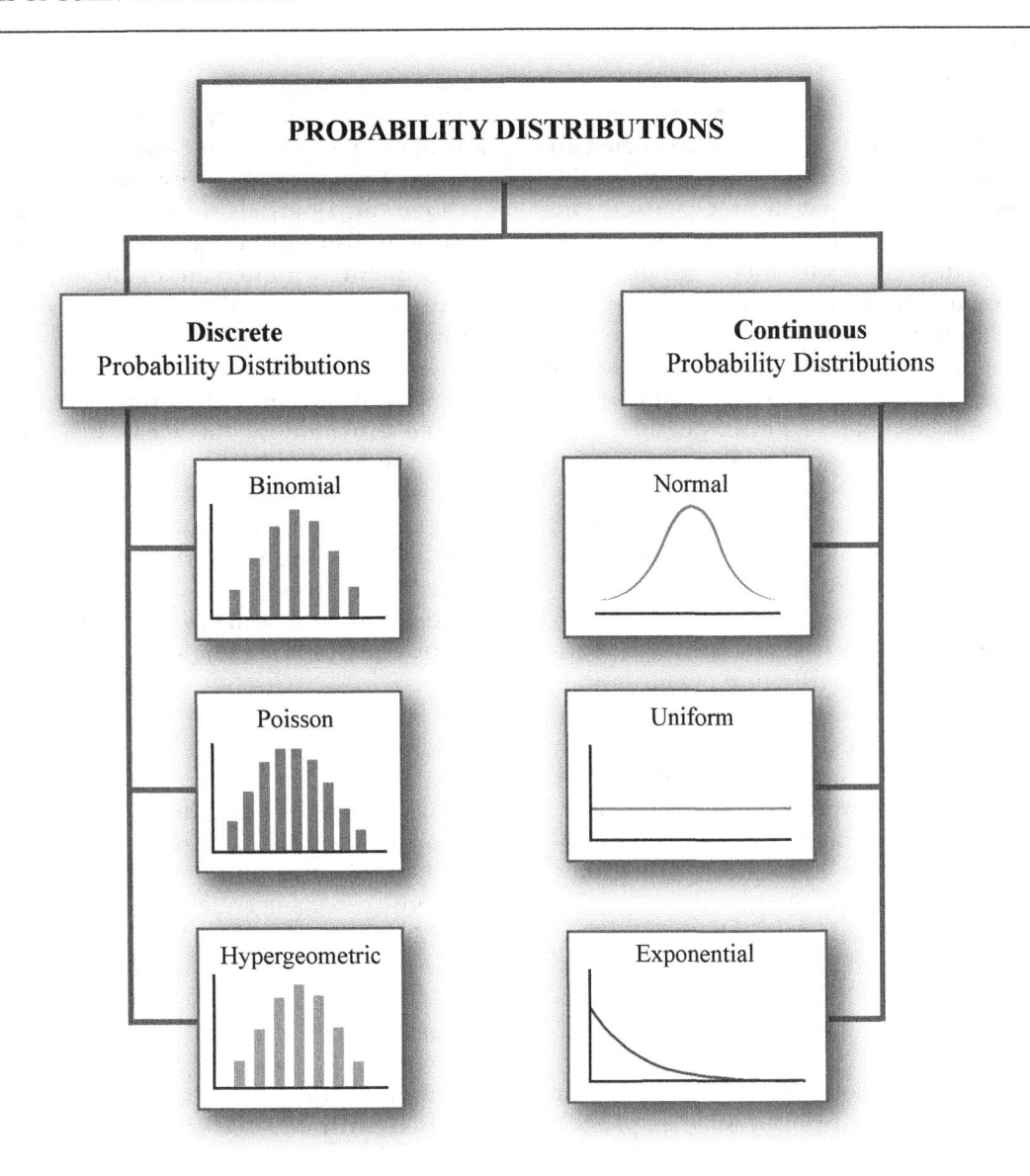

PROBABILITY DISTRIBUTIONS

Discrete Probability Distributions
- Binomial
- Poisson
- Hypergeometric

Continuous Probability Distributions
- Normal
- Uniform
- Exponential

PRINCIPAL TERMS

- **continuous variable:** a variable that can have any measurable value within a given range of values, such as age or temperature.
- **discrete variable:** a variable that can only have a finite, countable value, such as the result of a dice roll.
- **histogram:** a graphical representation of a distribution using the widths and areas of rectangles to demonstrate class intervals and frequencies, respectively.
- **normal distribution:** a probability or frequency distribution in which plotting the values contained in a data set, according to either the probability of their occurring or the frequency with which they occur, results in the appearance of a symmetrical bell-shaped curve, with the majority of values clustered around the middle; also called a bell curve.
- **skewed distribution:** a probability or frequency distribution in which plotting the values contained in a data set, according to either the probability of their occurring or the frequency with which they occur, results in the appearance of a nonsymmetrical curve, with one tail extending farther than the other.

Types of Variables

A distribution is a mathematical function that defines all possible values of a variable according to either the probability of their occurring or the frequency with which they occur in a data set. The type of data being represented by a variable can be continuous or discrete. Continuous variables are used to represent data that can potentially have an infinite number of values. For example, the weight of an apple, if measured to a sufficiently precise degree, can have any number of values: 3 ounces, 3.1 ounces, 3.14 ounces, and so forth. Variables that can only take on certain specific values are known as discrete variables. In this context, discrete means that the variable has clearly identifiable boundaries or characteristics. For example, a data set might include values of red, blue, or yellow. A variable is either red or it is not.

Distributions are often best analyzed and communicated visually. The distribution of continuous variables is often represented visually through the use of a line or curve. Because a line contains an infinite number of points, it is able to show all of the values that the variable may take on. The distribution of continuous variables can also be graphically represented with a histogram. Histograms are visually similar to simple bar graphs or charts. However, unlike a standard bar graph, they provide numerical data along both the x-axis and the y-axis. In order to show continuity, there are no gaps between the "bars" of a histogram, which are known as classes or bins.

Graphical representations of the distribution of discrete variables are often made using simple bar graphs. Each bar represents one value that the variable can take on. The length of the bar indicates how frequently the variable had that value in the distribution. So, if the value "red" occurred four times in a distribution, then the bar for "red" would be four units long.

Normal and Skewed Distributions

Distributions can be described based on the mean, median, and mode of the data and the shape of the graph that the data points form. For data collected from a random sample of a population, it is common for the graph to have the shape of a bell. This is called a bell curve or normal distribution, due to its common occurrence. In a normal distribution, values are symmetrically distributed: half of the values are greater than the mean and half of the values are less than the mean. The mean, median, and mode are equal in a normal distribution.

Some distributions do not follow this pattern. A skewed distribution has a larger number of data points falling on one side of the most commonly occurring variable (the mode). Left-skewed distributions (also called negatively skewed distributions) have a left tail that extends farther from the peak of the curve than the right tail. In a left-skewed distribution, the mean has a lower value than the mode. Right-skewed distributions (or positively skewed distributions) are less common than left-skewed distributions. Right-skewed distributions have a right tail that extends farther than the left tail. Distributions can also be bimodal or multimodal (appearing as a curve with two or more "bumps" instead of a single bell).

Predictive Value

In addition to representing values of data that have already been collected, distributions are also useful for their predictive value. By analyzing the patterns of distribution in a data set, one can make predictions about the values of data that have not yet been collected. This is known as a probability distribution. For example, consider a normal distribution that plots the heights of a population. By calculating the standard deviation, researchers can predict with confidence that 68 percent of the values will fall within one standard deviation of the mean height and 99.7 percent of values are within three standard deviations of the mean height.

Scott Zimmer, JD

Further Reading

Cooksey, Ray W. *Illustrating Statistical Procedures: Finding Meaning in Quantitative Data*. Tilde Publishing, 2014.

Geher, Glenn, and Sara Hall. *Straightforward Statistics: Understanding the Tools of Research*. Oxford UP, 2014.

Keller, Dana K. *The Tao of Statistics: A Path to Understanding (with No Math)*. SAGE, 2016.

Reinhart, Alex. *Statistics Done Wrong: The Woefully Complete Guide*. No Starch Press, 2015.

Smith, Gary. *Standard Deviations: Flawed Assumptions, Tortured Data, and Other Ways to Lie with Statistics.* Overlook Duckworth, 2014.

Wheelan, Charles J. *Naked Statistics: Stripping the Dread from the Data.* W. W. Norton, 2014.

SAMPLE PROBLEM

Imagine a research project in which the data being collected via random sampling include the number of pets owned by each person responding to the survey. Is the number of pets a discrete or continuous variable? How should the researchers graph the data? What type of distribution might you expect to find for this data set?

Answer:

All of the responses to this question would be either zero or a whole number. It is not possible to have a partial quantity of a living creature, such as one-half of a cat or three-fourths of a goldfish, or to have an infinite number of pets. Therefore, the data are discrete variables.

The researchers could graph their data showing the frequency of various responses with a bar chart. The x axis could plot the variables of 0, 1, 2, 3, 4, 5, 6, 7, and 8 or more pets. The y axis would plot the number of respondents for each discrete variable. For most populations, one would expect the graph to skew left, as a large number of people have less than three pets and a smaller number of people own four or more pets.

E

Einstein, Albert

Albert Einstein during a lecture in Vienna in 1921. Ferdinand Schmutzer [Public domain], via Wikimedia Commons

GERMAN PHYSICIST

Regarded as the most important scientist of the twentieth century, German physicist Albert Einstein received the Nobel Prize in Physics in 1921. While he made many contributions to the fields of physics, quantum mechanics, and statistical mechanics, he is best known for his theory of relativity.

- **Born:** March 14, 1879; Ulm, Germany
- **Died:** April 18, 1955; Princeton, New Jersey
- **Primary field:** Physics
- **Specialty:** Theoretical physics

EARLY LIFE

Albert Einstein was born on March 14, 1879, in Ulm, Württemberg, Germany, to Hermann and Pauline Einstein, both of whom were secular Jews. Einstein's family moved to Munich shortly after his birth and continued to move around throughout his childhood. They later moved to Milan, Italy, leaving Einstein behind to attend high school in Munich.

With the help of his scientifically minded father, Einstein became intrigued by complex scientific and mathematical concepts at a very early age. His curiosity and scientific ability clearly showed an advanced mind at work, though he disliked formal education. He wrote his first scientific paper on ether in magnetic fields when he was a teenager.

Dissatisfied with the level of education offered to students at his high school in Munich, Einstein joined his parents and sister in Milan in 1894. He renounced his German citizenship, thereby excusing himself from military service, and finished his schooling at the Swiss Federal Institute of Technology in Zurich. He earned his teaching certificate and became a Swiss citizen in 1901. That same year, unable to find work as a teacher, Einstein became a patent clerk, a job he would hold for the next seven years. It was during those years that he would complete his most profound work in physics.

In 1903, one year after the birth of their daughter (who either died in infancy or was given up for adoption), Einstein married Mileva Maric, a fellow student. Einstein and Maric spent a great deal of time discussing physics, and although the extent of her involvement in Einstein's work cannot be proved, there is a high probability that, at the very least, she helped him work out some of the ideas that would later make him famous. Einstein and Maric would have two more children, sons Hans Albert and Eduard, before divorcing in 1919.

In 1905, Einstein earned his doctorate at the Swiss Federal Institute of Technology. He left the patent office in 1908 to become a privatdozent (a professor

69

who is paid by his students). In 1909, he became professor extraordinary in Zurich, and in 1911 he moved to Prague, Czechoslovakia (now the Czech Republic), to become a professor of theoretical physics. He then returned to Germany as the head of the Kaiser Wilhelm Institute of Physics.

Life's Work

Albert Einstein's first important works in physics were the four papers, called the *Annus Mirabilis* (Latin for "year of wonder") papers, that he submitted to the German physics journal *Annalen der Physik* in 1905. The subjects of these papers include Brownian motion (which, among other things, allowed scientists to begin thinking of atoms as real, not theoretical, objects) and the photoelectric effect (which helped create the field of quantum mechanics).

In the third of these papers, entitled "On the Electrodynamics of Moving Bodies," Einstein laid out his theory for what would become known as the special theory of relativity, the first part of his theory of relativity. The special theory of relativity expresses the relationship between space and time. At a very basic level, it explains that two fundamental physics concepts, Newtonian mechanics and electromagnetism, are incompatible. Einstein's theory says that in order to arrive at answers to certain questions of physics, one must factor in the relativity of objects to each other and the constancy of the speed of light. A short time later, as an afterthought, Einstein added an essential element to the special theory of relativity: the equation $E = mc^2$, which expresses the equivalence between energy (E) and mass (m) by factoring in the velocity of light (c). This theory altered the very boundaries of physics and mathematics as they were understood at the time.

Einstein expanded his special theory of relativity into a more universal theory, the general theory of relativity. The general theory of relativity removes any restrictions of relative motion from the earlier theory and suggests a curvature in space and time. This general theory added to Einstein's previous work by proposing a theory of gravitation (simply put, the movement of masses toward each other, as opposed to gravity, which causes gravitation). Although his revolutionary ideas were controversial at the time of publication, the scientific world eventually came to accept them as the basis for all further work in physics.

Einstein presented his general theory of relativity in 1915 as a series of lectures and published it in 1916. Over the next few years, as physicists and others worked their own way through Einstein's theory, his extraordinary intelligence became widely known, first within the scientific and mathematics communities and then far beyond. In 1921, Einstein was awarded the Nobel Prize for his work on the photoelectric effect, described in his Annus Mirabilis papers of 1905.

Between 1915 and the mid-1930s, he spent much of his time defending his theories from his critics. Resistance to his work, he felt, was due to a failure among many of his peers to understand the mathematics involved. Opposition to Einstein and his work also stemmed from the anti-Semitism rampant throughout Germany and other parts of Europe during this period. In Germany, the Nazi Party began enforcing anti-Jewish laws soon after gaining power in 1933, and Einstein left Europe for the United States that same year. His second wife, his cousin Elsa Lowenthal, died soon after, in 1936. In 1940, he became a US citizen.

In addition to his continuing breakthroughs in physics, Einstein maintained a rich array of intellectual partnerships with other scientific luminaries of his age, including Niels Bohr, whom he famously debated on quantum mechanics, and the Indian physicist Satyendra Nath Bose. Physicist Leo Szilard, who worked on the Manhattan Project, helped Einstein draft a letter to President Franklin D. Roosevelt on the importance of nuclear development to counteract the Nazis' atomic-weapon ambitions. Einstein applied for the security clearance to work on the Manhattan Project but was denied. In another letter to Roosevelt a few years later, he asked the president to resist using atomic weapons against Japan, but the president never read this letter (it was found unopened after Roosevelt's death in April 1945, four months before the bombings of Hiroshima and Nagasaki, Japan).

During Einstein's tenure at the Institute for Advanced Study in Princeton, New Jersey, he continued his work in physics by investigating the possibility of a unified field theory and the generalized theory of gravitation, but no breakthroughs along the lines of his earlier work were ever achieved. He died on April 18, 1955, by which time he had become famous not only for his work as a physicist but also for his staunch pacifism, his advocacy for the Jewish state of Israel, and his opposition to the use of nuclear weapons.

Impact

Upon his death, Einstein's brain was removed by pathologist Dr. Thomas Harvey, who promised to study it and publish the results. He never did, and only relinquished the brain to Einstein's heirs several decades later.

Einstein's special and general theories of relativity corrected some of Sir Isaac Newton's theories, which had until that time been highly respected by many physicists. Central to Einstein's special theory, in particular, is his shattering of the Newtonian principle that space exists in three dimensions and that time exists in only one. Rather than thinking of space and time separately, Einstein saw space-time as a single four-dimensional system in which neither space nor time can exist without the other. These new ideas, once they became accepted, led other scientists to pursue studies that resulted in the splitting of the atom, the development of nuclear energy, space exploration, the development of a theory of superconductivity, and countless other accomplishments that have helped humankind explore the universe and understand it in much greater detail.

Einstein's later research led him to conclude that moving particles such as protons and electrons, basic building blocks of the atom, travel in waves and are closely connected with photons, the basic particles of electromagnetic energy. This conclusion led to the development of the field of wave mechanics in physics and was a fundamental element in Einstein's work with the photoelectric effect, for which he was awarded the Nobel Prize. Although the prize did not come in recognition of Einstein's work in relativity (his best-known achievement), his special theory established him as the most compelling and influential physicist of his day. It led to a total rethinking of the entire field of physics.

In addition to his relativity experiments, Einstein also became well known for his work in cosmology, in which he attempted to tackle the problem of how the universe itself is structured and discover its physical limits, if any.

—*Craig Belanger*

Further Reading

Einstein, Albert. *Relativity: The Special and General Theory*. Trans. Robert W. Lawson. New York: Penguin, 2006. Print. Explains Einstein's theories of special and general relativity in an accessible way for a general audience.

———. *The World As I See It*. New York: Open Road, 2010. Print. A collection of Einstein's writings on religion, politics, ethics, and other topics.

Isaacson, Walter. *Einstein: His Life and Universe*. New York: Simon, 2007. Print. A full biography of Einstein's life and work.

EINSTEIN'S THEORY OF RELATIVITY

Albert Einstein's special theory of relativity was first presented in an article in 1905. In it, he points out that time cannot be viewed as absolute. He states, rather, that time is relative. He also questions the idea of simultaneity—that is, viewing an event, such as a man throwing a baseball, at the exact moment it occurs. Because it takes an infinitesimal amount of time for light to travel from the ball to the observer, the event can be seen only after it happens.

Light always travels at the same speed, according to Einstein, and nothing can move faster than light. Einstein further theorized that as a moving object approaches the speed of light, its mass will increase dramatically. Experiments with high-speed electrons show that they often achieve ten thousand times their normal mass as their speed increases. An object's length also appears to change, depending on the rate at which it is moving.

In most cases, the physical changes that rapidly moving objects experience are very hard to observe. An object moving at a speed one-seventh that of the speed of light would change in mass, length, and time by only one percent. Drastic alterations would take place, however, if the object were to begin moving at nearly the speed of light. When the object's velocity reached six-sevenths of the speed of light, the mass would be twice what it was at rest, while the length and time measurements would be cut in half. Were the velocity of the object to equal the speed of light, its mass would become infinite. According to Einstein's theory of relativity, if a person were in a vehicle capable of reaching

> 99.995 percent of the speed of light, the mass of the person and the vehicle would be increased one thousand times.
>
> In addition, time would pass for that person at one one-thousandth of the normal rate; in other words, each year of such travel would be the same as one thousand years as time is normally measured. Einstein also theorized that a clock would run more slowly at the speed of light than would an identical clock at rest. This observation was proven in various ways, notably by physicists who carried atomic clocks on around-the-world commercial flights and, upon their return, checked them against comparable clocks in their laboratories; they found that the clocks they had carried with them had lost time. According to Einstein, if it were possible for an object to travel faster than the speed of light, that object would move backward in time.
>
> Einstein's theories on relativity led to his famous formula, $E = mc^2$, which states that the energy (E) of a particle of matter equals its mass (m) multiplied by the speed of light (c) squared. Matter itself is a form of concentrated energy. This equation would later suggest the possibility of creating very powerful explosions.
>
> The only major physical phenomenon that Einstein could not explain using his special theory of relativity was gravity. This problem led him to formulate his general theory of relativity in 1915. Gravity needed to be explained separately because space and time are curved; therefore, when scientists are trying to draw up mathematical calculations having to do with space and time, their findings are affected by this curvature.

EXPERIMENTAL RESEARCH

FIELDS OF STUDY

Hypothesis Testing; Experimental Design; Research Design

ABSTRACT

Experimental research involves the manipulation of certain variables in a controlled environment in order to better understand the effect of the change. Most experiments investigate what causal relationships, if any, exist between certain variables. Researchers try to determine whether and to what degree manipulating an independent variable results in a change to dependent variables. To do this, they select a sample group that is representative of the larger population and then divide its members into an experimental group and a control group. The experimenters manipulate certain variables for the experimental group, while the control group does not receive the intervention or receives a placebo or sham intervention. The groups are then compared to document any changes.

PRINCIPAL TERMS

- **control group:** the subjects in an experiment who do not receive any intervention or who receive a placebo or sham intervention.
- **null hypothesis:** a prediction that no significant differences will be found between the control and experimental groups.
- **randomization:** the use of methods that are based on chance to assign subjects to the experimental or control groups.
- **sample group:** the portion of a larger population that is selected to participate in an experiment.
- **variable:** in research or mathematics, an element that may take on different values under different conditions.

WHAT IS EXPERIMENTAL RESEARCH?

Experimental research is widely used to determine whether and to what degree causal relationships exist between various phenomena. The process begins when scientists observe certain phenomena and wish to understand in greater detail the mechanisms that affect them. To do so, they develop a hypothesis about what factors are influencing certain observed outcomes. Scientists must then test their hypothesis by conducting an experiment.

When the research question concerns a large group of subjects, the entire group is referred to as a population. All residents of a city could be the target population, for example. Populations can also be defined by several characteristics, such as gender, age, occupation, health status, income level, or any other category. For example, the target population could be all American men between the ages of sixty-five and eighty years old who have lung cancer. When researchers want to study an issue that affects a large population, it is impractical and even impossible to enlist every member of the population in the experiment. Instead, researchers select a sample group of subjects who are representative of the larger population. The sample group's characteristics are selected to statistically represent those of the overall population. Members of the sample group should be selected through a process of randomization. Through this sampling process, every member within the target population has an equal chance of being included in the sample group. Randomization reduces the likelihood that researchers will introduce bias to the study. If subjects are not randomly selected, the results of the study cannot be generalized to the overall population with confidence.

Randomization is also used to divide subjects into an experimental group and a control group. The experimental group receives an intervention from the researchers. The experimental group may undergo a treatment, take a drug, participate in a program, eat a particular diet, or be exposed to a certain environment. The control group either receives no intervention or receives a placebo or sham intervention. Following the intervention, the researchers compare the outcomes of the experimental and control groups to draw conclusions about the effect of the intervention. The intervention is the independent variable. Any outcomes are the dependent variables. For example, a particular drug treatment would be the independent variable. The participants' reported pain level, side effects, and recovery rate would be the dependent variables to be measured and compared.

Applications of Experimental Research

Experimental research is common in many scientific fields, including medicine, chemistry, physics, biology, psychology, and sociology. Companies that create medicines rely heavily on experimental research in order to ensure that new drugs are effective and have limited side effects. Once a drug under development reaches the testing stage, it is first tried on animals. If the drug proves safe and effective, it can be tested on human beings. Researchers recruit a sample of people who would likely use the medication. For a medication designed to help with headaches, for example, subjects who frequently suffer headaches would be randomly selected to participate.

The researchers would then divide the sample group into a control group and an experimental group. Subjects typically do not know which group they are in, as this knowledge can influence the outcomes they report. The subjects of the control group do not take the new drug, or they take a placebo. On the other hand, the experimental group takes the drug. The initial research question is often based on the prediction that there will be no relationship between the drug and the frequency of headaches. This prediction is called the null hypothesis. Researchers must then disprove the null hypothesis and establish that the drug does have a measurable effect on treating headaches.

To do this, researchers collect information from both groups about the frequency and severity of their headaches. They also collect information on any possible side effects. At the end of the experiment, the researchers analyze the data they have collected to see if there are any differences between the two groups. If there are differences, researchers use statistical methods to determine if the differences are significant. Through statistical analyses, they can determine if these differences were likely caused by the drug or if they were simply minor variations that can be attributed to chance.

Maintaining Control

A crucial component of experimental research is controlling for all of the variables. This ensures that any changes observed during the experiment are attributable to the treatment itself and not to other factors. In the example above, participants would need to stop taking other pain medications for the duration of the experiment. In this way, researchers can be sure that any improvement in the participants' headaches is attributable to the experimental drug and not to other medications or their interaction.

Subjects will also be asked to avoid making any major changes to their lifestyle during the study. For example, altering their diet or level of physical activity could also influence their perceptions of pain. Other potentially confounding variables, such as the dosage or the medication schedules of the new drug, must be standardized across all participants. Failing to standardize or eliminate other variables can invalidate the results of an experiment.

Scott Zimmer, JD

FURTHER READING

Creswell, John W. *Research Design: Qualitative, Quantitative, and Mixed Methods Approaches.* 4th ed., Sage Publications, 2014.

Feinberg, Fred M., Linda Court Salisbury, and Yuanping Ying. "When Random Assignment Is Not Enough: Accounting for Item Selectivity in Experimental Research." *Marketing Science,* vol. 6, 2016, p. 976.

Hemminki, Elina. "Experimental Research in Public Health." *Scandinavian Journal of Public Health,* vol. 43, no. 5, 2015, pp. 445–46.

Horakova, Tereza, and Milan Houska. "On Improving the Experiment Methodology in Pedagogical Research." *International Education Studies,* vol. 7, no. 9, 2014, pp. 84–98.

Kirk, Roger E. *Experimental Design: Procedures for the Behavioral Sciences.* 4th ed., Sage Publications, 2013.

Montgomery, Douglas C. *Design and Analysis of Experiments.* 8th ed., John Wiley and Sons, 2013.

Explanatory Research

FIELDS OF STUDY

Hypothesis Testing; Research Theory

ABSTRACT

Explanatory research seeks to explain the causes of a phenomenon or the relationships between variables. It is sometimes referred to as causal research, because a significant percentage of explanatory research studies seek to determine whether or not a causal relationship exists between two or more variables, and, if it does, what type of causation it is and how strong the causal relationship is.

PRINCIPAL TERMS

- **causation:** a relationship between two events, actions, or other phenomena in which one event is determined to have directly caused or otherwise affected the other.
- **dependent variable:** in experimental research, a measurable occurrence, behavior, or element with a value that varies according to the value of another variable.
- **independent variable:** in experimental research, a measurable occurrence, behavior, or element that is manipulated in an effort to influence changes to other elements.
- **quantitative data:** information that expresses attributes of a phenomenon in terms of numerical measurements, such as amount or quantity.

OVERVIEW

Explanatory research tends to follow descriptive research. Descriptive research seeks to gather information about a phenomenon and present findings to the scientific community. Descriptive research is intended to produce an accurate description of the current state of a phenomenon or population. However, descriptive research does not generate conclusive evidence about when, why, or how the current state came to be. After descriptive research studies have been conducted, explanatory research projects attempt to explain some of the relationships between variables. The purpose of explanatory research is to build upon previous research by testing the hypothesis that causation exists between a dependent variable and an independent variable.

In most cases, researchers test the hypothesis by performing some type of experiment. Researchers then analyze the data generated by that experiment. Explanatory research usually generates quantitative data. However, qualitative data can also play a role. Explanatory research is crucial in helping researchers move beyond discovering what is happening. It is conducted to determine why and how a phenomenon occurs. Understanding whether a causal relationship exists and determining its degree of influence allow researchers to manipulate the relationship between variables. For example, descriptive research may uncover a potential relationship between a particular behavior and lower susceptibility to a disease. To expand upon this finding, explanatory research attempts to find out if the relationship actually exists and, if it does, how it works. By understanding how the relationship between the behavior and the disease process works, researchers can develop strategies to reduce the incidence of the disease.

To Control or Not Control

Explanatory research can take different forms. However, governments and other institutions that provide research funding often prefer rigorous, scientific research designs. Therefore, the traditional approach for explanatory inquiry is to use a case control study. In this paradigm, researchers randomly assign subjects to experimental and control groups. The subjects in the experimental group receive an intervention, such as a treatment or drug. The subjects of the control group do not receive the intervention, or they receive a placebo or a sham intervention. Researchers then compare both groups to determine if there are any differences that might have been caused by the treatment. The use of the control group, random selection of subjects and random assignment to groups help to ensure that any observed changes are unlikely to have been caused by external factors. However, these steps make the process of conducting a study much more complex.

To avoid this complexity, researchers may opt to conduct an uncontrolled before-and-after study. This type of research is less rigorous than case control studies. However, before-and-after studies can still be persuasive in providing evidence for a causal relationship between variables. In an uncontrolled study, subjects are not divided into groups. Instead, a single group of subjects is measured, given a treatment, and then measured again to see if any change can be detected. Both controlled and uncontrolled studies can be viable options for explanatory research. However, controlled studies generate more definitive conclusions.

Pet Ownership and Longevity

Studies of the effects of owning a pet on the owner's life expectancy are examples of explanatory research. In a descriptive study, researchers found that people who own pets tend to report a higher quality of life than people who do not have pets. Even more startling, however, was that pet owners tend to have significantly longer life expectancies than people who do not own pets.

These findings generated a great deal of interest in identifying the root cause of these longer lifespans. Toward that end, researchers have conducted explanatory studies to build upon the earlier, descriptive findings. These explanatory studies seek to clarify the reason why owning a pet results in prolonging the owner's life. The common assumptions are that the companionship of a pet makes people happier and lowers their levels of stress, and that this accounts for their longer lives. Other researchers have hypothesized that owning a pet gives owners a sense of purpose and self-worth that is beneficial. No concrete answers have emerged so far. However, the findings of such explanatory research studies could help researchers find ways to create similar benefits for people who are unable to own or uninterested in owning pets.

The Heart of the Matter

Descriptive research is a vital first step toward understanding the current state of a phenomenon. But explanatory research is better suited to satisfying the deep sense of curiosity that underlies all scientific pursuits. In descriptive research, recording observations about the world is essential. However, these observations almost always leave researchers wanting to know more. It is not enough to simply know that certain types of butterflies migrate south for the winter or that nerve cells do not regenerate. Researchers seek to understand how and why these things happen. Understanding the factors influencing a particular phenomenon allows researchers to determine whether it can be influenced or changed.

Scott Zimmer, JD

Further Reading

Grange, Louis, et al. "A Logit Model with Endogenous Explanatory Variables and Network Externalities." *Networks & Spatial Economics*, vol.15, no. 1, 2015, p. 89.

Holgersson, H. E. T., L. Nordström, and Ö. Öner. "Dummy Variables vs. Category-Wise Models." *Journal of Applied Statistics*, vol. 41, no. 2, 2014, pp. 233–41.

Ingham-Broomfield, Rebecca. "A Nurses' Guide to Mixed Methods Research." *Australian Journal of Advanced Nursing*, vol. 4, 2016, pp. 46–52.

Noorossana, R., S. Fatemi, and Y. Zerehsaz. "Phase II Monitoring of Simple Linear Profiles with Random Explanatory Variables." *International Journal of Advanced Manufacturing Technology*, vol. 76, no. 5, 2015, pp. 779–87.

Seawright, Jason. "The Case for Selecting Cases That Are Deviant or Extreme on the Independent Variable." *Sociological Methods & Research*, vol. 45, no. 3, 2016, pp. 493–525.

Thoemmes, Felix. "Empirical Evaluation of Directional-Dependence Tests." *International Journal of Behavioral Development*, vol. 39, no. 6, 2015, pp. 560–69.

Exploratory Research

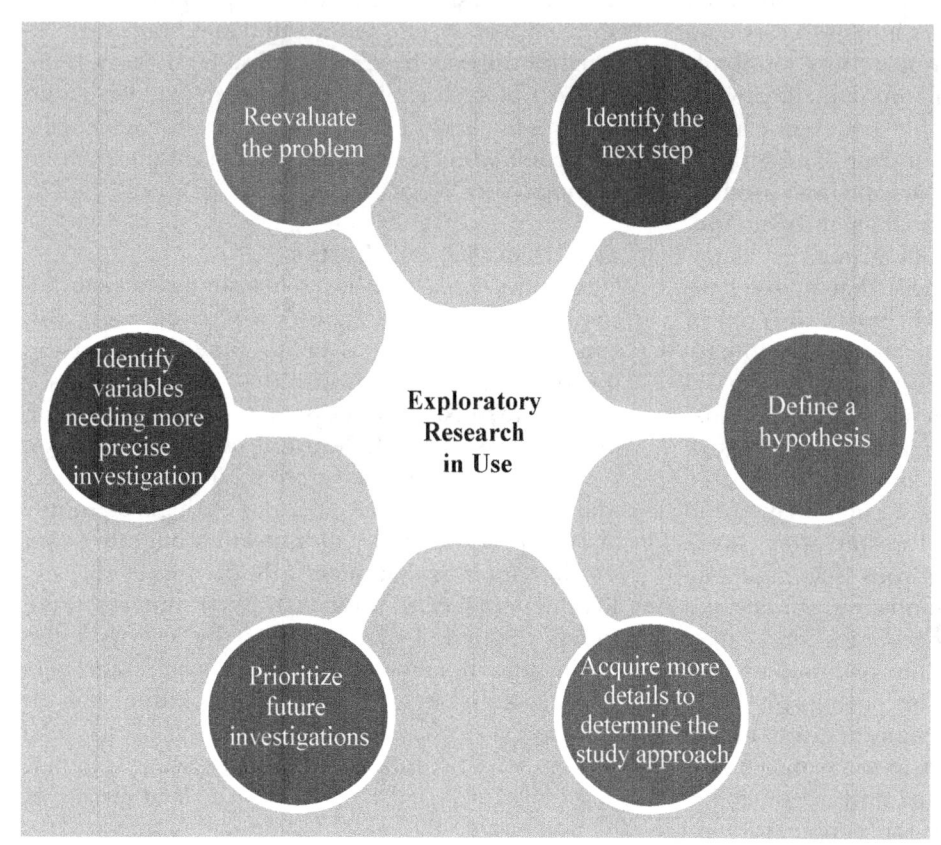

FIELDS OF STUDY

Research Design; Hypothesis Testing; Research Theory

ABSTRACT

Exploratory research examines a topic that has not yet been studied in depth. It ordinarily precedes a larger and more focused study on a particular topic or problem. This type of research helps researchers determine the scope of the topic, estimate how much work will be involved and what tasks will be necessary, prioritize the questions to ask about it, and so forth.

PRINCIPAL TERMS

- **descriptive research:** research intended to produce an accurate description of the current state of its subject, whether a phenomenon or a population, without addressing when, why, or how the current state came to be.
- **observational research:** a research method in which researchers objectively examine and analyze phenomena in their natural setting, rather than under experimental conditions.

WHY CONDUCT EXPLORATORY RESEARCH?

There are two reasons why a researcher may choose to conduct exploratory research. The first is when the researcher observes a new development or unknown phenomenon. Before an in-depth investigation can begin, it is necessary to understand the basic topic, any related phenomena, and the most urgent areas of inquiry. The second is when the subject is known to science but unknown to the researcher. In the latter situation, the researcher can usually study the topic by consulting with other researchers and reviewing the scholarly literature.

The former situation is much more challenging. Because the subject matter has not been studied yet, there is no one with whom the researcher may confer. In this case, the researcher's initial work will attempt to understand the phenomenon and its context, compare it with related phenomena, and develop basic research questions to direct further study. This type of preliminary, open-ended research differs from many other forms of research, such as, experimental research, which is designed to draw definitive conclusions or prove a specific hypothesis. Some researchers prefer exploratory research models due to their adaptability.

Benefits and Drawbacks

Exploratory research is unusual in its broad focus. Instead of seeking to provide definitive answers to research questions, exploratory research is aimed at opening up additional areas of inquiry and facilitating future research. Some researchers find exploratory research to be frustrating because they are anxious to gain a complete understanding of a phenomenon. However, exploratory research is essential to developing appropriate research questions.

Exploratory research can provide valuable insights about what type of research design should be used in future studies. It can also suggest which methodology would be best suited to the subject, whether a case study, a hypothesis-based study, or some other approach. Exploratory research can help researchers make good use of limited time, money, and other resources by refining their research question and improving their research design before they launch a study.

The primary disadvantage of exploratory research is due to its open-ended nature. Exploratory research typically generates qualitative data, which is more susceptible to bias than quantitative data. Because researchers often do not know where to begin with exploratory research, they are more prone to make guesses about the phenomenon they are studying based on their past experiences and personal beliefs. This can cause them to pursue fruitless lines of inquiry or to dismiss other sources of information, the merit of which they do not recognize.

Applications

One field that relies heavily on exploratory research is business, particularly marketing and sales. Exploratory research is often used when a company sets out to design, manufacture, and distribute a new product. The company will want to know as much as possible about the potential demand for the product, the characteristics of existing products, consumer desires and spending habits, and even social trends that may influence the demand for the product. Gathering all of this information involves exploratory research.

A company that plans to create a new deodorant will research the marketplace to find what brands are already available and how popular each one is. The company will research the characteristics of each brand. It may explore further to find out what, if any, connections there are between these characteristics and the brand's popularity. Doing this will give the company a better idea about what features consumers want a deodorant to have (no residue, mild scent, packaging, branding, etc.). The company is then able to include these features in their own design. As the company's research and development progress, the research and statistical data will become more refined and focused than the initial surveys conducted for the exploratory phase.

An Intermediate Step

Exploratory research is often conducted between an initial stage of observational research and a later stage of explanatory and experimental research. As such, it performs a crucial function. Descriptive research attempts to accurately describe the phenomenon to be studied. Exploratory research then allows the researcher to learn more about the phenomenon and to narrow down which research questions are most pressing. Finally, explanatory and experimental research use controlled experiments and trials to find answers to these questions.

Many people think of research as primarily consisting of the explanatory and experimental types. These approaches to research tend to be the most visible, as they are the types of studies whose findings are most often reported in popular media. However, without the groundwork laid by descriptive and exploratory research, the findings and conclusions reported by explanatory and experimental research would be difficult and costly to attain.

Scott Zimmer, JD

FURTHER READING

Boyko, Edward J. "Observational Research—Opportunities and Limitations." *Journal of Diabetes and Its Complications*, vol. 27, no. 6, 2013, pp. 642–48. *Academic Search Complete*, search.ebscohost.com/login.aspx?direct=true&db=a9h&AN=91792644&site=ehost-live. Accessed 31 Mar. 2017.

Davies, Pamela. "Exploratory Research." *The Sage Dictionary of Social Research Methods*, edited by Victor Jupp, Sage Publications, 2006, pp. 110–11.

Hair, Joseph F., Jr., et al. *Essentials of Business Research Methods*. 2nd ed., M. E. Sharpe, 2011.

McNabb, David E. *Research Methods for Political Science: Quantitative and Qualitative Approaches*. 2nd ed., M. E. Sharpe, 2010.

Stebbins, Robert A. *Exploratory Research in the Social Sciences*. Sage Publications, 2001.

Wrenn, Bruce, et al. *Marketing Research: Text and Cases*. 2nd ed., Haworth Press, 2007.

EXTERNAL VALIDITY

> **Outside the Study Frame**
> (External Validity)
> Are the results repeatable in other studies?
>
> > **Inside the Study Frame**
> > (Internal Validity)
> > Can the conclusions accurately be assessed from the study?

FIELD OF STUDY

Experimental Design; Research Theory; Sampling Design

ABSTRACT

External validity is one of several factors that are considered when evaluating the soundness, or overall validity, of a research study. It measures the degree to which the results of a study can be generalized to or across populations or situations other than those studied. It is possible for a study to uncover valuable information about its subjects but for that information to apply only to those subjects. Such a study would have low external validity.

PRINCIPAL TERMS

- **causation:** a relationship between two events, actions, or other phenomena in which one event is determined to have directly caused or otherwise affected the other.
- **generalizability across populations:** the degree to which experimental findings can be applied to populations other than the population from which the sample being studied has been drawn.
- **generalizability across situations:** the degree to which experimental findings can be applied to situations that are not subject to experimental controls.
- **internal validity:** the extent to which the results of a study can confidently be attributed to the quality being tested, as determined by how well the study was designed to minimize systematic errors.
- **sampling bias:** a form of statistical bias in which the selection of participants for a study is not truly random, meaning that certain members of the target population are more likely to be chosen than others.

EXTERNAL VERSUS INTERNAL VALIDITY

Because of time and funding constraints, most research studies must balance efficiency with meaningful results. For example, if a researcher wants to answer a question about a large population group, the most accurate method is to collect information from each and every member of that group. However, in most cases, this method is impractical because it takes too long and costs too much. The most common solution to this problem is to study a small sample of the larger population in order to draw a conclusion about that population. This method is more efficient, but because it involves generalizing attributes of a smaller group in order to apply them to the larger group, it is also less accurate. In order to produce meaningful results, the study must be designed in a way that minimizes inaccuracies. "Validity," in this context, is a function of research design that measures how accurately a study's results reflect the population or phenomenon being studied.

The two main forms of validity are internal validity and external validity. Internal validity refers to the accuracy with which a study establishes causation between variables. For example, in a trial of a medical treatment, internal validity describes how successful the research design was in eliminating or accounting for any factors other than the treatment that could affect the outcome. External validity, on the other hand, refers to the accuracy with which a study's results can be generalized to populations or situations beyond the ones being studied.

Generalizability

External validity is a measure of how generalizable a study is. A study that cannot be generalized is only relevant to its participants. It is important to note that the ideas of "generalizability across" and "generalizability to" are distinct. If a sample group is randomly selected from a larger population group and the findings from the sample group can be confidently attributed to the larger group, those findings are said to be generalizable to a target population. The same can be said if findings from one population group (such as youth) can be confidently attributed to another (such as the elderly). On the other hand, generalizability across populations, or population validity, means that study findings can be generalized to multiple population groups or to the whole population. One way to improve generalizability is to eliminate sampling bias as much as possible by using random sampling methods.

Another form of generalizability is generalizability across situations, or ecological validity. If ecological validity is lacking, the study findings depend on the conditions established in the study. For example, researchers might find that a new drug to prevent headaches is 100 percent effective, but only if the people who take it sleep three hours per night or less and eat only fruit. These findings would lack ecological validity, because outside of a lab setting, very few people would follow the conditions required for the drug to be completely effective.

The Price of Expediency

To understand the importance of external validity, consider a hypothetical study investigating the effects of a new health supplement on young people. After safety testing, the researchers design a study in which subjects receive a physical exam to obtain a baseline. Members of the experimental group then consume the supplement daily for six months, and those in the control group take a placebo. At the end of the six months, both groups again receive physical exams, and the results are compared with the baseline measurements to look for differences between the groups.

If, however, the researchers have trouble finding young people to participate for the full six months, they may decide to use older adults instead. Obviously, this study would be fatally flawed due to its lack of external validity, as there is no evidence that the results from the sample group (older adults) can be generalized to the target population (young people). If the study is to have any meaning, it would have to be reconfigured to use subjects that are representative of the target population.

Apples and Oranges

The idea behind external validity is to compare like entities, rather than ones that are fundamentally different. As in the old saying, one must compare apples to apples, because apples and oranges are too different for a meaningful comparison to be made. The more similar given populations or situations and the population or situation examined in a study, the more the study results can be generalized to those populations or situations.

While no study is ever generalizable across all populations and situations, several methods can be used to increase external validity. Perhaps the simplest method, though not the most convenient, is to repeat the study with different populations in different situations in order to determine whether and how the study's findings change.

Scott Zimmer, JD

Further Reading

Braver, Sanford L., and Melinda E. Baham. "External Validity." *Encyclopedia of Educational Psychology*, edited by Neil J. Salkind, vol. 1, Sage Publications, 2008, pp. 386–87.

Brewer, Marilynn B., and William D. Crano. "Research Design and Issues of Validity." *Handbook of Research Methods in Social and Personality Psychology*, edited by Harry T. Reis and Charles M. Judd, 2nd ed., Cambridge UP, 2014, pp. 11–26.

Dyrvig, Anne-Kirstine, et al. "Checklists for External Validity: A Systematic Review." *Journal of Evaluation in Clinical Practice*, vol. 20, no. 6, 2014, pp. 857–64. *Academic Search Complete*, search.ebscohost.com/login.aspx?direct=true&db=a9h&AN=100766188&site=ehost-live. Accessed 20 Mar. 2017.

Ferguson, Linda. "External Validity, Generalizability, and Knowledge Utilization." *Journal of Nursing Scholarship*, vol. 36, no. 1, 2004, pp. 16–22. *Academic Search Complete*, search.ebscohost.com/login.aspx?direct=true&db=a9h&AN=12682026&site=ehost-live. Accessed 20 Mar. 2017.

Gliner, Jeffrey A., et al. *Research Methods in Applied Settings: An Integrated Approach to Design and Analysis.* 3rd ed., Routledge, 2017.

Marcellesi, Alexandre. "External Validity: Is There Still a Problem?" *Philosophy of Science*, vol. 82, no. 5, 2015, pp. 1308–17. *Academic Search Complete*, search.ebscohost.com/login.aspx?direct=true&db=a9h&AN=111483429&site=ehost-live. Accessed 20 Mar. 2017.

F

FARADAY, MICHAEL

Michael Faraday, (22 September 1791 – 25 August 1867) in his late thirties. Painted by H.W. Pickersgill (1782-1875), Engraved by John Cochran (1821-1865) [Public domain], via Wikimedia Commons.

BRITISH PHYSICIST AND CHEMIST

In the 1820s, British physicist and chemist Michael Faraday sprinkled iron filings on a piece of paper and guided an electromagnet beneath it to illustrate lines of magnetic force. Since then, generations of students have learned about the principle of magnetic fields and other basics of electromagnetism from repeating this simple exercise. Faraday's discovery of magnetic fields remains one of the most significant contributions to science and provided the foundation for the development of the telegraph and other important innovations.

- **Born:** September 22, 1791; London, England
- **Died:** August 25, 1867; London, England
- **Primary fields:** Physics; chemistry
- **Specialty:** Electromagnetism

EARLY LIFE

Michael Faraday was born on September 22, 1791, in Newington, London, England. His family was poor; his father worked as a blacksmith. With three other children in the family, young Faraday sometimes went hungry. When he was twelve years old, Faraday took a job as an errand boy for a local bookseller, marking the end of his formal education. Two years later, he was apprenticed to a bookbinder, a position he would hold for the next seven years. A naturally curious youth with access to many books, Faraday read everything that looked engaging, including books on electricity and chemistry.

During this period, Faraday attended four of the weekly lectures given by Sir Humphry Davy, a professor of chemistry and director of the Royal Institution of Great Britain. Faraday took detailed notes at the lectures and began to conduct simple experiments by himself. When he was twenty-one years old, Faraday was hired as Davy's assistant and began working in the lab at the prestigious Royal Institution.

Davy was impressed with Faraday's enthusiasm, and offered to take him along on a cultural and scientific journey to France, Italy, Belgium, and Switzerland. This was Faraday's first trip outside of London, and it broadened his horizons immensely. He climbed Mount Vesuvius, visited a French laboratory, and experimented with glowworms. While in Florence, Davy and Faraday burned a diamond and concluded that it was composed of pure carbon, much to the dismay of many diamond owners.

After returning from his trip, Faraday was promoted to assistant in the Laboratory and Mineral Collection and Superintendent of the Apparatus at the Royal Institution. Faraday and Davy continued to collaborate, and in 1816 they invented a safety lamp for miners.

LIFE'S WORK

By 1819, Faraday had distinguished himself as one of the most prominent chemists in England. He was

81

the first to liquefy certain gases, such as carbon dioxide, which were previously thought to be incapable of undergoing change. Some of these experiments resulted in the discovery of tetrachloroethene, also called perchloroethene and used in dry cleaning as a water repellent, and as an industrial cleaning solution. Faraday's lab work was quite dangerous, and he suffered minor injuries from accidents with chlorine and other substances.

In the early 1820s, the British Admiralty requested the assistance of Davy and Faraday in finding a way to prevent the corrosion of copper-plated ship bottoms. After solving this problem, the chemists sought to improve optical glass used in telescopes, which would allow for more accurate navigation.

In 1825, Davy retired and Faraday was selected to replace him as the director of the laboratory at the Royal Institution. He discovered benzene and naphthalene, which later became important to the development of the pharmaceutical industry. He researched the composition and manufacture of alloy steels, and his work resulted in the first steel razors.

Despite his success, Faraday was not interested in obtaining patents or working in the manufacturing sector, which would have made him quite wealthy. Instead, he used his talents to seek out scientific truths and to share his findings freely with the world.

In December 1826, Faraday began offering a series of lectures for children at the Royal Institution, intending to spark an early interest in science. Referred toas the Christmas Lectures, they demonstrated his interest in education and his talent for communicating with children. Topics included magnetism, electricity, and gravitation. Many of these lectures became standard curricular materials in schools around the world. The most well-known lecture, "The Chemical History of a Candle," was first published in 1861 and remains in print today. The popular lecture series was continued throughout the twentieth century by Faraday's successors, and in later generations it was transmitted to millions of children via television.

In addition to his lectures for children, Faraday began the Friday Evening Discourses for the general public and also lectured to members of the Royal Institution. He also published dozens of scholarly articles. In the late 1820s, Faraday was appointed professor of chemistry at the Royal Military Academy in London, where he spent the next twenty-five years preparing and delivering lectures.

Faraday's interest in electromagnetism began in the early 1820s, when he surveyed all available research on the topic for a paper entitled "Historical Sketch of Electro-magnetism."

Following in the footsteps of Hans Christian Ørsted, who discovered electromagnetism, and André-Marie Ampère, who determined that the force around a wire is circular, Faraday created a device that is considered the first rudimentary electric motor. He subsequently discovered the phenomenon known as electromagnetic rotation, or the movement of the circular magnetic force.

Faraday was distracted by other research, and it wasn't until the 1830s that he returned to electromagnetism. On August 29, 1831, a date widely considered the birth of the electric industry, Faraday took an iron ring and wrapped separate coils of wire around the opposite sides. He connected one coil to a galvanometer, which used a needle to detect electric current. He attached the other coil to a battery. When the battery was connected, the galvanometer's needle moved, proving that an electric current had been induced from the second coil into the first. This experiment demonstrated the principle of electromagnetic induction, and it gave the world its first transformer. Later that year, Faraday built and demonstrated the first electric generator.

From this and other experiments performed in 1831, Faraday devised three laws of electromagnetic induction: a changing magnetic field will produce an electric current within a circuit; the electrical force is proportional to the rate of change in the field; and the direction of the rate of change in the field will affect the force.

In 1833, Faraday was named professor of chemistry at London's Royal Institution. For the rest of the decade he continued to conduct experiments that would prove his theory that all forms of electricity (voltaic, electromagnetic, and static) contained similar properties and produced the same effects. His research provided the foundation for the new discipline of electrochemistry.

By the end of the decade, Faraday was exhibiting symptoms of stress. He took a long vacation to Switzerland and limited his work to publishing and lecturing. He returned to the lab rejuvenated, and in 1845 he made another important discovery.

In a series of experiments involving light and magnetism, Faraday demonstrated what is now called the Faraday effect, which states that the direction of the plane of polarization of light waves is dependent upon the direction of the magnetic field. The study of this phenomenon became known as magneto-optics. Faraday also discovered that glass and all other materials—not just iron, cobalt, and nickel—had the potential to be affected by magnetism. He termed this new study of magnetic materials diamagnetism.

Faraday's health began to decline in the 1850s. In 1858, he moved into Hampton Court, a palace presented to him by Queen Victoria for his lifetime of achievement. Faraday did not accept the queen's offer of a knighthood, nor did he accept the presidency of the Royal Institution, which was offered to him in 1864.

IMPACT

Faraday's work had a significant impact on various scientific fields and lead to several modern technologies, including generators, electric motors, benzene, and Bunsen burners. Two units of measure are named in his honor: the farad, which is used to determine the storage potential of a capacitor (a device used to store an electrical charge), and the faraday, a unit of electricity. The Faraday constant, which represents the amount of electrical charge carried by one mole of electrons, is an important constant in chemistry, physics, and electronics and is commonly symbolized by an italicized uppercase *F*.

In addition to his achievements in electromagnetism, electricity, and physics, Faraday also contributed to analytical and organic chemistry and engineering. His work carried considerable commercial applications as well. Among Faraday's last projects was his research on electric lights for lighthouses. In 1861, due to his deteriorating memory, he resigned his position with the Royal Institution. He died at his home on August 25, 1867.

—*Sally Driscoll*

FURTHER READING

Hirshfeld, Alan. *The Electric Life of Michael Faraday.* New York: Walker, 2006. Print. A biography of Michael Faraday and descriptions of his discoveries.

James, Frank A. J. L. *Michael Faraday: A Very Short Introduction.* New York: Oxford UP, 2010. Print. A brief introduction to Faraday and his work. References and further reading included.

Klein, Maury. *The Power Makers: Steam, Electricity, and the Men Who Invented Modern America.* New York: Bloomsbury, 2008. Print. Narrative history describing the introduction of steam and electric power to the United States. Faraday's contributions to the study of electricity are detailed.

FARADAY'S LAW OF ELECTROMAGNETIC INDUCTION

During the early nineteenth century, several scientists were studying the relationship between electricity and magnetism, first discovered by Hans Ørsted in 1820. Using the failed experiments of Davy and British chemist and physicist William Wollaston as a guide, Faraday successfully produced electromagnetic rotation, the basis of the electric motor, in 1821. After neglecting to acknowledge their work in his paper, Faraday was assigned to work on other projects at the laboratory. However, he continued experimenting with electromagnetism in his spare time.

By the end of the 1820s, French physicists Charles-Augustin de Coulomb and André-Marie Ampère, as well as Ørsted had shown that stationary charges produce electric fields on other stationary charges. They also found that moving charges similarly create magnetic fields. Faraday believed that if electricity produced magnetic fields, that magnetic fields should somehow induce a current. In 1831, Faraday placed two coils of copper wire within the magnetic field created when a current was passed through the first coil. The coils were insulated from each other to prevent the transfer of electricity. Faraday was disappointed that he saw no effect on a magnetic needle placed by the second coil.

Faraday continued experimenting using larger coils of wire, a more sensitive galvanometer, and a more powerful battery. He was surprised by the results of his third experiment: a fast, slight movement on the galvanometer when contact was made or when contact was broken with the battery. However, Faraday noticed no effect when the current flowed steadily.

Next, he replaced the galvanometer with a nonmagnetized iron needle within a solenoid. He found similar results—the

> needle moved when the current through the first coil was altered. The needle was left magnetized with a polarity opposite of the first coil's field. Faraday had discovered electromagnetic induction, later known as Faraday's law.
>
> The nonmathematical version of the law is commonly stated as the induced electromotive force (emf) across a closed-circuit of thin wire equals the time rate of change of the magnetic flux through the circuit. The emf is measured in volts and is generated by the magnetic field. Magnetic flux is basically the strength of a magnetic field across a given surface area, measured in volt-seconds. Faraday published his findings in his 1839 work Experimental Researches in Electricity. Scottish physicist James Maxwell later reinterpreted the law mathematically; that equation became one of four known as Maxwell's equations that explain field theory.
>
> Faraday continued his research, rotating a copper disc on edge between the poles of a horseshoe magnet. He found a constant voltage was produced across the disc. Faraday's disc, as it became known, was the first simple generator.

FEYNMAN, RICHARD

Richard Feynman's Los Alamos ID badge. By United States Army (Atomic Heritage) [Public domain], via Wikimedia Commons

AMERICAN PHYSICIST

Nobel Prize winner Richard Feynman was a groundbreaking physicist who helped to combine classical electrodynamics and quantum physics into a theory that guided the formation of modern physics.

- **Born:** May 11, 1918; New York, New York
- **Died:** February 15, 1988; Los Angeles, California
- **Primary field:** Physics
- **Specialties:** Quantum mechanics; theoretical physics

EARLY LIFE

Richard P. Feynman was born on May 11, 1918, in Far Rockaway, Queens, New York. His father, Melville, was originally from Minsk and worked as a sales manager. He influenced Feynman greatly by teaching him to question everything. Feynman's mother Lucille, a homemaker, helped to mold the sense of humor for which he was noted. His parents were of Russian and Polish descent and Jewish, although not devout. He had a younger brother, Henry, who died as an infant, and a younger sister, Joan, who also went on to become a physicist.

Feynman showed a talent for electronics and engineering from an early age. He dismantled radios and repaired electronics and was remarkably gifted in mathematics. Feynman learned both integral and differential calculus by the time he was fifteen, and he experimented with math and reinvented mathematical topics that he had not yet learned. He was an honor student in high school and won the New York University Math Championship in his senior year. His final score was so far ahead of the rest of the participants that the judges were reportedly openly amazed.

After high school, Feynman attended the Massachusetts Institute of Technology in Cambridge, MA. While still an undergraduate, he took a graduate-level course in physics and published an article in *Physical Review*. He received his bachelor's degree with honors in 1939, and he was named a Putnam Fellow, which placed him as one of the top five scorers in a mathematics competition for colleges in the United States and Canada. Feynman earned a perfect score in mathematics and physics on the graduate school entrance exams for Princeton University—a feat never before accomplished.

Feynman's grades in history, literature, and fine arts were, however, close to the bottom of the class.

Feynman's first seminar that he gave at Princeton was attended by such luminaries in physics as Albert Einstein, Wolfgang Pauli, and John von Neumann. Feynman's PhD thesis explored the relationship between quantum mechanics and classic electrodynamics and laid the groundwork for his future Nobel Prize. His mathematical solutions, applied to different aspects of physics, were seen as an important step in the development of physics. In 1941, he married his high school sweetheart, Arline Greenbaum. Feynman received his PhD in 1942.

LIFE'S WORK
While at Princeton, Feynman was asked to work on the Manhattan Project—the research program for developing the atomic bomb. He was persuaded to join the program after learning the goal of developing the bomb before Nazi Germany could develop the weapon first. He worked in the theory division and helped develop the formula to determine the yield of a fission bomb. Feynman's sense of humor came into play during the secluded stay at the Los Alamos National Laboratory in New Mexico. He picked the locks of other scientists' filing cabinets and left notes, leading some to believe that there was a spy in their midst. He also went to an isolated section of the mesa to pound a drum, creating a rumor about a Native American drummer. His wife Arline died from tuberculosis a month before the Trinity test of the first atomic bomb, at which Feynman was present.

Another of his important contributions at Los Alamos was the discovery of how to calculate how close a mass was to criticality—that is, the point where fissionable material can sustain a chain reaction by itself. This helped establish safety procedures for storing fissile materials. After Feynman saw the devastation at Hiroshima, he expressed regret at not having left the project when Germany was defeated.

After the war, Feynman turned down an offer from the Institute for Advanced Study despite the prestige of the position. He instead went to Cornell University, where he taught theoretical physics from 1945 to 1950. He experienced a bout of depression after the bombing of Hiroshima and worked on solving physics problems to relax rather than to further research. He developed an explanation for a balancing, spinning dish that would later feature in his Nobel Prize work.

Other universities offered him professorships, and in 1950 he accepted one from the California Institute of Technology (Caltech), which promised a milder climate than that of New York. He married Mary Lou Bell in 1952, but they divorced in 1954.

At Caltech, Feynman constantly strove to take complex issues in physics and make them understandable at a layperson's level, which would earn him the nickname "The Great Explainer." Memorization of facts with no discovery through experiment was abhorrent to him, a position he strongly advocated. Feynman married his third wife Gwyneth Howarth in 1960. They had a son and a daughter.

At Caltech, Feynman conducted research on the behavior of super-cooled helium, which allowed physicists to see quantum behavior on an observable level, and he developed a model for weak decay. He also developed Feynman diagrams, which can be used to note and calculate the interactions between particles in space-time. Additionally, Feynman's undergraduate lectures were compiled into a physics textbook, *The Feynman Lectures on Physics* (1963). The book has sold millions of copies and is considered the best introduction to college-level physics. His memoir, *Surely You're Joking, Mr. Feynman!* (1985), spent fourteen weeks on the *New York Times* best-seller list.

Feynman also helped to develop the first massively parallel computer, and in 1986, he served on the presidential commission to determine the cause of the *Challenger* space shuttle explosion; despite having been diagnosed with cancer, he persevered and found the design flaw in the rubber rings used in the joints of the fuel boosters. He died in Los Angeles on February 15, 1988, from complications from his disease.

IMPACT
Feynman was a groundbreaking physicist who helped to usher in the modern age of physics. He helped to develop the atomic bomb, and he solved the mathematics problems that united the disparate elements of quantum and electrodynamic physics. In 1965, Feynman won a Nobel Prize in Physics for his work in quantum electrodynamics, and he also received the Albert Einstein Award (1954) and the E. O. Lawrence Award (1962). He developed Feynman diagrams, which physicists use as a tool to track particle interactions.

He was an educator who inspired his students to learn through experiment and experience—what he considered to be the true ideals of science. His unique personality and strong drive to question everything brought physics to the attention of the public instead of keeping it sequestered in a research lab. His breakthroughs and discoveries in the fields of mathematics and physics are still used today.

—*James J. Heine*

FURTHER READING

Feynman, Richard Phillips, and Laurie M. Brown. *Feynman's Thesis: A New Approach to Quantum Theory*. Hackensack, NJ: World Scientific, 2010. Print. Describes the development of Feynman's thesis and explains the thought process that led to his groundbreaking theories.

———, Ralph Leighton, and Edward Hutchings. *"Surely You're Joking, Mr. Feynman!" Adventures of a Curious Character*. New York: Bantum, 1989. Print. A series of anecdotes relating Feynman's life in his own words. His views on learning through understanding and questioning everything come through the humor.

Gleick, James. *Genius: The Life and Science of Richard Feynman*. New York: Pantheon, 1992. Print. Biography of Feynman covering his scientific achievements and his personality.

Krauss, Lawrence M. *Quantum Man: Richard Feynman's Life in Science*. New York: Norton, 2011. Print. A biography of Feynman that discusses the physicist's life and personality, as well as his scientific legacy. Includes information from Feynman's lectures.

Ottaviani, Jim, and Leland Myrick. *Feynman*. New York: First Second, 2011. Print. Biography in graphic novel format provides insight into Feynman's life from childhood through his work on the *Challenger* disaster.

RICHARD FEYNMAN INVENTS THE FEYNMAN DIAGRAM

In the years during World War II much work in theoretical physics was suspended with the exception of that required by the atomic bomb project. Theoretical work resumed after the war. By 1947 and 1948, physicists understood that the interaction of subatomic particles (such as electrons, positrons, and photons) was not entirely predictable. When a collision or encounter between particles took place, the outcome could be any physically possible path; the actual paths taken are determined by the laws of probability. Mathematical analysis was unavailing because the processes of integration, which the mathematics required, often gave infinity as the end result as they were extended past the first order approximation. Thus the extended analysis yielded results that were physically meaningless.

Although many physicists studied the problem, the three leading figures were Richard Feynman, Julian Schwinger of Harvard University, and Sin-Itiro Tomonaga of Tokyo University. The key to the solution of the problem was twofold: first, the mathematical technique of renormalization of the equations, and second, adding in the result of each particle's effect on itself—its self-interaction. The formal mathematical techniques for doing this were published first by Schwinger. His results were consistent with those of Feynman and Tomonaga, but the mathematics were so formal and complex that their connection to physical reality could not be easily understood.

Feynman's great contribution was first to understand that some of the complications introduced by Schwinger were unnecessary. He and Freeman Dyson, a British-born mathematician and physicist, devised simpler mathematics for the analysis. Feynman also provided visual representations of quantum interactions, making the interactions more clear and comprehensible. The graphic depictions of quantum interactions came to be known as Feynman diagrams.

Nearly all subsequent development of quantum electrodynamics rely on Feynman diagrams, in part because of Dyson's insistence among physicists that Feynman's techniques were superior to those of Tomonaga and Schwinger. These diagrams and their associated mathematics provide images of the movement of subatomic particles and can successfully account for the movement of electrons and positrons backward and forward through time.

Feynman published his fullest description of these techniques in a 1949 paper. Feynman diagrams depict time moving forward in the vertical dimension and the interactions of the particles with time. But the diagrams are more than an aid to visualization. Each is obtained from a calculated complex number derived from Lagrangian equations. These equations, originally springing from classical mechanics, describe the motion of particles over time. This derived number represents the amplitude of the reaction. The square of the amplitude gives the probability of the interaction. Quantum solutions obtained from Feynman's equations and diagrams have been experimentally verified consistently over the last half-century.

As it turned out, Feynman diagrams also have extensive applications in statistical mechanics and solid-state theory. In October, 1965 Feynman, Schwinger, and Tomonaga were awarded the 1965 Nobel Prize in Physics for "their fundamental work in quantum electrodynamics."

Field Experiment

Laboratory Experiment vs. Field Experiment

Laboratory Experiment	Field Experiment
Few Extraneous Variables	Many Extraneous Variables
Low Realism	High Realism
High Control	Low Control
Low Cost	High Cost
Short Duration	Long Duration

FIELDS OF STUDY

Research Theory; Experimental Design

ABSTRACT

Field experiments are conducted in a natural setting where the phenomena under study can be observed in real-world conditions. This deprives experimenters of the precise control over the environment that the laboratory offers and involves a greater risk of contamination. Yet it also provides greater confidence in the accuracy and relevance of observations compared to those derived from simulated conditions.

PRINCIPAL TERMS

- **confounding variable:** an unexpected, external element in an experiment that may interfere with the results.
- **control group:** the subjects in an experiment who do not receive any intervention or who receive a placebo or sham intervention.
- **dependent variable:** in experimental research, a measurable occurrence, behavior, or element with a value that varies according to the value of another variable.
- **experimental group:** the subjects in an experimental research study who receive the intervention or treatment being tested.
- **independent variable:** in experimental research, a measurable occurrence, behavior, or element that is manipulated in an effort to influence changes to other elements.
- **observational research:** a research method in which researchers objectively examine and analyze phenomena in their natural setting, rather than under experimental conditions.

EXPERIMENTING IN THE FIELD

The scientific method consists broadly of observing phenomena, proposing a hypothesis that could explain those phenomena, designing and carrying out experiments to test that hypothesis, and evaluating the results of the experiment. Unlike laboratory experiments, which take place in artificial settings where most conditions can be controlled, field experiments take place in "the field" of the real world. An independent variable is manipulated so that effects on a dependent variable can be observed. Subjects are randomly assigned to the experimental group or to the control group.

The classic field experiment is the natural experiment. Other types of field experiments include artefactual field experiments and framed field experiments. Such experiments are more lab-like but test nonstandard subjects or include real-world tasks, knowledge, or exchange to make them more naturalistic.

Observational research does not typically randomize subjects for treatment or intervention. Such studies are termed "quasi-experiments." Natural experiments may be considered quasi-experiments rather than true experiments when it is impossible or unethical to randomize the subjects.

BENEFITS AND DRAWBACKS
Some experiments depend on observing phenomena in as close to natural circumstances as possible or by their nature cannot be conducted in an artificial setting like a lab. Such experiments may require that subjects not know that they are being observed, or there may be necessary variables that cannot be manipulated in a lab. While some social science lab experiments are designed such that participants believe they are being tested on one thing when actually something else entirely is what is being observed, this is not always practical or possible. However, field experiments also raise ethical concerns around subjects' consent. Field research must then be done in a less intrusive way than lab research can be.

The chief drawback to field experiments is diminished control. It is harder to establish groups that differ only in the key variable being tested, as can ideally be done in the lab. Field experiments are more prone to contamination. Contamination refers to the unwanted effects of confounding variables, external factors that may skew the results. Experiments with such unaccounted-for variables can involve significant bias and show false correlation.

The benefits of field experiments are also considerable. They avoid the sample selection bias to which lab experiments are vulnerable. This bias arises because of the demographic difference between the population of volunteer participants (typically skewed in favor of college students) and the general population. Causal relationships can be established. Finally, it is easier to establish external validity in field experiments and generalize results to the wider population.

POLICE RAIDS AS CRIME DETERRENT
In a 1995 study of the effects of police raids of crack houses, researchers conducted a field experiment in which the crime levels in both experimental and control blocks were observed. Experimental blocks were those in which court-authorized uniformed police raids were made on crack houses on the block. Raids were only allowed on blocks where five or more police calls had been made regarding a crack house during the preceding month. This provided researchers with a norm against which to compare. Furthermore, different blocks were observed in winter and spring to account for the possibility of seasonal impacts.

The experiment was designed to test whether these raids were effective at "harm reduction" by deterring crime in the immediate surrounding area. What the researchers found, however, was that although there was a small drop in police calls and offense reports right after a raid, that reduction disappeared within two weeks. There was also no overall difference between a raid resulting in arrest and one without arrests.

The police raid study is just the sort of experiment that could not be redesigned for a lab setting. It requires observing real-world effects of real-world behaviors. In this case, those phenomena are not easily modeled. A lab experiment would require participants in the roles of police and at least three civilian roles (drug user, narcotics criminal, and non-criminal) and indirect methods of observing the effects of modeled "raids."

THE NECESSITY OF FIELD EXPERIMENTS
Field experiments offer opportunities that the lab cannot. They are necessary to a wide array of fields of study, from economics and politics to social sciences to epidemiology and meteorology. Testing real-world conditions and behaviors greatly enriches research in those subject matters. The outcomes of field experiments continue to inform real-world solutions and policies, such as the recommendation against using police raids to deter crime.

Bill Kte'pi, MA

FURTHER READING
Abdi, Hervé, et al. *Experimental Design and Analysis for Psychology.* Oxford UP, 2009.
Benz, Matthias, and Stephan Meier. "Do People Behave in Experiments as in the Field? Evidence from Donations." *Federal Reserve Bank of Boston Working Paper,* no. 06-8, Jan. 2006, pp. 1–20.
Christensen, Larry B., et al. *Research Methods, Design, and Analysis.* 12th ed., Pearson, 2014.
Frankfort-Nachmias, Chava, et al. *Research Methods in the Social Sciences.* 8th ed., Worth Publishing, 2015.
Harrison, Glenn W., and John A. List. "Field Experiments." *Journal of Economic Literature,* vol. 42, no. 4, Dec. 2004, pp. 1009–55. *EBSCOhost,* search.ebscohost.com/login.aspx?direct=true&db=ecn&A

N=0770719&site=ehost-live&scope=site. doi:www.aeaweb.org/jel/index.php. Accessed 17 Mar. 2017.

Kirk, Roger. *Experimental Design: Procedures for the Behavioral Sciences.* 4th ed., SAGE Publications, 2013.

Sherman, Lawrence W., et al. "Deterrent Effects of Police Raids on Crack Houses: A Randomized, Controlled Experiment." *Justice Quarterly*, vol. 12, no. 4, 1995, pp. 755–81.

FRACTIONAL FACTORIAL DESIGNS

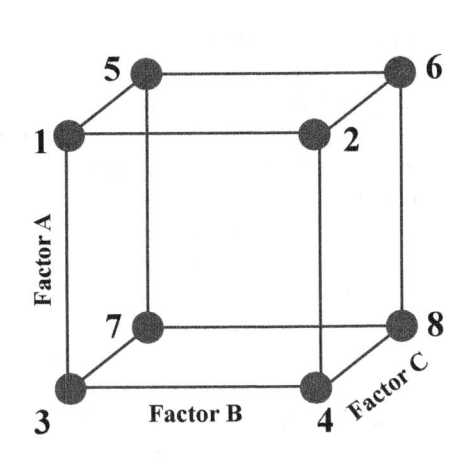

	Factors		
Run	A	B	C
1	+	−	−
2	+	+	−
3	−	−	−
4	−	+	−
5	+	−	+
6	+	+	+
7	−	−	+
8	−	+	+

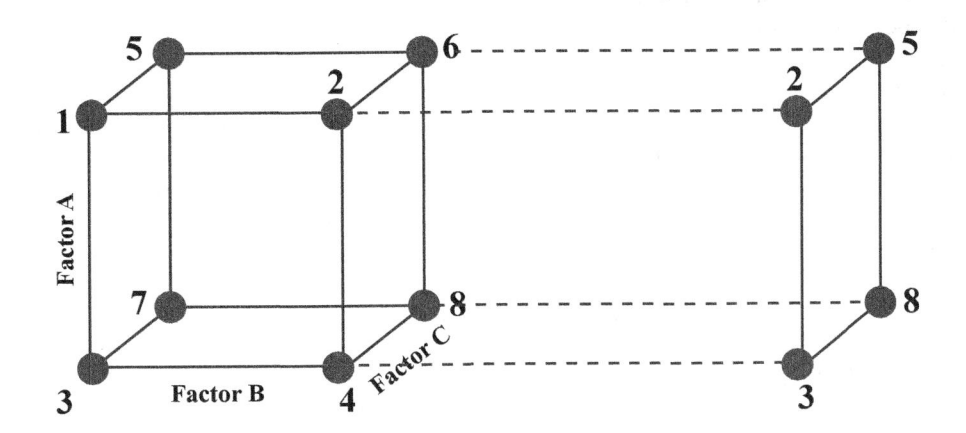

FIELDS OF STUDY

Research Design; Experimental Design; Research Theory

ABSTRACT

The fractional factorial design is a type of experimental research design used to screen factors. This type of design condenses the large number of runs needed for a full factorial study using strategies that maintain a balanced experiment. For that reason, it is employed across many scientific fields of study.

PRINCIPAL TERMS

- **confounding:** in quantitative research, describes an extraneous variable that affects both the depen-

dent and the independent variables, the effects of which must be suppressed in order to accurately determine the relationship between the variables being studied.
- **experimental condition:** the level of an independent variable being tested.
- **experimental run:** a combination of experimental conditions that are measured at a time during the execution of an experiment; one component of an experimental design.
- **factor:** an independent variable or a confounding variable that affects the outcome, or dependent, variable.
- **interaction:** in statistics, an effect by which the impact of an independent variable on the dependent variable is altered, in a nonadditive way, by the influence of another independent variable or variables.
- **Plackett-Burman design:** an experimental research design used to identify main effects among large numbers of factors.

Fractionalize an Experiment

The fractional factorial design is a type of true experimental research design used to screen several factors. Factorial designs are most often used for quantitative data but can also be used for qualitative factors. The fractional factorial design is essential in conducting studies where a researcher can only examine a subset or fraction of all the factors involved. These factors are also subject to confounding. The design nevertheless allows for comparison of main effects along with interactions.

Note that the design and data collection in a fractional factorial design stems initially from a full factorial design. The standard notation used to describe a fractional factorial design is L^{k-p}, where L is the number of levels; k, the factors; and p, the confounded interactions. Although the number of factors can theoretically be higher in a fractional factorial design, two-level factors are easiest to work with and therefore a common setup. In two-level designs, the levels are referred as "high" and "low," or +1 and −1. There must be an equal number of observations across all combinations of levels, and the design must be orthogonal, meaning the effects of each factor must sum to zero across those of other factors.

The Most Information with the Fewest Runs

A full factorial design has two or more levels per factor and examines all the combinations of the factor levels. As the number of factors rises, so do the experimental runs needed to measure the experimental conditions. Factorial designs compare the means of various *combinations* of conditions that include a given main effect to combinations that exclude it. Subjects are randomly assigned to groups by experimental condition. In effect, each group serves as an experimental group for analyses of certain effects and as a control for others. Thus, the overall sample size can be based on the number of subjects needed for the smallest expected effect to have statistical power.

When a researcher can investigate only a predetermined fraction of runs, such as one-half or one-quarter, from a full factorial design, the fractional factorial design is best utilized. It is well suited when resources, including time and funds, are limited or there are a large number of factors. However, because the fractional factorial design studies a subset, the main effects and interactions can be confounded, where the interaction cannot be separated from the effects. Confounding can lead the results to being misinterpreted.

Further, when a researcher is designing a fractional factorial study, the first step is choosing an alias structure. The alias structure aids in choosing which factors are confounded. For example, a full factorial experiment with four two-level factors would need sixteen runs, so confounding one factor limits the runs by half ($2^{4-1} = 2^3 = 8$). Using the formula L^{k-p} aids in determining the right number of runs needed to attain an effective analysis. Researchers generally assume that higher-order interactions (that is, interactions between a greater number of factors) have negligible effects on the outcome and select those for aliasing.

Similarly, the Plackett-Burman design is an experimental quantitative research design that aims to screen a large number of factors in order to identify the most important main effects through fewer runs. This design relies on a base of four, rather than two, for determining the runs allowed and can handle even more factors than fractional factorial design. It is important to such real-world situations as vehicle accident analysis. However, while Plackett-Burman designs can narrow down the factors to the most significant ones, confounding is more pronounced. Thus, factorial designs and other research design types may be needed later in order to analyze the main effects more closely.

Fractional Factorial Designs in Practice

Among other real-world applications, fractional factorial designs can be useful in manufacturing settings where several factors can influence quality outcomes in a product. Such studies can help industrial quality assurance engineers determine where defects are coming from with a minimum of disruption and cost. Analyzing the results of a fractional factorial study lets them create models that can be further tested in order to correctly identify the cause of the defect and address it.

Significance

The fractional factorial design is useful in fields ranging from psychology and medicine to manufacturing. For example, in a clinical trial, it may be impossible or undesirable to test all combinations of a drug treatment regimen. In such cases, a fractional factorial design may be a more feasible alternative to the full factorial design and still provide meaningful results.

Pamelyn Witteman, PhD

Further Reading

Creswell, John. *Research Design: Qualitative, Quantitative, and Mixed Approaches.* 4th ed., SAGE Publications, 2014.

Field, Andy. *Discovering Statistics Using IBM SPSS Statistics.* 4th ed., SAGE Publications, 2013.

"Fractional Factorial Designs." *Engineering Statistics Handbook.* NIST SEMATECH, www.itl.nist.gov/div898/handbook/pri/section3/pri334.htm. Accessed 14 Feb. 2017.

Jaynes, Jessica, et al. "Application of Fractional Designs to Study Drug Combinations." *Statistics in Medicine,* vol. 32, no. 2, 2013, 307–18. *PMC,* www.ncbi.nlm.nih.gov/pmc/articles/PMC3878161. Accessed 4 May 2017.

"Lesson 8: 2-Level Fractional Factorial Designs." *Design of Experiments.* Pennsylvania State U, onlinecourses.science.psu.edu/stat503/node/48. Accessed 4 May 2017.

Orcher, Lawrence T. *Conducting Research: Social and Behavioral Science Methods.* 2nd ed., Routledge, 2017.

Swanson, Richard A., and Elwood F. Holton III, editors. *Research in Organizations: Foundations and Methods of Inquiry.* Berrett-Koehler Publishers, 2005.

SAMPLE PROBLEM

An engineer wants to investigate several factors that affect a new car design for mileage: engine size in liters, drag, weight, shape, tire pressure, suspension, transmission type, number of cylinders, and tank size. The factors are denoted as *A, B, C, D,* and so forth. However, the engineer only has resources for a single replication of seven factors at two levels. Recall that for a fractional factorial experiment, the notation is L^{k-p}. How many runs could the engineer complete during the experiment?

Answer:

The engineer had to drop two factors that would be required to do a full factorial study. To find how many runs they could do, substitute the value of the number of levels, the number of factors, and the number of confounders into the formula above and solve:

$$L^{k-p} = 2^{7-2} = 32$$

Thus, the engineer can only perform thirty-two experimental runs.

Full Factorial Design

Factors		
Run	A	B
1	+	−
2	+	+
3	−	−
4	−	+

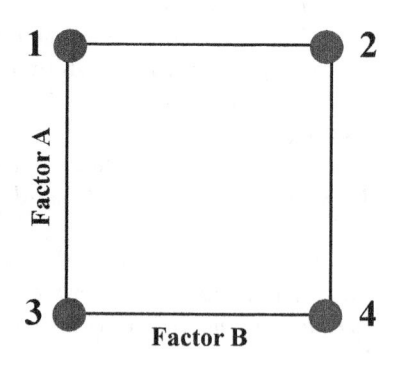

Factors			
Run	A	B	C
1	+	−	−
2	+	+	−
3	−	−	−
4	−	+	−
5	+	−	+
6	+	+	+
7	−	−	+
8	−	+	+

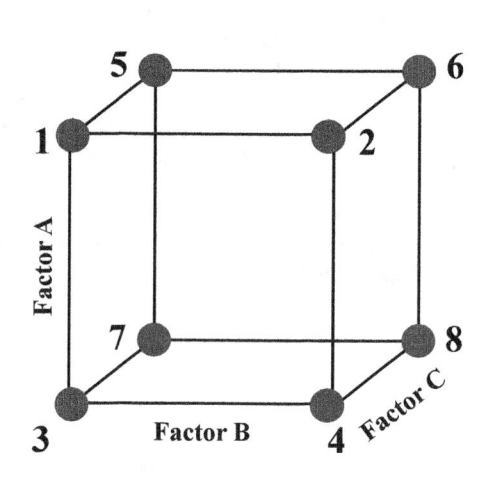

FIELDS OF STUDY

Research Design; Experimental Design

ABSTRACT

The full factorial design is a flexible experimental research design that analyzes all possible combinations of factors in an experiment. Because it enables researchers to consider many factors in a single study, this research design is important across scientific fields.

PRINCIPAL TERMS

- **experimental condition:** the level of an independent variable being tested.
- **experimental run:** a combination of experimental conditions that are measured at a time during the execution of an experiment; one component of an experimental design.
- **factor:** an independent variable or a confounding variable that affects the outcome, or dependent, variable.

- **interaction:** in statistics, an effect by which the impact of an independent variable on the dependent variable is altered, in a nonadditive way, by the influence of another independent variable or variables.

A Thorough Design

The full factorial design is a thorough quantitative experimental design. This design type aims to investigate the main effects of and the interactions between important multilevel factors through a controlled experiment. A key aspect of the full factorial design is understanding that independent variables are the factors that a researcher is most interested in examining. Each factor can have two or more levels.

Unlike other types of research designs, the full factorial design does not compare the means of individual experimental conditions. Instead, it compares the means of various *combinations* of conditions that include a given main effect to combinations that exclude it. Study subjects are divided into groups by experimental condition; ideally, this is done randomly. In effect, each group serves as an experimental group for analyses of certain effects and as a control group for other effects. This means that smaller sample sizes are needed than in traditional randomized controlled trials. The overall sample size need is based on the number of subjects needed for the smallest expected effect to have statistical power.

There are two types of full factorial designs: one contains only two-level factors and the other contains factors with more than two levels. The more factors are involved in a full factorial design, the more experimental runs are needed. An experimental run of a prior set of levels or factors is called a "replicate." Depending on the needs of the study, multiple replications may be performed.

Design Types Compared

A full factorial experiment offers several benefits. First, it addresses all possible combinations of factor levels. The two-level design aids researchers in examining the interactions between the independent variables in terms of their effect on the dependent variable. Because of its ability to determine the most important effects among multiple factors involved in an experiment, the full factorial design has been used extensively in scientific study.

However, because the full factorial design analyzes all combinations of factors, this type of design can be costly and time-consuming. Involving more factors increases the experimental runs needed exponentially. In some cases, such as in certain drug trials, it may not be possible or desirable to combine treatments or interventions. Lastly, it can be difficult to interpret the results from a full factorial design due to the large number of interactions that may exist.

By contrast, the fractional factorial design addresses only a subset or fraction of the total factors involved in an experiment. Consequently, the researcher chooses only a fraction of the experimental runs to do, making this design useful when there are a large number of factors and resources are limited. The chief drawback to this approach is that certain effects merge, leading to some degree of confounding.

Setting Up Independent Variables

In 1995, researchers conducted a study on the effects of a type of smile displayed by a transgressor had on leniency. The types of smiles (levels) were felt, miserable, fake, and neutral. The dependent variable was the degree of leniency exhibited toward the smiling person. Thus, in this experiment, there was one factor, facial expression, at four levels with possible multiple outcomes.

The number of experimental conditions is computed using the number of factors involved and the number of levels to be assessed. A full factorial design using three factors at two levels would be described as having $2 \times 2 \times 2$, or $2^3 = 8$ experimental conditions. The base of 2 in the notation 2^3 represents the levels, and the exponent of 3 shows the number of factors. For two-level factorial designs, the standard notation is 2^k, where k is the number of two-level factors under study.

Applications

Full factorial designs are a commonly used design in experimental research due to their thoroughness. Although an experiment using the full factorial design can become very large quickly, the design assesses all possible combinations of factors and makes it nearly impossible to miss any interactions. The full factorial design is the best way to examine interaction effects. In addition, it is efficient because researchers do not need to conduct many studies to examine the effects of different factors. The full factorial design is flexible and well suited for diverse fields from medicine to agriculture.

Pamelyn Witteman, PhD

FURTHER READING

Collins, Linda, et al. "Design of Experiments with Multiple Independent Variables: A Resource Management Perspective on Complete and Reduced Factorial Designs." *Psychological Methods*, vol. 14, no. 3, 2009, pp. 202–24. doi: 10.1037/a0015826.

Creswell, John W. *Research Design: Qualitative, Quantitative, and Mixed Approaches*. 4th ed., SAGE Publications, 2014.

Field, Andy. *Discovering Statistics Using IBM SPSS Statistics*. 4th ed., SAGE Publications, 2013.

"Fractional Factorial Designs." *Engineering Statistics Handbook*. NIST SEMATECH, www.itl.nist.gov/div898/handbook/pri/section3/pri334.htm. Accessed 14 Feb. 2017.

"Introduction to Factorial Designs." *Design of Experiments*. Pennsylvania State U, onlinecourses.science.psu.edu/stat503/node/26. Accessed 16 Feb. 2017.

"Introduction to Factorial Experimental Designs." *The Methodology Center*. Pennsylvania State U, methodology.psu.edu/ra/most/factorial. Accessed 16 Feb. 2017.

Swanson, Richard A., and Elwood F. Holton III, editors. *Research in Organizations: Foundations and Methods of Inquiry*. Berrett-Koehler Publishers, 2005.

FULL FACTORIAL DESIGN SAMPLE PROBLEM

A scientist seeks to examine the main effects of how two types of ice cream respond to two different temperatures, hot (30 degrees Celsius) and cold (0 degrees Celsius). Assuming a single replication, determine the number of experimental runs required in this full factorial design.

Answer:

Substitute the number of levels and the number of factors into the formula for experimental runs, and compute:

$$\text{levels}^{\text{factors}} = \text{experimental conditions}$$

$$2^2 = 4$$

In this study, there are four experimental conditions. Each condition would need its own experimental run; therefore, there would be four runs total.

Geller, Margaret

American Astrophysicist

American astrophysicist Margaret Geller's research into the structure of galaxies has yielded a new understanding of their shape and distribution. With colleague John Huchra, Geller discovered and named the Great Wall, a superstructure cluster of galaxies.

- **Born:** December 8, 1947; Ithaca, New York
- **Primary fields:** Physics; astronomy
- **Specialties:** Astrophysics; cosmology

Early Life

Margaret Joan Geller was born in Ithaca, New York, on December 8, 1947. Her father, a chemist who specialized in crystallography, encouraged her early interest in science and mathematics. She also learned from his example that science could be a useful and rewarding field. Often bored with her curriculum at school, Geller became an ambitious autodidact. It was unusual in the 1950s for a girl to be interested in mathematics and science, but the reactions she drew prepared her for a future in a male-dominated field.

Geller's family moved to Los Angeles during her high-school career. Geller began attending the University of California, Berkeley in 1966 and earned her undergraduate degree in physics four years later. While deciding on her field of study for graduate and doctoral work, she was advised by one of her physics professors to enter a field that would evolve over the next decades into a major discipline, and thus she chose astrophysics.

The physics department at Princeton University, New Jersey, in the 1970s was still male dominated, and Geller was only the second woman to earn a place in the program. She obtained her master's degree in 1972 and her PhD in 1974. Her doctoral thesis, entitled "Bright Galaxies in Rich Clusters: A Statistical Model for Magnitude Distributions," employed statistics to discern the patterns of brightness and motion given off by galaxy clusters, with the ultimate goal of mapping the structure of the universe.

Life's Work

Between 1974 and 1980, Geller undertook two fellowships, one as a postdoctoral fellow at the Harvard-Smithsonian Center for Astrophysics in Cambridge, Massachusetts, and the other as a senior visiting fellow at the Institute of Astronomy at Cambridge University in England. She also worked as a research associate and lecturer at Harvard University, which led to an offer in the 1980s to join the faculty as an untenured assistant professor. She became a full professor of astronomy at Harvard in 1988.

At Harvard, Geller began collaborating with fellow American astrophysicist John Huchra. Over the course of the 1980s, they performed major research and published results that detailed newly found structures of the universe. In 1983, Geller began working as an astronomer in the Smithsonian Astrophysical Observatory at the Harvard-Smithsonian Center. She was promoted to senior astronomer in 1991.

Before Geller began her astrophysics research, portions of the universe had been two-dimensionally mapped. Geller wanted to map the circumference of the space around Earth, extending five hundred million light-years away in three dimensions. Thus, not only the galaxies' positions would be understood, but so would their large-scale structures and distance from Earth.

The American astronomer Edwin Hubble posited in 1929 that the distance of galaxies can be inferred from the speed with which they move away from Earth; those moving the fastest are farthest away. Using what was later named Hubble's law, a scientist can estimate the age and size of the universe, and thus calculate its expansion rate. Geller and Huchra were not the first astrophysicists to apply Hubble's law to create maps of the universe, but they went further than anyone had before.

Geller and Huchra began mapping their proposed five-hundred-million-light-year space around Earth in large wedges, until they had completed the circumference; one of the wedges was later named after them. To do this, they used the 1.5-meter telescope of the Smithsonian Observatory located in Arizona, and later a 6.5-meter telescope, to obtain spectrographic images in what are called redshift surveys. Redshift indicates the speed at which a galaxy is moving away from Earth because of the universe's expansion; when this movement occurs, light waves appear to lengthen and thus move toward the red end of the visible spectrum. Though redshift does not yield exact measurements of distance, it gives a close approximation: the faster a galaxy is moving away, the greater its redshift and thus the greater its distance from Earth. Redshift surveys allow astrophysicists to view the universe in its current state, adjusted for the amount of time it takes light to travel, and understand the form the universe took historically. The initial aim of Geller and Huchra's project was to establish the specifics of fifteen thousand galaxies in this manner, with each galaxy measured individually.

In 1989, Geller and Huchra discovered an enormous cluster of galaxies that they deemed the "Great Wall." Also called the Coma Wall, Coma Cluster, or CfA2 Great Wall, this superstructure measures more than five hundred million light-years in length, three hundred million light-years in width, and fifteen million light-years in thickness. It was the largest known structure in the universe at the time.

Geller is the author of numerous academic articles in collaboration with other scientists, as well as articles written for general audiences. Her commitment to educating the public on astronomy has made her a popular figure in lecture halls and on radio, television, and film. Geller has popularized her discoveries and her working methods in two educational films, both made with the cinematographer Boyd Estus. The first, *Where the Galaxies Are* (1991), demonstrates the results of her redshift surveys; the second, *So Many Galaxies . . . So Little Time* (1992), shows a team of scientists collaborating on an astronomy project.

With only a small percentage of the observable universe mapped thus far, Geller has continued her explorations. She has studied distant parts of the universe with the aim of better understanding its history, evolution, and geometric structures. She has also researched the distribution of dark matter and mapped the Milky Way Galaxy as a means of understanding its relation to other galaxies.

Geller resigned from Harvard in 2001. She has remained at the Smithsonian Astrophysical Observatory, continuing her work as a senior astrophysicist there. During her career, she has earned many awards and honorary degrees, including the Newcomb-Cleveland Prize from the American Academy of Arts and Sciences in 1989 and a MacArthur Fellowship in 1990. She also received the Magellanic Premium award from the American Philosophical Society (APS) in 2008.

IMPACT

Geller's redshift surveys and her discovery of the Great Wall created a new image of the universe, which led to a series of metaphors to explain its structure. One of the metaphors is that the universe resembles bubbles of soap in a sink, with each bubble representing a galaxy. Another is that the universe resembles a sponge, in which the pockets of the sponge are the walls of the universe surrounding large voids. The clusters that emerged from Geller's maps form a pattern called a "stick man" for the form it appears to represent.

Before Geller's groundbreaking research, scientists held a different perspective on the structure of the universe. Galaxies, it was thought, were uniformly distributed. Moreover, scientists believed that the size of galaxies was limited by gravity and that massive structures therefore could not exist. Geller and Huchra, among others, proved these assumptions wrong through their spectrographic maps. Contrary to expectation, they discovered the Great Wall superstructure, and their maps revealed that galaxies are not evenly distributed, but rather take on the form of clusters and exist amongst huge voids.

With the advances that have been taking place in the field of astronomy, scientists have been able to further broaden their perspective of the universe as they examine galaxies at ever-more-distant reaches. In 2003, evidence for an even larger structure emerged from the redshift surveys of another team of astrophysicists. Called the Sloan Great Wall (SGW), it is at least 80 percent larger than the Great Wall and is based on surveys of a million galaxies.

—*Michael Aliprandini*

Further Reading

Kanipe, Jeff. *Chasing Hubble's Shadows: The Search for Galaxies at the Edge of Time.* New York: Hill, 2007. Print. Chronicles astronomers' efforts to map the universe farther and farther away from Earth. Covers such topics as dark matter, redshifts, the Milky Way Galaxy, and other clusters of galaxies. Illustrations, bibliography, index.

Rossiter, Margaret W. *Women Scientists in America: Forging a New World Since 1972.* Vol. 3. Baltimore: Johns Hopkins UP, 2012. Print. Biographies of American women scientists and their significant research from 1972 onward. Includes Margaret Geller. Illustrations, bibliography, index.

Weinberg, Steven. *Cosmology.* New York: Oxford UP, 2008. Print. An in-depth introduction to modern cosmological research. Discusses numerous topics, including those that Geller has worked on. Illustrations, glossary, index.

MARGARET GELLER DISCOVERS A GREAT WALL OF GALAXIES

With her research partner, astronomer John P. Huchra, Margaret Geller discovered what was at the time the largest structure in the known universe: the first Great Wall. The Great Wall is a system of thousands of galaxies that stretches across 5 percent of the known universe. The Wall is peculiar in its sheetlike shape; it is very thin, about fifteen million light-years, but stretches at least five hundred million light-years across. The finding changed scientists' understanding of large-scale structures in the universe and posed important questions about the nature of the big bang theory.

Technology for measuring the distance between galaxies became available in the 1970s, and early surveys using three-dimensional mapping techniques suggested the existence of large-scale structures. Geller and Huchra began their collaboration in 1980. Using the mapping technology, they set out to prove what cosmologists had long theorized: that galaxies are uniformly distributed in space. But their research, which provided the first visual conception of the universe and was published in 1986, showed something different. The surveys showed that galaxies are not, in fact, distributed uniformly; instead, there exist huge voids in space, described among astrophysicists as bubbles, and large sheetlike structures containing clusters of thousands of galaxies among them. Geller and Huchra reported their largest finding, the Great Wall, in 1989.

The existence of the Great Wall and similar large-scale structures challenges the predominant big bang theory of the creation of the universe. The big bang theory states that the universe was created 13.7 billion years ago in a state of nothingness, when an infinitely dense, infinitely hot singularity (so called because scientists do not understand its origin) began to cool and expand to the size of the universe known today. The universe continues to expand, which is known because galaxies appear to be moving away from Earth at the same rate over time. Large-scale structures challenge this theory simply because their sheer size defies these calculations. Geller and Huchra estimated that the Great Wall would have taken at least 100 billion years to form, far surpassing the big bang theory's estimate for the age of the universe.

Additionally, large-scale structures present the problem of lumps. Cosmologists have long theorized that the big bang distributed matter uniformly throughout the universe, as evidenced by the microwave background radiation considered to be the "echo" of the event. However, for the Great Wall to exist, there would have had to be a number of what scientists call lumps in the cosmos at the moment of expansion. Gravity presents another problem, as large-scale structures are too large to have been formed by the mutual gravitational attraction of their galaxies. Though answers regarding their origins remain scarce, the discovery of large-scale structures presents an opportunity to better understand the universe in which we live. Geller has suggested that large-scale structures are a clue to what might be a fundamental missing piece of that understanding.

H

Hawking, Stephen

NASA StarChild image of Stephen Hawking. By NASA [Public domain], via Wikimedia Commons

ENGLISH PHYSICIST

Many consider Stephen Hawking to be the greatest physicist of the late twentieth and early twenty-first centuries. His work combined the two primary developments of early twentieth-century physics, general relativity and quantum mechanics, to explain the origins and structure of the universe. His 1988 book A Brief History of Time was a best seller that introduced studies of the universe to nonscientists.

- **Born:** January 8, 1942
- **Birthplace:** Oxford, England

EARLY LIFE

Stephen Hawking was born on the three hundredth anniversary of the death of one of the greatest physicists of all time, Galileo Galilei, who is generally credited with proving that the earth revolves around the sun. Hawking's birth also came only four days after the three hundredth birthday of another great physicist, Isaac Newton, who developed a mathematical model to explain the structure of the universe that was essentially unchallenged for over two hundred years. Hawking was born in Oxford, England, and both of his parents had attended Oxford University. However, the Hawkings had only recently returned to Oxford to escape the likelihood of London being bombed during World War II.

Hawking's father, Frank, was a physician who had the same ambition for his son. After the war, the family returned to London, where Frank directed the parasitology division of the National Institute for Medical Research. The family moved again in 1950 to St. Albans in Hertfordshire, where, at age eight, Hawking enrolled in St. Albans School. He was an unexceptional student there, although his mathematics teacher proved to be an early inspiration to him.

Despite his father's encouragements, Hawking did not find medicine and biology theoretically rich enough and instead decided to major in physics at University College, Oxford. By his own admission, Hawking averaged barely one hour per day of studying at Oxford and decided to concentrate on theoretical physics as a way to avoid the busywork of memorizing facts. After graduating from Oxford, Hawking had hoped to study at Cambridge University with Fred Hoyle, who had developed steady-state cosmology, which argues that the structure of the universe remains relatively constant over time. However, Hawking's acceptance to Cambridge was contingent on his receiving honors from Oxford. Because of his lack of studying, his final examination scores at Oxford were only borderline for an honors degree. In an interview, he then told his examiners that if they gave him honors, he would go to Cambridge. Otherwise, he would stay at Oxford. They awarded him a first-class honors degree in natural science in 1962.

Upon reaching Trinity Hall, Cambridge, Hawking was disappointed to learn that he would not study with Hoyle but with Dennis Sciama, another steady-state cosmologist, but one unfamiliar to Hawking. However, Sciama turned out to be much more available and open to students developing their own alternative perspectives than Hoyle would have been. Within a few months of arriving at Cambridge, Hawking faced a far more serious disappointment. Although he was coxswain on a university crew team, he was never robust or athletic, and he was becoming increasingly clumsy. At his parents' insistence, he saw a doctor. He was diagnosed with amyotrophic lateral sclerosis, or ALS (also known as Lou Gehrig's disease in the United States), a degenerative condition causing the gradual loss of control of all muscles, including those necessary to move, gesture, speak, and swallow. Hawking was told he would live perhaps another two years and entered a deep depression as he contemplated the futility of trying to complete his doctorate. Sciama persuaded him to continue but refused to lower his standards.

In 1965, a turning point in Hawking's outlook for the future came when he married Jane Wilde, also a university student. As his condition worsened, he came to need a cane, then a wheelchair, and then an artificial speech synthesizer. However, he overcame the depression and survived despite the diagnosis. After completing his doctorate, he became a research fellow at Cambridge.

Through Sciama, Hawking met Roger Penrose, a mathematician who had developed the idea of a "singularity," a point at which the laws of mathematics and science break down. Hawking earned his doctorate by proposing that singularities could be used to understand the structure of the universe. He came to rethink the concept of singularity and argued that the laws of physics are continuous throughout the universe.

Life's Work

Early in the twentieth century, Newton's physics faced two challenges: relativity, conceived by Albert Einstein, and quantum mechanics, which had several founders. Although Einstein contributed to the development of quantum mechanics, he was uncomfortable with it. What most disturbed him was the uncertainty principle of Werner Heisenberg, which suggests that not everything is knowable and measurable. Because Einstein's relativity insists that everything can ultimately be determined, a position that it shared with Newtonian mechanics, Hawking calls it a "classical theory." One implication of relativity that Einstein himself shunned was that large stars could collapse into black holes, single points with such overwhelming gravitational strength that nothing, including light, can escape. Penrose considered black holes to be singularities and thought that since they emitted no signals, they were unknowable by the laws of physics as scientists understood them.

In the 1920s, astronomers began to believe that the universe was expanding. Hoyle's steady-state cosmology was one model of an expanding universe. However, a rival cosmology emerged to account for the expanding universe, which Hoyle disparagingly dismissed as the "big bang." The big bang theory proposes that billions of years ago the entire universe was compressed into a single point that exploded. The energy released from that explosion produced the four forces that govern the universe (gravity, electromagnetism, strong nuclear force, and weak nuclear force), as well as elementary particles, atoms, galaxies, stars, and planets. The initial explosion was so strong that it continues to cause the universe to expand to the present day. The big bang theory suggests that residual radiation from that explosion should pervade the cosmos even now. Around the time Hawking wrote his doctoral dissertation, the radiation was discovered. This caused most physicists to reject Hoyle's steady state. In his dissertation, Hawking suggested that the entire universe was originally a singularity like a black hole and that the big bang could be understood by comparing the universe to a star collapsing into a black hole. Hawking's analogy reversed the time sequence of a compressing star: the universe explodes from a singularity rather than imploding into one.

One of the most serious problems in twentieth-century physics is that its two main theories, relativity and quantum mechanics, appear to be incompatible. Black holes are predicted by relativity, which implies that they should emit no energy. In the 1970s, Hawking began to wonder if he could reconcile relativity and quantum mechanics by applying quantum mechanics to black holes. He found that according to quantum mechanics, black holes would indeed emit energy, a phenomenon since known as Hawking radiation, and would eventually explode. Hence, if the universe prior to the big bang was analogous to a black hole, it would one day burst like a black hole; therefore, the big bang could be explained by combining relativity and quantum mechanics. This means that the primordial universe as well as black holes can be understood by laws of mathematics and science and neither are really singularities. In a 2005 journal article, Hawking modified his theory of black holes to allow that information is never truly lost inside them and is eventually released into the universe again, if in garbled form.

Hawking came to believe that the universe was continuous and governed by a single set of laws. This would not mean that all events were predictable because a comprehensive theory of the universe would still contain Heisenberg's uncertainty principle. Hawking cautions that even if humans knew all the laws that underlie the operation of the universe, an account of all possible occurrences would require knowing the history of every particle, something clearly impossible within finite time.

Modern physics believes that the universe is governed by the four forces of gravity (which binds the planets, stars, and galaxies together), electromagnetism (which binds the atom together), the strong nuclear force (which holds the atomic nucleus together), and the weak nuclear force (which is necessary to account for radioactive decay). Einstein spent his later years unsuccessfully searching for a "grand unified theory," a single set of equations that would account for all four forces. Hawking was convinced that Einstein failed because he did not incorporate quantum mechanics. According to Hawking, the place in time to search for the grand unified theory is the time of the big bang, when, according to Hawking and other physicists, the four forces were one.

The universe assumed the shape it did because of the particular way energy happened to have been distributed at the time of the big bang. With slight differences, the galaxies and stars might never have developed. Indeed, the big bang may have produced regions where space, matter, and energy assumed different forms. Hawking proposed that the result could be an infinite number of "baby universes" and that the universe in which humans live may be only one of many possibilities.

The ultimate fate of the universe may depend upon how much matter it contains. If it contains only the matter that astronomers can see with visible light, then the universe should expand forever. If there is more matter, the universe will eventually contract and ultimately collapse into a single point. Hawking hypothesizes that black holes may have formed not only out of imploding stars but also from residues of the big bang. If that is true, black holes may pervade the universe, and there may be enough invisible matter to one day reverse its expansion. Long after the universe again congeals into a single point, it will again explode and expand. Hence, the universe would have a continual history whose broad outline is predictable. If Hawking is correct, then the dream of physics, a single set of laws to explain the development of the universe and all its contents, is indeed attainable.

In 2012, another of Hawking's long-held assumptions was proven false, when researchers Peter Higgs and Francois Englert discovered evidence for the existence of the Higgs boson (colloquially called the "God particle"), which, first theorized in 1964, is believed to give all elementary particles their mass. Hawking had wagered that the Higgs boson would never be found and that physicists would be required to reexamine the fundamental laws of the universe in light of unexpected experimental results. Consequently, he was rather disappointed at the 2012 discovery and encouraged the physics community to pursue other, little-explored areas of research, such as experiments on the M theory of the universe.

Hawking became the Lucasian Professor of Mathematics at Cambridge in 1979, a professorship once occupied by Newton, and he is a fellow of Gonville and Caius College. Believing that the public should be kept informed about scientific theory, in 1988, he published *A Brief History of Time*, a popular book about the intellectual history of cosmology. To the surprise of all, it remained a best seller for 237 weeks, and it sold more than nine million copies.

Other popular books followed: *Black Holes and Baby Universes* (1993), *The Universe in a Nutshell* (2001), *A Briefer History of Time* (2005), *God Created Integers: The Mathematical Breakthroughs That Changed History* (2005), and *The Grand Design* (2010). Hawking also cowrote an illustrated children's book series with his daughter Lucy—*George's Secret Key to the Universe* (2009), *George's Cosmic Treasure Hunt* (2011), and *George and the Big Bang* (2013)—and penned an autobiography, *My Brief History* (2013).

In 1985, Hawking had to have an emergency tracheotomy after suffering from pneumonia. He then required round-the-clock nursing care, and he lost his ability to speak. A computer system was built for him that enabled him to speak with an electronic voice with an American English accent. He could control the system with his little finger, by blinking to activate an infrared sensor, or by twitching his cheek. He writes of his condition,

> I am quite often asked: How do you feel about having ALS? The answer is, not a lot. I try to lead as normal a life as possible, and not think about my condition, or regret the things it prevents me from doing, which are not that many.

Many awards and honors have come to Hawking for his work in science. He was made a commander of the Order of the British Empire in 1982 and companion of honor in 1989. He is a member of the Royal Society, an honorary fellow of the Royal Society of Arts, and a member of the National Academy of Sciences in the United States. He received the Eddington Medal (1975), the Hughes (1976) and Copley medals (2006) of the Royal Society, the Albert Einstein Medal (1979), the Wolf Prize in Physics (1988), Prince of Asturias Awards in Concord (1989), the Julius Edgar Lilienfeld Prize of the American Physical Society (1999), the Michelson Morley Award (2003), and the Smithson Bicentennial Medal (2005), the Presidential Medal of Freedom (2009), and the Fundamental Physics Prize (2012). Hawking was also awarded a dozen honorary degrees.

Hawking also was a popular public speaker. In addition to physics, he often discussed his worries about the dangers to humanity's future from disease, nuclear weapons, and global warming. By 2006, he feared that global warming might already have proceeded too far to reverse and that humanity's chance for long-term survival might require colonizing other planets. In 2013, he urged further space exploration, stating that he believed humanity could not survive another thousand years on earth.

Hawking considers his own work incomplete. However, as he suggests, science is never complete: it is supposed to be self-critical and forever subject to revision.

Hawking is the director of research at Cambridge's Centre for Theoretical Cosmology and was the Sherman Fairchild Distinguished Scholar at the California Institute of Technology (Caltech). He continues to travel the world giving lectures, actively pursues his research at Cambridge, and mentors a group of graduate students.

Stephen and Jane Hawking divorced in 1991. He was married Elaine Mason, a nurse, from 1995 to 2006. He has three children and three grandchildren. To celebrate his sixty-fifth birthday in 2007, he took a trip on a specially designed plane that allowed him to float in zero-gravity, the first paraplegic to do so. In 2011, Hawking participated in a trial of the iBrain, a single-channel brain monitor mounted on a headband that might one day help Hawking and other ALS patients communicate more easily. In 2014, he wrote about artificial intelligence, specifically noting the dangers of such technology.

In July 2015, Hawking, along with tech billionaire Yuri Milner, launched a $100 million program, the Breakthrough Initiative, to search for extraterrestrial life. In May 2016, he hosted a PBS program, *Genius by Stephen Hawking*, which purported to "teach normal people to think like a genius." At the 2016 Pride of Britain Awards, he received a lifetime achievement award for his contribution to science and British culture.

SIGNIFICANCE

Hawking's theories led to a reassessment of Einstein's reworking of Newton's theory. Newton's theory had served as the foundation of physics for more than two centuries. Hawking, in turn, would not have developed his model had Einstein not preceded him, nor would Einstein have produced relativity without Newton's prior framework. The ideal for scientists is to build on each other's work. Modern scientific truths could become the scientific "errors" of the future, but these truths (and errors) nevertheless are foundational for future work in the sciences. As Newton famously admitted, "I can see far because I stand on the shoulders of giants." Hawking has proved to be another "layer" in a pyramid of giants.

There are critics who charge that Hawking simply was one of a number of scientists who tried to unify physics and to explain the development of the universe and that his disability is the focus of the media. Even if true, the claim cannot take away Hawking's actual accomplishments in physics and astronomy while living with ALS. Einstein, Bertrand Russell (who revised mathematics and worked for world peace), and Linus Pauling (who won Nobel Prizes in chemistry and peace) are remembered not only as great scientists but also as great humanitarians. Hawking, too, has been a great scientist and a great humanitarian. His life story inspires those of all physical ability levels.

—*Yale R. Magrass*

Further Reading

Benford, Gregory. "Leaping the Abyss." *Reason* 4 (2002): 24–31. Print.

Criss, Doug. "Stephen Hawking Says We've Got About 1,000 Years to Find a New Place to Live." *CNN*, 18 Nov. 2016, www.cnn.com/2016/11/17/health/hawking-humanity-trnd. Accessed 6 Jan. 2017.

Ferguson, Kitty. *Stephen Hawking: An Unfettered Mind.* New York: Macmillan, 2012. Print.

Hawking, Jane. *Music to Move the Stars: A Life with Stephen Hawking.* New York: Macmillan, 1999. Print.

Hawking, Stephen. *A Brief History of Time.* New York: Bantam, 1988. Print.

———. *A Briefer History of Time.* Rev. ed. New York: Bantam, 2005. Print.

———. *Black Holes and Baby Universes.* New York: Bantam, 1993. Print.

———. *My Brief History.* New York: Bantam, 2013, Print.

———. *The Universe in a Nutshell.* New York: Bantam, 2001. Print.

Hawking, Stephen, ed. *The Dreams That Stuff Is Made Of: The Most Astounding Papers of Quantum Physics and How They Shook the Scientific World.* Philadelphia: Running, 2011. Print.

Hawking, Stephen, and Roger Penrose. *The Nature of Space and Time.* Princeton: Princeton UP, 1996. Print.

Hern, Alex. "Stephen Hawking: AI Will Be 'Either Best or Worst Thing' for Humanity." *The Guardian,* 19 Oct. 2016, www.theguardian.com/science/2016/oct/19/stephen-hawking-ai-best-or-worst-thing-for-humanity-cambridge.

Kolbert, Elizabeth. "'I Think We Have It': Is the Higgs Boson a Disapppointment?" *New Yorker.* Condé Nast, 5 July 2012. Web. 24 Dec. 2013.

Mailet, Hélène. *Hawking Incorporated: Stephen Hawking and the Anthropology of the Knowing Subject.* Chicago: U of Chicago P, 2012. Print.

McCoy, Terrence. "How Stephen Hawking, Diagnosed with ALS Decades Ago, Is Still Alive." *The Washington Post,* 24 Feb. 2015, www.washingtonpost.com/news/morning-mix/wp/2015/02/24/how-stephen-hawking-survived-longer-than-possibly-any-other-als-patient. Accessed 6 Jan. 2017.

McEvoy, J. P., and Oscar Zarate. *Introducing Stephen Hawking.* New York: Totem, 1997. Print.

Merali, Zeeya. "Stephen Hawking: 'There Are No Black Holes.'" *Scientific American.* Scientific American, 27 Jan. 2014. Web. 15 Aug. 2014.

Singh, Simon. *Big Bang: The Origin of the Universe.* New York: Fourth Estate, 2004. Print.

"Stephen Hawking Talks of Impact of Higgs Boson Discovery." *UPI.* United Press International, 12 Nov. 2013. Web. 24 Dec. 2013.

White, Michael, and John Gribbin. *Stephen Hawking: A Life in Science.* New York: Penguin, 1993. Print.

STEPHEN HAWKING: HAWKING DISCOVERS BLACK HOLE RADIATION

One of the most mysterious objects in the universe, the black hole has been the subject of heated debate since the 1960s. It is theorized that a black hole can form in two ways. The first is through the collapse of a large star that is at least three times bigger than our sun. In the 1970s, physicist Stephen Hawking argued for the existence of much smaller, ancient black holes. These black holes were formed even before the stars through fluctuations in the primordial soup of the universe.

Hawking theorized that in the seconds following the Big Bang, the incredible explosion of matter and energy would have been a swirling, chaotic mess. In the midst of this maelstrom, slight fluctuations in density could have solidified into small black holes. Unlike their star-formed peers, which are so dense and massive that almost no energy can escape their pull, these primordial black holes are smaller, weaker, and hotter.

According to Hawking, the heat and weakness of small black holes allows them to radiate energy at a much higher rate than bigger black holes. As they radiate energy into space, they shrink, and as they shrink, they heat up. Eventually, small black holes evaporate faster and faster until they disappear completely. Physicists calculate that the smallest primordial black holes are likely gone already, evaporating into bursts of gamma rays.

Hawking's calculations are based on quantum mechanics, which often conflicts with Einstein's theory of general relativity. Black holes are used by physicists like Hawking as models for extreme gravity, where quantum mechanics and classical physics clash. While the mathematics involved in the study of general relativity and quantum mechanics is some of the most advanced in the world, scientists are at a loss to combine the two theories. The study of black holes may one day uncover the key to uniting all branches of physics into one "theory of everything."

As a testament to Hawking's work on black hole equations, the energy emitted from black holes was named Hawking radiation. Normal massive black holes do theoretically emit radiation but not much more than the surrounding radiation of space, known as the cosmic microwave background. This lack of radiation makes black holes very hard to detect. They are "black" because they do not emit light, but they can be detected by their gravitational effect on nearby objects.

Primordial black holes could be easier to find. In June 2008, NASA launched the Fermi Gamma-ray Space Telescope into low orbit. Part of its mission is to detect the evaporation of a primordial black hole based on Hawking's calculations. According to NASA, a rapidly deteriorating small black hole could be detected with the Large Area Telescope, which attempts to scan as much of the sky as possible in order to detect elusive gamma-ray bursts. Detecting Hawking's primordial black holes through their gamma radiation could be the first step to developing the theory of everything.

HIGGS, PETER

BRITISH PHYSICIST

British physicist Peter Higgs proposed in 1964 the existence of a subatomic particle to account for the origin of mass in other subatomic particles such as protons and neutrons. In July 2012, his predictions were verified, providing an experimental basis for a complete revolution in unified field theory.

- **Born:** May 29, 1929
- **Birthplace:** Newcastle upon Tyne, United Kingdom

EARLY LIFE

Peter Ware Higgs was born in Newcastle upon Tyne, England, on May 29, 1929. His father, Thomas Higgs, was a sound engineer for the British Broadcasting Company (BBC). From 1930 through 1941, Higgs lived and attended elementary schools in Birmingham. Britain's involvement in World War II interrupted his formal early schooling, and he was taught at home by his mother, Gertrude Higgs, in Bristol. From 1940 to 1941, he attended Halesowen Grammar School in Worcestershire, then Cotham Grammar School in Bristol from 1941 to 1946 and the City of London School from 1946 to 1947. While at Cotham Grammar School, he learned of and was inspired by the work of Paul Dirac, a former Cotham student and one of the founders of the field of quantum mechanics. In 1947, he began studies in physics at King's College London. In 1950, Higgs was awarded the bachelor of science degree, with first class honors in physics. Studying under Charles Coulson and Christopher Longuet-Higgins, Higgs focused on symmetry in physical systems during graduate school. He completed his master of science degree in 1951 and his doctorate in 1954.

LIFE'S WORK

Following the completion of his PhD, Higgs took a position as a senior research fellow at the University of Edinburgh. An ICI Research Fellowship took him back to London in 1956, where he worked and lectured in mathematics at University College and then Imperial College. In 1960 he returned to Edinburgh to lecture in mathematical physics at the university's

Tait Institute. He was promoted to professor of theoretical physics in 1980, a position he held until his retirement in 1996, when he became professor emeritus. At Edinburgh he met Jody Williamson, a linguist. They were married in 1962 and had two sons. Christopher Higgs, born in 1966, is a computer scientist; Jonathan Higgs, born in 1969, is a musician. The couple divorced in 1972, and Jody Williamson Higgs died on February 3, 2008, from leukemia.

When Peter Higgs began his career, theoretical physicists were already working to develop what Albert Einstein called a unified field theory that explains how the universe works in terms of physical matter and all the forces that affect it. The most important theoretical model that twenty-first-century physicists use is called the Standard Model of particle physics. It shows the basic subatomic particles and how they comprise matter as we know it. For example, protons and neutrons are made of different combinations of a variety of particles called quarks. Electrons and neutrinos are composed of particles called leptons. The Standard Model also shows bosons, which are particles that carry forces, such as electromagnetism and the wear nuclear force. Bosons include photons, gluons, Z bosons, and W bosons.

One of the greatest problems in the development of the unified field theory has been the relationship between energy and mass, which is the primary characteristic of physical matter. Higgs's work demonstrates the principles that relate energy, mass, and matter. The Higgs mechanism, a process described by Higgs and five other physicists in 1964, states that through interaction with a universal field, a subatomic particle (later named the Higgs boson) decays and transforms massless energy into the mass of other fundamental particles, particularly protons and neutrons. In other words, the Higgs boson explains how other particles have the physical property of mass.

Particle physicists began to search for evidence of the Higgs boson in high-energy particle collision experiments in 1975, following the formalization of Higgs's theories into a testable hypothesis over the intervening years since 1964. On July 4, 2012, researchers at the Large Hadron Collider at the European Organization for Nuclear Research (also known as CERN, Conseil Européan pour la Recherche Nucléaire) announced that they had identified results consistent with the existence of the Higgs boson. Although the results were not definitive proof that the particle exists, they still provided strong verification of Higgs's work and represent one step closer to the development of the unified theory that has so far been elusive. By 2014, scientists who had continued to research the findings declared that, because they had been able to observe the decay of the particle considered necessary to truly identify it as the Higgs boson, the particle discovered by CERN was indeed the Higgs boson.

Higgs's work has garnered him numerous awards of recognition from his peers and the scientific community, including the Saltire Society & Royal Bank of Scotland Scottish Science Award (1990), the Royal Society of Edinburgh James Scott Prize Lectureship (1993), the Paul Dirac Medal and Prize of the Institute of Physics (1997), and the Stockholm Academy of Sciences Oskar Klein Memorial Lecture and Medal (2009). He has received honorary degrees from the Universities of Bristol (1997), Edinburgh (1998), Glasgow (2002), Swansea (2008), Kings College London (2009), University College London (2010) and Cambridge (2012). In 2013, he and Belgian theoretical physicist François Englert shared the Nobel Prize in Physics for their contribution to the theoretical existence of the significant Higgs boson.

Retired from his position with the University of Edinburgh since 1996, Higgs has stated in interviews since his reception of the Nobel Prize that he does not believe that were he still actively working that he would have been able to make the same kind of discovery that he made in the 1960s in today's more fast-paced, high-pressure academic science climate.

Impact

Philosophers and scientists have sought an explanation for the existence of the physical world, and of the universe in general, for thousands of years. Ancient Greek philosophers, from whom came the concept of atoms, tried to perceive the physical world as being constructed from properties within an all-pervasive "ether." Alchemists, natural philosophers, and mystics since that time have also tried to explain the universe in a similar way. Their works and studies eventually produced the modern atomic theory based on the principles of quantum mechanics developed by Paul Dirac, Albert Einstein, and many other physicists. At the heart of this ongoing effort is the desire for a unified "theory of everything" that describes the essence of matter. The major stumbling block for such a theory has been in identifying the fundamental relationship between matter, mass, and energy. Higgs's work has provided a viable means of defining and identifying that relationship, with the Higgs field (which gives rise to the Higgs boson) perhaps fulfilling the role of the "universal ether," and the Higgs boson providing the means whereby energy becomes mass and matter. Higgs's work holds the promise of revolutionizing the science of physics with the development of a complete unified theory.

—*Richard M. Renneboog, MS*

FURTHER READING

Close, Frank. *The Infinity Puzzle: Quantum Field Theory and the Hunt for an Orderly Universe*. Basic, 2011. An account that does not rely on mathematics to give the reader an appreciation of the intricacies of quantum field theory and its development to the present day.

Martin, Victoria. "A Layperson's Guide to the Higgs Boson." *The University of Edinburgh*. School of Physics and Astronomy / University of Edinburgh, 2 July 2012, www.ph.ed.ac.uk/higgs/laypersons-guide. Accessed 24 July 2012.

Overbye, Dennis. "A Pioneer as Elusive as His Particle." *The New York Times*, 15 Mar. 2014, www.nytimes.com/2014/09/16/science/a-discoverer-as-elusive-as-his-particle-.html. Accessed 20 Mar. 2017.

Sample, Ian. *Massive: The Missing Particle That Sparked the Greatest Hunt in Science*. Basic, 2010. Presents the history of the search for the Higgs boson in a highly readable format that describes the development of Higgs's theories in the context of the history of subatomic and elementary particle physics.

Tully, Christopher G. *Elementary Particle Physics in a Nutshell*. Princeton UP, 2011. Explains the concepts of elementary particle physics and Higgs interactions in an understandable way, but requires familiarity with the mathematics of quantum mechanics.

PETER HIGGS AND THE SEARCH FOR THE HIGGS BOSON

Since Peter Higgs and others conceived of the Higgs boson, physicists have been trying to prove that the particle actually exists. If they can do so, they think that the Higgs boson will provide the missing piece of unified field theory, a single theory that explains how the universe works in terms of physical matter and the forces that affect it.

In 1964, Peter Higgs and others suggested that there must be a relationship between mass and masslessness. Higgs stated that this relationship is detectable through the interaction of the Higgs boson and an omnipresent field, the Higgs field. During the interaction, the Higgs boson theoretically decomposes into protons and neutrons and gives them mass in the process. Higgs's theoretical work became integral to the Standard Model of particle physics after 1971, by which time Tini Veltman and Gerard t'Hooft had worked out some of the mathematical problems with Higgs's theory.

In the 1970s, different quark particles began to be observed and identified and the pieces of the Standard Model puzzle started to come together. In 1975, experimental work toward identifying the Higgs boson began. W bosons and Z bosons were discovered in the 1980s, though the mechanism by which they acquired their mass was unidentified, and was believed to be dependent on the Higgs boson. By the end of the year 2000, the Higgs boson had not itself been detected, but each subsequent observation refined the range of masses within which it should exist. The masses of such particles are given in terms of how much energy they have relative to other particles, and in 2001 the mass for the Higgs boson had been narrowed to a range between 115 and 158 gigaelectron volts (GeV; one billion electron volts).

The search for the Higgs boson continued at CERN, located in Geneva, Switzerland. CERN houses the Large Hadron Collider (LHC), which, with a ring of superconducting magnets some 27 kilometers in diameter and situated about 100 meters below the border between France and Switzerland, is the largest and most powerful particle collider ever constructed. The LHC operates at energies of up to one teraelectron volt (TeV; one trillion electron volts) to accelerate particle streams to 99.99998% of the speed of light before bringing them into collision. Particle physicists measure the interactions of the accelerated particles and the particles that they produce in the collisions to verify the mathematical predictions of quantum mechanics theory, including that of the Higgs boson. On July 4, 2012, CERN reported having identified results consistent with the formation of the Higgs boson, a particle with a mass of 125.3 GeV, which is about 125 times the mass of a single proton.

HISTOGRAMS

Seasons in Which Students' Birthdays Fall

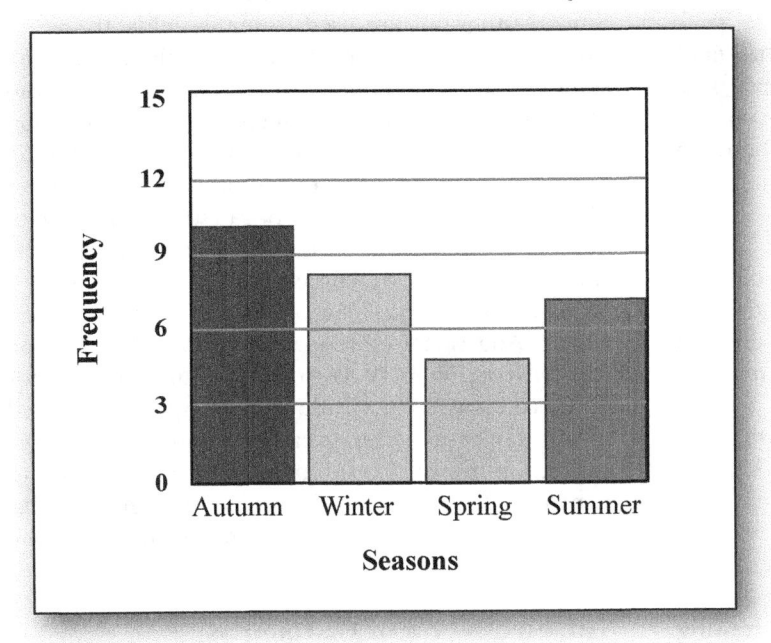

FIELDS OF STUDY

Statistical Analysis

ABSTRACT

Histograms present data sets as simple, easily understood graphics explaining how often a particular member of a group falls within a selected category. They also visually summarize trends, such as how one group differs from another in the depicted population.

PRINCIPAL TERMS

- **class frequency:** the number of times that a particular value occurs when data are sorted into designated classes of values on a histogram.
- **class interval:** the size of a designated class of values represented on a histogram.
- **continuous variable:** a variable that can have any measurable value within a given range of values, such as age or temperature.
- **discrete variable:** a variable that can only have a finite, countable value, such as the result of a dice roll.
- **frequency:** the number of times that an event occurs within a data set; may either absolute, representing actual totals, or relative, given as percentages.
- **qualitative data:** information that describes but does not numerically measure the attributes of a phenomenon.

A Simple Visual Aid

A histogram is a statistical graphic that allows researchers to present large amounts of data in a simple, easily understood way. It looks much like the bar chart, which has long been used for a similar purpose. The histogram and the bar chart follow different rules, however. One major difference is that histograms depict continuous variables, which are quantitative data, while bar charts depict discrete variables, which are often qualitative data. Thus, gaps may appear between bars on a bar chart but not between classes, or bins, on a histogram.

Another difference is that bar charts only represent frequency as length along one axis, usually the *y*-axis. Histograms present numerical data along both the *x*- and *y*-axes. The *x*-axis gives each class interval, and the *y*-axis represents class frequency. Thus, the area, not just the length, of the bin indicates the size of the class. Most intervals on a histogram are equal in width, but if these are unequal in size, some intervals must be broader or narrower than others.

The peaks on a histogram provide an immediate explanation of distribution. A unimodal histogram, for instance, depicts a lone peak that represents the mode, or most common value. By contrast, a bimodal histogram has two peaks and thus two modes. In multimodal histograms, data peaks on at least three points.

Data that follow a normal distribution tend to cluster toward and peak in the middle of the histogram. That area of clustering is called the "center" of distribution. Such data generally produce symmetrical histograms that depict the familiar bell-shaped curve.

Data that vary from the normal distribution are asymmetric and said to be "skewed." distributions may be skewed either left (negative) or right (positive). The direction of skew reflects where the "tail," or values decreasing from the center toward zero, lies.

Histograms also show the spread of the data and any outliers, or data points that lie outside the main body of data.

Advantages and Disadvantages

The chief advantage of histograms is their ability to make large amounts of data easily understood. They are also fairly simple in design and flexible enough to be used across fields. However, that very simplicity may lead to statistical errors if researchers are not careful to follow the basic rules governing histogram creation. Five steps are generally followed when creating histograms. First, the uppermost and lowermost values in the data set are found. Next, one chooses the number of bins that will be represented, and sets limits on those bins by applying a rule. Once data is sorted into the bins, the final step is drawing the histogram.

Various statisticians have attempted to establish rules for bin creation and sorting. The rule most commonly used in software is Sturges's rule, developed by statistician Herbert A. Sturges in 1926. Sturges's rule for the number of intervals is

$$C_{intervals} = 1 + 3.322 \log_{10}(n)$$

where n is the sample size. That value is then rounded to the nearest whole number. For instance, a sample size of 16 would call for five bins, a sample size of 64 would call for seven bins, and a sample size of 1,024 would call for 11 bins. To find the class interval, that is, the width of each bin, divide the range of the data, R, by that same formula. Sturges's rule works best for normally distributed data; skewed data call for more bins. An alternate rule is the Rice rule:

$$C_{intervals} = 2n^{1/3}$$

Here n is the sample size. The Rice rule divides large sample sizes into smaller class intervals.

Very small samples can lead to errors when plotted on a histogram. Small data sets may appear to have more or less variability in distribution. They may appear more skewed than they really are.

Histograms in the Real World

Since histograms are used in many fields, there are numerous real-world examples of their use. For instance, the US Centers for Disease Control and Prevention examined the effects of introducing mosquito control in New York City on the number of reported cases of West Nile encephalitis in the summer of 1999. Control measures began the first week of September. Data were sorted into nine bins, by week: July 25 (zero cases), August 1 (one case), August 8 (four cases), August 15 (six cases), August 22 (ten cases), August 29 (three cases), September 5 (one case), September 12 (zero cases), and September 19 (zero cases). The resulting histogram was skewed to the left because the greatest number of cases occurred earlier in the summer, before the mosquito control efforts began. Thus, the histogram helps show the effectiveness of the intervention on the spread of disease.

The Importance of Histograms

Histograms are considered a major research tool across disciplines because they allow researchers to present large data sets in a simple manner. Data sets that seem overwhelming when taken individually fall into recognizable patterns when placed in histograms. At a glance, histograms allow readers to identify normal distributions, anomalies, and trends. It does not require special knowledge of a field to understand a histogram, although further understanding may be essential to understanding why distributions follow a particular pattern.

Elizabeth Rholetter Purdy, PhD

Further Reading

Aldrich, James O., and Hilda M. Rodriguez. *Building SPSS Graphs to Understand Data.* SAGE Publications, 2013.

Doane, David, and Lori Seward. *Applied Statistics in Business and Economics.* McGraw-Hill Higher Education, 2015.

"Histograms." *Student Learning Development,* University of Leicester, www2.le.ac.uk/offices/ld/resources/numerical-data/histograms. Accessed 12 Apr. 2017.

Jackson, Daniel. *Statistics for Quality Control.* Industrial Press, 2015.

Lane, David. "Histograms." *Online Statistics Education: A Multimedia Course of Study,* onlinestatbook.com/Online_Statistics_Education.pdf. Accessed 12 Apr. 2017.

Moore, David S., William Notz, and Michael A. Fligner. *The Basic Practice of Statistics.* 7th ed., W. H. Freeman, 2013.

Schumacker, Randall, and Sara Tomek. *Understanding Statistics Using R.* Springer Science and Business Media, 2013.

SAMPLE PROBLEM

In an American government class of sixty-four students, scores on a recent test ranged from 14 to 102. A score of 60 was needed to pass. Describe the histogram that could be created using the following data sample and Sturges's rule.

Recall that Sturges's rule is

$$C_{intervals} = 1 + 3.322 \log_{10}(n)$$

Where *n* is sample size.

Point Range	Number of Students
0-14	1
15-29	2
30-44	3
45-59	4
60-74	28
75-88	20
88-102	6

Answer:

Substitute the value of the sample size (*n* = 64) into the formula and solve:

$$C_{intervals} = 1 + 3.322 \log_{10}(n)$$
$$= 1 + 3.322 \log_{10}(64)$$
$$= 1 + 5.99$$
$$= 6.99$$

Based on Sturges's rule, seven bins would be created.

The histogram would show that the majority of students in the class scored in the 59–74 and 75–88 intervals. Six students scored at least 88 or higher, and ten students failed the exam. Thus, the histogram would be asymmetrical, with a longer tail on the left (negative side) than on the right (positive).

Holistic Study (Ethnography)

Dos and Don'ts of a Holistic (Ethnographic) Study

DOs	DON'Ts
Do observe discreetly	Don't be obtrusive and obvious
Do watch for nonverbal behaviors	Don't miss observations due to note-taking
Do listen for verbal cues	Don't focus on only one individual
Do preserve objectivity	Don't be biased
Do identify themes or patterns among behaviors	Don't conduct a study for validation
Do work with other researchers	Don't generalize observations

FIELDS OF STUDY

Research Design; Research Theory; Experimental Design

ABSTRACT

A holistic study is the study of a system as a whole, taking into account how the parts of the system work together to form a functioning unit. In anthropology, "holistic study" is typically synonymous with ethnography, the systematic and comprehensive study of different peoples and cultures. The researcher lives with and among the people being studied for a predetermined period of time and observes them, taking daily field notes. From these notes, interview subjects, called "informants," are identified and interviewed to contribute to a holistic view of the people being studied.

PRINCIPAL TERMS

- **case study:** a research method in which selected individuals or small groups are observed and examined thoroughly.
- **participant observation:** a method of research in which the researcher observes a group of individuals by participating in their activities, typically over an extended period of time.
- **realist ethnography:** the practice of writing about the subjects of an ethnographic study in the first person, from the researcher's point of view, while taking into account the ways in which the researcher's presence might have influenced the observed behaviors.

HOLISM AND ETHNOGRAPHY

A holistic study is a study based on the philosophy of holism (that is, the idea that systems should be studied and understood as whole, complex entities rather than as the sums of their parts). While this philosophy can be applied to any given system in any field, holistic studies are most often practiced by anthropologists in the study of different cultures. When applied to people, as in anthropology, the term "holistic study" is typically synonymous with ethnography. "Ethnography" means both the practice of conducting a holistic case study and the account produced by such a study.

Ethnography looks at subjects in their full context, taking into account their families, societies, and individual and group histories. It observes objects, behaviors, and rituals, using both qualitative and quantitative data, although it is fundamentally qualitative. A classic approach to ethnography is participant observation, in which the researcher observes a culture and conducts interviews with various members while also participating in the daily activities of that culture. It is an iterative process, as the review of daily field notes may prompt new lines of inquiry. With very little control of variables, the researcher may need to tweak or completely change their research design as circumstances change.

There are various approaches one may take when writing an ethnography (that is, a report on the results of an ethnographic study). The traditional approach, which remains the most common, is realist ethnography, in which the researcher reports their observations from their own first-person perspective. Such studies acknowledge the ethnographer's role in the study and attempt to identify ways in which the ethnographer may have influenced the events observed. They strive for objectivity with the understanding that true objectivity is ultimately impossible to achieve.

Ethnographies are often supplemented by governmental or cultural data, such as birth, death, and baptismal records, as well as photographs and recordings. Quantitative analysis may be performed on qualitative data, such as analyzing the frequency of word usage. Replication is difficult with ethnographic studies, as each researcher brings their own perspective, and ethnography is not an objective science. In addition to anthropology, ethnographic studies are often conducted in sociology, psychology, and city planning.

CONSIDERATIONS OF ETHNOGRAPHY

Ethnography does not begin with a hypothesis or prediction. Rather, it is an open-ended, multifaceted exploration that is intended to contribute to general knowledge and may identify questions that might be answered with further study. Ethnography cannot be conducted in a laboratory, because culture occurs within context.

There are some drawbacks to ethnography. The researcher's own culture might differ so much from the culture being studied as to prevent useful conclusions from being drawn. Bias is inevitable in ethnography, both from preconceptions and from interactions between the researcher and the host culture. Time constraints may prevent researchers from identifying cultural cycles that last longer than their stay. Despite these limitations, however, ethnography provides invaluable information that cannot be obtained by other methods of research.

Ethnography in Practice

An open-ended, exploratory question regarding a culture or society requires ethnographic study. For example, if a researcher wants to understand the daily workings of an indigenous Arctic culture, they conduct an ethnographic study. The researcher requests and, ideally, is granted the opportunity to live among the people of that culture. They make field notes each day, including all of the events that took place and their personal observations and impressions about those events. The researcher constantly reviews these notes during the course of the study. After a period, when they have gathered enough information, they identify potential interview subjects, called "informants," whom they will ask questions based on previous observations. These informants are not selected using random sampling, but rather are carefully selected based on the depth of information and the particular perspective they may be able to provide. Cultural records, photographs, and recordings are also added to the data when possible.

On the other hand, if the researcher wants to answer a particular question, such as how an indigenous Arctic culture finds enough food in a barren climate, they would not choose an ethnographic study to answer this question. The researcher might still live among the people and observe their practices firsthand, but it would not be an ethnographic study, as the researcher would focus one particular aspect of the society rather than the society's function as a whole. The researcher might instead decide that interviews, historical research, or some other research design is a better approach to answering this question than firsthand observation.

The Importance of Ethnography

Traditionally, ethnography was the purview of anthropologists. While still a common practice in anthropology, ethnography is now used in many other fields, with a broader definition of culture that includes microcosms of societies, such as a town, a neighborhood, or a workplace. A company may be considered a culture, and a researcher in human interaction or sociology might complete an ethnographic study by working there, taking field notes, and interviewing key players. A political science researcher might consider a campaign to be a culture and volunteer for the campaign as a way of living among the members of that culture. To study a factory town, multiple researchers might live in the town, some working at the factory and some not. In all of these situations, obtaining a complete picture requires submersion in the culture and the daily lives of its inhabitants.

Maura Valentino, MSLIS

Further Reading

Boellstorff, Tom, et al. *Ethnography and Virtual Worlds: A Handbook of Method.* Princeton UP, 2012.

De Munck, Victor C. *Research Design and Methods for Studying Cultures.* AltaMira Press, 2009.

Fetterman, David M. *Ethnography: Step-by-Step.* 3rd ed., SAGE Publications, 2010.

O'Reilly, Karen. *Key Concepts in Ethnography.* SAGE Publications, 2009.

Otto, Ton, and Nils Bubandt, editors. *Experiments in Holism: Theory and Practice in Contemporary Anthropology.* Wiley-Blackwell, 2010.

Hammersley, Martyn, and Paul Atkinson. *Ethnography: Principles in Practice.* 3rd ed., Routledge, 2007.

Wolcott, Harry F. *Ethnography: A Way of Seeing.* 2nd ed., AltaMira Press, 2008.

HOOKE, ROBERT

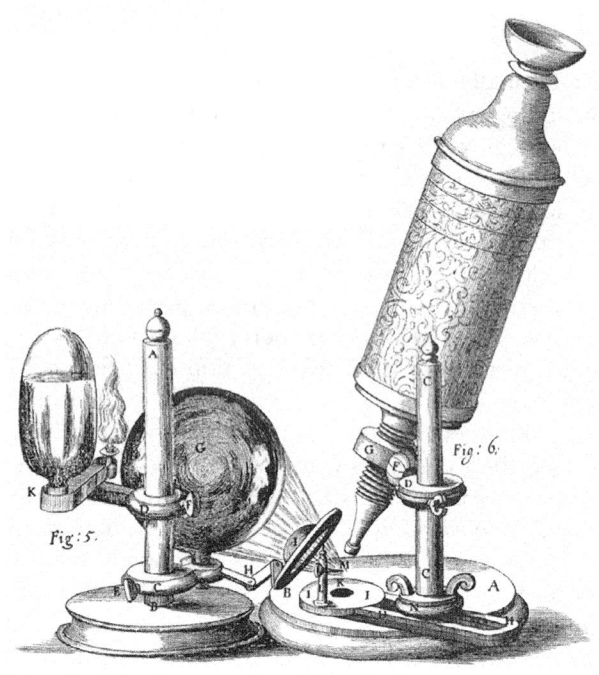

Robert Hooke's microscope. From Scheme I. of his 1665 Micrographia. On permanent display in "The Evolution of the Microscope" exhibit at the National Museum of Health and Medicine, in Washington, DC. By Robert Hooke (w:en:Image:Hooke-microscope.png) [Public domain], via Wikimedia Commons

ENGLISH PHYSICIST

As curator of experiments for England's Royal Society, Robert Hooke proved to be one of the most influential experimentalists and inventors of the seventeenth century, contributing to a wide range of scientific fields.

- **Born:** July 18, 1635; Freshwater, England
- **Died:** March 3, 1703; London, England
- **Primary field:** Physics
- **Specialties:** Mechanics; theoretical physics

EARLY LIFE

Robert Hooke was born in 1635 into the household of John Hooke, a minister, and his second wife, Cecelie. He was so sickly as a child that his parents did not expect him to survive, and frequent headaches later kept him from attending school. As a result, Hooke was schooled at home and largely left to pursue his own interests. These included drawing and studying the inner workings of machines, which he disassembled and used as guides for making his own devices; for instance, he constructed wooden clocks and a working model of a warship, complete with firing cannons.

Hooke's father died in 1648, and after collecting his inheritance, the thirteen-year-old Hooke moved to London. At first he intended to become the apprentice of Peter Lely, a celebrated painter of miniature portraits, but he soon changed his mind and entered Westminster School, a premier preparatory school. There, Hooke impressed the headmaster, Richard Busby, with his talent for language and geometry and his mechanical skills. With Busby's special tutelage and support, Hooke entered Christ Church College, Oxford, in 1653.

At Oxford, Hooke eventually joined a group of natural philosophers who viewed nature as a vast mechanism and sought to determine how it worked. Some members of this group, including Christopher Wren and Robert Boyle, became Hooke's lifelong friends and collaborators. In 1657, Boyle hired Hooke to construct laboratory equipment and help with experiments; Hooke's vacuum pump, a famous instrument of its day, permitted Boyle to explore the properties of air, part of the research that led to the discovery of Boyle's law (published 1662). At the same time, Hooke began to explore his lasting interest in chronometers as he attempted to construct a timepiece that would remain accurate enough aboard a ship to use in determining longitudes.

LIFE'S WORK

Hooke's first solo publication was a 1661 pamphlet explaining capillary action. The work so impressed contemporary scientists that, with the help of Wren and Boyle, Hooke was hired as curator of experiments for the newly founded Royal Society, an organization of scientists and natural philosophers, in 1662. He was elected a full member the following year. His duties for the society were onerous; for the benefit of members, he was to perform several experiments of his own, as well as any that members themselves suggested at each weekly meeting. This he did with ingenuity and gusto, setting a high intellectual standard for the Royal Society. During the next twenty-six years, in hundreds of experiments, Hooke investigated the nature of light, air, gravity, magnetism, gunpowder, comets and other celestial phenomena, optics, chronometers (particularly the use of springs and pendulums), lightning, earthquakes, respiration, circulation, fossils, and medical treatments while also inventing carriages, the iris diaphragm, meteorological instruments, watches, and a wide variety of scientific tools. Hooke also took on a number of scientific posts, each with a considerable workload. In 1664, John Cutler, a wealthy merchant, endowed a lecture series especially for Hooke in which he was to discuss the practical sciences and trades. The following year, Hooke became Gresham College's professor of geometry; the appointment included an apartment in the college's London premises, which remained his home from then on. Amid these manifold duties, Hooke published one of the masterpieces of seventeenth-century science literature, *Micrographia* (1665). It quickly became a best seller and was admired for the wide range of topics discussed, including new theories of light and combustion, as well as for its beautiful drawings of the objects and creatures that Hooke had examined under his improved microscope. The book established the importance of the microscope as a scientific instrument and embodied Hooke's guiding principle, drawn from the ideas of Francis Bacon, that philosophers must base their understanding of the world on rigorous observation and experimentation; that is, theory must arise from demonstrable fact.

In subsequent publications, Hooke formulated the law of elasticity, later known as Hooke's law, which states that stress in springs is directly proportionate to strain; in other words, the force applied to and released by coiling and uncoiling a spring is directly proportionate to the amount of deformation undergone by that spring. Hooke also conducted an early analysis of harmonic motion, helped found the fields of meteorology and crystallography, proposed an explanation for celestial dynamics, and advanced broadly correct explanations for the origin of fossils and the evolution of species during environmental change.

Following the Great Fire of London in 1666, Hooke helped to rebuild the devastated central city. He was appointed city surveyor in 1667 and became responsible for laying out the new streets, designing many new public buildings and overseeing their construction, enforcing building codes, and settling property disputes. He executed these many tasks in coordination with Wren, who had been appointed royal surveyor. This partnership broadened when Hooke became an assistant and virtual partner in Wren's architectural firm. Hooke assisted Wren in projects such as designing the new Saint Paul's Cathedral and also designed a number of private houses and public buildings on his own, including Bedlam Hospital, the Royal College of Physicians, and the towering monument memorializing the fire. In 1677, Hooke began a five-year term as secretary of the Royal Society, and he later served on its council while also caring for the society's library and its collection of rarities.

During the 1670s and 1680s, Hooke quarreled with fellow scientist Isaac Newton over the nature of light and the theory of universal gravitation. He held that light was composed of waves, while Newton insisted it was made up of tiny particles; both were later shown to be correct. Hooke also claimed to have first stated that gravity was centered in bodies and decreased in strength by the square of the distance between those bodies (the inverse-square law), but he could not demonstrate it mathematically; Newton did so in *Philosophiae Naturalis Principia Mathematica* (1687; *The Mathematical Principles of Natural Philosophy*, 1729) and ridiculed Hooke's claim to the discovery. These and other disputes, as well as failing health, led Hooke to become somewhat reclusive in his later life. He died at Gresham College in London in 1703.

Impact

During his lifetime, Hooke had an international reputation as the premier experimentalist in England, and his biographers have called him the first professional scientist in an era when most people interested in research were aristocratic dabblers. Despite his experimental genius and mechanical talent, however, Hooke's energies were scattered, and he seldom had the time or the mathematical skill to investigate topics fully. His insights were usually remarkably accurate and his suggested solutions to problems basically correct; however, he left it to others to supply proof. Accordingly, Hooke failed to receive credit for many ideas that others pursued and developed largely or solely on the basis of his work.

Hooke's rivalry with Newton further contributed to his decline in recognition. After Newton became president of the Royal Society in 1703, Hooke's influence quickly waned, and Newton's system of rigorous mathematical demonstration in scientific investigations became the intellectual standard, displacing Hooke's style of hypothesis and experimentation. Until a revival of interest in Hooke around the three hundredth anniversary of his death, he was known almost solely for Hooke's law. Nevertheless, Hooke's scientific methods and ideas were pervasively influential during his lifetime, affecting the work even of such rivals as Newton.

—*Roger Smith*

FURTHER READING

Cooper, Michael. *"A More Beautiful City": Robert Hooke and the Rebuilding of London after the Great Fire.* Stroud: Sutton, 2003. Print. Offers a brief sketch of Hooke's life and work in science, while discussing in detail his architectural work and role in rebuilding London.

'Espinasse, Margaret. *Robert Hooke.* Berkeley: U of California P, 1956. Print. Proposes that Hooke's conflict with Newton changed how science was pursued in England, from broad, practical empirical studies to narrower mathematical induction.

Hooke, Robert. *The Diary of Robert Hooke, 1672–1680.* Ed. Henry W. Robinson and Walter Adams. London: Wykeham, 1968. Print. A detailed record of eight years in Hooke's life that testifies to the variety, intensity, and burden of his workload.

Inwood, Stephen. *The Forgotten Genius: The Biography of Robert Hooke, 1635–1703.* San Francisco: MacAdam, 2005. Print. Explains Hooke's varied scientific achievements and discusses the attendant controversies, suggesting that a tendency in Hooke to overstate his claims led him into conflicts.

Jardine, Lisa. *The Curious Life of Robert Hooke: The Man Who Measured London.* New York: Harper, 2005. Print. Discusses Hooke's grueling schedule of work for the Royal Society, his partnership with Wren, and his official duties for London and how these affected his research, professional standing, and health.

ROBERT HOOKE PUBLISHES THE FIRST SCIENTIFIC BEST SELLER

Robert Hooke's primary duty while serving as curator of experiments for the recently chartered Royal Society was to design and carry out experiments. Nevertheless, his colleagues also recognized his ability to observe, describe, and draw objects he studied with the rudimentary microscopes of his time, and he was asked to present drawings from his microscopic observations during the meetings of the membership. Throughout 1663 and the beginning of 1664, he shared drawings of objects ranging from a ribbon to a gnat. Impressed by the quality of his drawings, his colleagues urged him to publish his work in book form. Hooke completed his drawings by mid-1664; the work was published in January 1665.

Micrographia; or, Some Physiological Descriptions of Minute Bodies Made by Magnifying Glasses with Observations and Inquiries Thereupon, or simply Micrographia, was an instant success among Hooke's scientific colleagues and, later, with the public at large. The first major scientific text to be written in English, the book contains Hooke's descriptions of sixty diverse objects and phenomena, including mold, the head of a pin, and the sparks generated by a flint, and accompanying drawings rendered in minute detail. Perhaps the most impressive drawings are the microscopic views of insects; the book includes fold-out drawings of a flea and a louse that magnify each insect to about the size of a cat.

The book begins with a dedication to the reigning monarch, Charles II, followed by a dedication to the Royal Society. Hooke then presents a long preface that describes his methods in preparing his specimens as well as procedures for using the microscope itself—no simple task for even his scientific colleagues. Hooke's initial descriptions are those of mundane, common objects: a razor, snowflakes, cloth, and even gravel found in urine. He later goes on to describe his observations regarding air, stars, and the moon. The first known use of the term cell as applied to the functional unit of life occurs almost in passing, as Hooke describes the honeycomb appearance of "pores" or "cells" found in a slice of cork.

Hooke's purpose was not only to provide entertainment for his scientific colleagues but also to enable members of the general public to enjoy the field that came to be known as science. In this he was partially successful, as many members of the public who had the opportunity to read his book became intrigued by the microscopic world. The book was not inexpensive, even for the time, costing thirty shillings and therefore somewhat limiting its readership. However, the significance of the work was far reaching, in that it inspired a new generation of microscopists.

Hypothesis Testing

FIELDS OF STUDY

Hypothesis Testing; Research Theory; Experimental Design

ABSTRACT

Hypothesis testing is the process of formulating an experiment that will enable analysis of the validity of a proposition about observed phenomena. This is the essential function of the scientific method, which typically involves repeated hypothesis testing. Each experiment produces a new body of evidence, allowing researchers to reject or refine the original hypothesis.

PRINCIPAL TERMS

- **control group:** the subjects in an experiment who do not receive any intervention or who receive a placebo or sham intervention.
- **null hypothesis:** a prediction that no significant differences will be found between the control and experimental groups.
- **parameter:** in statistics, a quantity that defines a certain characteristic of a population, such as mean, median, mode, or standard deviation.
- **variable:** in research or mathematics, an element that may take on different values under different conditions.

The Scientific Method

The classic scientific method follows the structure of collecting data based on observations of phenomena. Researchers first propose a hypothesis to explain those phenomena. Then they design an experiment to test that hypothesis. Lastly, they make observations during and after the experiment to collect more data bearing on the hypothesis. It is important to distinguish between a hypothesis, a working hypothesis, and a scientific theory. A hypothesis is essentially an educated guess. A working hypothesis is a hypothesis that has been accepted for the time being, usually because some amount of evidence has been collected to support it. A scientific theory, on the other hand, has been tested and confirmed repeatedly. It represents the most rigorous form of scientific knowledge in its subfield. The journey from hypothesis to scientific theory requires collecting data in order to repeatedly test a specific hypothesis.

In a typical experiment, the hypothesis is formulated based on observations of existing data, including the results of previous experiments. The hypothesis is usually a specific statement about observed phenomena. It can often be phrased in the form of the proposition, "If p, then q." This conditional statement represents a cause-and-effect relationship that the researchers aim to prove. For example, researchers might propose the hypothesis that if loud music is played while participants are taking a test, then participants will perform worse than the control group taking the same test in a quiet room. In this example, test performance and music are experiment variables. Researchers would also specify parameters for choosing the sample groups to study. The proposition that the control group and experimental group will have the same average test results is the null hypothesis. The researcher hopes to reject the null hypothesis through the testing.

Testing and Retesting

Hypothesis testing is what theorist Thomas Kuhn (1922–96) called "normal science." It is the day-to-day activity of scientists who work within an explanatory framework consisting of the many established and well-tested scientific theories of their field. They gather evidence for and test hypotheses within that framework that gradually reinforce the truth of some elements of the framework while revealing the inconsistencies of others. According to Kuhn, the accumulation of anomalies as the result of normal science leads to a paradigm shift. Scientific revolutions occur not because of a single breakthrough in science but because of the decades of normal science that came first. Countless experiments result in anomalies that need to be explained.

A hypothesis should be as simple as possible and should lay out clear variables that can be tested and measured in a standardized manner. The experiment designed to test the hypothesis should be carefully set up to exclude or reduce any bias on behalf of the researcher or subjects, as well as errors that could influence the authenticity of the results. If performed successfully, the evidence collected from the experiment should allow the researcher to reject the null hypothesis in favor of their alternative hypothesis. The researcher can also use the findings to rework the hypothesis if needed. If the data support the hypothesis, others must then replicate the experiment and consistently get the same results in order to establish the hypothesis as an accepted theory.

COMPUTERS READING COMICS
A 2016 student study conducted at the University of Maryland tested the hypothesis that artificial intelligence (AI) machines could not yet replicate the human ability to read comic-book narratives. Like many visual media, comic books use both graphics and text. However, where a comic book differs is in its closure-driven narrative. This refers to the cognitive process by which readers infer the logical connections between one panel and the next. Every panel portrays only one moment. Therefore, visual and narrative aspects such as motion, the passage of time, spatial relationships, and causality must be inferred. The researchers wanted to test whether the information contained in the accompanying text (captions, dialogue, and sound effects) is sufficient to convey this closure to an artificial intelligence. To test this, they had machines try to predict what would happen in the next panel.

For the experiment, a dataset of 3,948 publicly available comic books dating from between 1938 and 1954 were used. They included around 1.2 million total panels, which were broken into basic elements in order to analyze the closure. Three tasks were designed to test a software model's capacity for closure when fed these panels. Each task asked the model to predict some element of an upcoming panel based on preceding panels. This is a common way of testing comprehension in deep-learning models. Human readers were used for a baseline comparison. The human readers performed the easy-rated tasks without error and the hard-rated tasks with success rates from 84 to 88 percent. In comparison, the software models lagged well behind in every category of task and difficulty rating. The most successful performed at 83 percent on easy-rated tasks of one type (most success rates were closer to 60 percent). In other applications, deep-learning models are often expected to outperform human baseline models, not the other way around. The poor performance of the models, at least provisionally, offered strong support for the original hypothesis.

In the comic-book experiment, the human readers served the same purpose as a control group. They provided the benchmark against which the deep-learning models in the experimental group were compared.

THE IMPORTANCE OF HYPOTHESIS TESTING
Hypothesis testing is the basic function of the scientific method. It provides direction for experiments and, by extension, shapes most pursuits in professional sciences. However, not all research is experimental or concerned, at least at first, with the explanatory framework of a hypothesis. Testing hypotheses improves experiment design, which in turn refines hypotheses. It is this cycle that contributes to the accumulation of scientific knowledge, and especially to the well-reasoned predictive power of that knowledge.

Bill Kte'pi, MA

FURTHER READING
"AI Machine Attempts to Understand Comic Books . . . and Fails." *MIT Technology Review*, 23 Nov. 2016, www.technologyreview.com/s/602973/ai-machine-attempts-to-understand-comic-books-and-fails/. Accessed 10 Mar. 2017.

Barnard, Chris, et al. *Asking Questions in Biology: A Guide to Hypothesis Testing, Experimental Design and Presentation in Practical Work and Research Projects.* 4th ed., Pearson, 2011.

Iyyer, Mohit, et al. "The Amazing Mysteries of the Gutter: Drawing Inferences between Panels in Comic Book Narratives." *ArXiv.org*, Cornell U Library, arxiv.org/pdf/1611.05118.pdf 2016. Accessed 10 Mar. 2017.

Lehmann, E. L., and Joseph P. Romano. *Testing Statistical Hypotheses.* 3rd ed., Springer, 2010.

Weisberg, Herbert I. *Willful Ignorance: The Mismeasure of Uncertainty.* John Wiley & Sons, 2014.

Wilcox, Rand R. *Introduction to Robust Estimation and Hypothesis Testing.* 4th ed., Elsevier, 2017.

Hypothesis-Based Study

FIELDS OF STUDY

Hypothesis Testing; Experimental Design; Research Theory

ABSTRACT

Hypothesis-based studies are common to most fields of research. Hypothesis-based studies clearly identify a problem, predict an outcome, and either confirm or reject that prediction through data collection and analysis. Such studies are considered valid if they follow basic rules of objectivity and the scientific method and can be replicated by other scholars.

PRINCIPAL TERMS

- **control group:** the subjects in an experiment who do not receive any intervention or who receive a placebo or sham intervention.
- **null hypothesis:** a prediction that no significant differences will be found between the control and experimental groups.
- **parameter:** in statistics, a quantity that defines a certain characteristic of a population, such as mean, median, mode, or standard deviation.
- **variable:** in research or mathematics, an element that may take on different values under different conditions.

What Is a Hypothesis-Based Study?

In many fields, the chief purpose of research is to form a hypothesis that makes a clear prediction about a particular phenomenon and then to test if the hypothesis is true. A hypothesis makes assumptions about what factors are likely influencing a particular phenomenon. A hypothesis typically predicts that there is a relationship between two or more variables. Researchers must also form a null hypothesis. A null hypothesis states that there is no relationship between the variables. The null hypothesis encompasses all other possible outcomes not stated in the research hypothesis. The null hypothesis is therefore assumed to be true until it can be disproven. Research parameters serve as guidelines for the selection of a sample group to be studied.

The independent variable is the factor that researchers believe causes a particular outcome. Such variables may be the result of various attributes. For example, sociology or demographic research may focus on gender, age, race, or ethnicity. Independent variables may also be manipulated by the researcher in order to compare outcomes for when the condition is present or absent. For example, the independent variable may be a particular treatment, drug, diet, program, or environment. Researchers must clearly define terms and determine how variables will be measured. The hypothesis holds that a dependent variable will be affected by the independent variable. After a research question or problem has been identified, the researcher conducts a literature review to see what other researchers have found about the issue.

Once researchers have identified a problem and reviewed previously conducted studies on the subject, the next step is to develop a hypothesis to state in simple declarative sentences what they expect to find through the study. The hypothesis must clearly define all variables and explain the suspected relationships between variables. It must also be open to statistical testing and to replication by other scholars. Researchers then select experiment participants from a larger population. The participants are typically divided between an experimental group, which receives the factor under investigation, and a control group. Other methods may also be used to control the experiment and reduce bias. These include double-blind studies in which neither the researcher nor the subject is made aware of the study's hypothesis.

Impacts

Hypothesis-based studies are common to almost all fields of research. They are considered essential by natural scientists and mathematicians. The social sciences also make frequent use of hypothetical research when possible. In some fields, such as political science, there is an ongoing battle over the legitimacy of qualitative or descriptive-based research versus quantitative research. Hypothesis-based studies can help bridge this divide. Hypotheses formation and testing help to identify questions and find answers in ways that may be measured in various ways.

Proponents of hypotheses-based studies frequently argue that research that does not use hypothesis formation and testing is less legitimate because it is more subjective and more open to different interpretations. Another criticism of non-hypothesis-based research is that it is less likely to be replicated by other scholars. However, hypothetical research is not always possible or effective at providing in-depth examination of certain issues. For example, studies not based on hypotheses may be necessary when the subject being studied is rare, such as a disease or an idiosyncratic group.

Hypothesis in Action

Real-world examples of hypothesis-based studies are common in the world of biomedicine. Scientists commonly test the effectiveness of a drug, treatment, or device by using it with one or more groups and comparing those results with data from a control group that did not receive the experimental factor. This method continues to be successful in finding new drugs and determining new courses of treatment. For instance, if members of Group A and B improve after receiving a nasal spray containing salmon calcitonin as treatment for osteoporosis, and Group C receives no such treatment and does not improve, the hypothesis that salmon calcitonin is a valid alternative to estrogen as a treatment for osteoporosis is upheld.

Significance

Hypothesis-based studies continue to dominate research in many areas. It is considered imperative that researchers learn the basics of forming, developing, and testing hypothesis. No other method of research is considered to be as viable or as significant to rigorous scientific study. It is the primary method for answering research questions with data that is both verifiable and replicable.

Elizabeth Rholetter Purdy, PhD

Further Reading

Cargan, Leonard. *Doing Social Research.* Rowman and Littlefield, 2012. Print.

Gravetter, Frederick J., and Larry B. Wallnau. *Statistics for the Behavioral Sciences.* 9th ed., Wadsworth Cengage Learning, 2013.

Leedy, Paul D., and Jeanne Ellis Ormrod. *Practical Research: Planning and Design.* Pearson, 2013.

Martin, William E., and Krista D. Bridgmon. *Quantitative and Statistical Research Methods: From Hypothesis to Results.* John Wiley and Sons, 2012.

Richey, Rita C., and James D. Klein. *Design and Development Research: Methods, Strategies, and Issues.* Routledge, 2014.

Supino, Phyllis G., and Jeffrey S. Borer. *Principles of Research Methodology: A Guide for Clinical Investigators.* Springer Science and Business Media, 2012.

Toledo, Alexander H., Robert Flikkema, and Luis H. Toledo-Pereyra. "Developing the Research Hypothesis." *Journal of Investigative Surgery,* vol. 24. no. 5, 2011, pp. 191–94.

Independent Rariable Manipulation

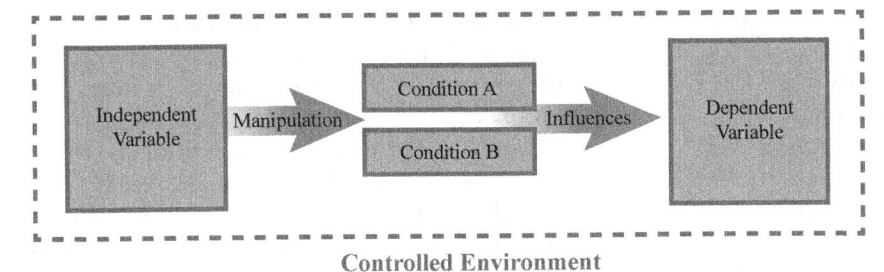

Controlled Environment

FIELDS OF STUDY

Research Theory; Experimental Design

ABSTRACT

Independent variable manipulation is the first step in running an experiment, when the variable is changed so that its effects on one or more dependent variables may be observed. Data may then be collected and examined for evidence in support or rejection of the hypothesis. The challenge is to reduce the effects of extraneous variables so that the results of the research reflect only the impact of the introduced change.

PRINCIPAL TERMS

- **covariate:** a continuous variables other than the primary independent variable that is observed because it is hypothesized to affect the outcome of the study.
- **elimination:** a study design in which the effects of covariates are removed from the model so that only the effects of the independent variable are observed.
- **extraneous variable:** a variable in a study that is not being studied explicitly and is undesirable because of its potential impact on the dependent variable.
- **inclusion:** a study design in which the effects of covariates are purposely observed, measured, and accounted for when observing the effects of the independent variable.
- **statistical control:** the reduction or removal of the effects of extraneous variables through experimental design to the point that they have no significant impact on the effect of the independent variable on the dependent variable.

MANIPULATING VARIABLES

In the scientific method, prior observations of a phenomenon, possibly from previous experiments, are used to formulate a hypothesis. A hypothesis proposes a specific change will occur when an independent variable is manipulated in order to effect change in dependent variables. An experiment is designed to test a hypothesis about this suspected cause-and-effect relationship. During the running of the experiment, data are collected and then analyzed in order to evaluate the accuracy of the hypothesis.

Dependent variables are those whose values are affected by the value of another variable. Independent variables are those whose values are not affected. The dependent variable represents the expected change in outcome that results from the change in the independent variable introduced by the experimenter's manipulation. Experimenters typically limit the number of independent variables to ensure that the relationship between the variables and the results are clear. In order for evidence from the experiment to be reliable, statistical control is critical. This helps to distinguish between covariates and extraneous variables through inclusion and elimination.

THE CHALLENGES OF STATISTICAL CONTROL

The difficulty that manipulation presents is in assuring that the results of the experiment reflect only the influence of the manipulation, and not of extraneous variables. Care in manipulation is itself important in achieving this. However, much of experimental design also concerns statistical control and various strategies for dealing with the effects of extraneous variables.

Approaches vary based on the nature of the experiment, the phenomenon being studied, and the independent variable. Inclusion deliberately involves extraneous variables as covariates in the experimental design. As data are collected, they include data on the effects of those covariates. This way, the effect of an extraneous variable on a dependent variable will not be mistaken for the effect of the independent variable. For instance, studies on human behavior often strive to select a subject group that is demographically diverse. Therefore, no demographic variable such as sex, race, age, or education that is unrelated to the topic of the study is overrepresented in the experimental or control groups.

Elimination takes the opposite approach. The effects of extraneous variables are removed by controlling them and making them constant. For instance, instead of using a diverse group, elimination would call for using an identical group. In that case, the effect of demographically linked variables is identical for every subject. Depending on the study, however, the research results may not be generalizable.

The Street Lighting Experiment

In 1991, London crime prevention researchers studied the effect of street lighting on crime. Regularly spaced street lights often make people feel safe around public spaces like roads and parking lots. They have long been advocated as a crime prevention device. However, both the London study and later studies have found that there is no correlation between nighttime lighting and drops in crime rates. The London study was conducted over one year in the borough of Wandsworth. It studied the effects of manipulating a single variable: 3,500 improved street lights introduced at the beginning of that period. The daylight crime rate was used as the control group, since no change had been introduced to the daylight environment. The researchers worked for the Home Office, the UK department that oversees police departments. They had full access to the database of crimes reported in the period under study. During that period, there was no statistically significant change in either the number of nighttime crimes, their frequency, or their character (such as violent crimes). Those results were borne out in other studies conducted in other parts of the world. Criminologists have suggested that while people feel safer when they can see their surroundings, the perceived benefits of nighttime lighting are offset by the benefit they offer to criminals, who also need to see their victims.

The Heart of Experimental Design

Manipulation of an independent variable, along with statistical control to reduce the impact of extraneous variables on experimental results, is at the heart of a well-wrought experiment. It is attention to these aspects of experimental design that best serves the testing of the experiment's hypothesis. The introduction of a change to the system being observed is what distinguishes experimental research from non-experimental research. The level of control that a researcher has over the experiment comes directly from that manipulation and their ability to protect the results from contamination by extraneous variables.

Bill Kte'pi, MA

Further Reading

Atkins, Stephen, et al. *The Influence of Street Lighting on Crime and Fear of Crime.* Crime Prevention Unit Paper no. 28, United Kingdom Home Office Crime Prevention Unit, 1991. *Center for Problem-Oriented Policing,* www.popcenter.org/library/scp/pdf/07-Atkins_Husain_Storey.pdf. Accessed 15 Mar. 2017.

Creswell, John W. *Qualitative Inquiry and Research Design: Choosing among Five Approaches.* 3rd ed., Sage Publications, 2013.

Creswell, John W. *Research Design: Qualitative, Quantitative, and Mixed Methods Approaches.* 4th ed., Sage Publications, 2014.

Goodwin, C. James, and Kerri A. Goodwin. *Research in Psychology: Methods and Design.* 8th ed., John Wiley & Sons, 2017.

Kazdin, Alan E. *Research Design in Clinical Psychology.* 5th ed., Pearson, 2017.

Smith, Gary. *Essential Statistics, Regression, and Econometrics.* 2nd ed., Academic Press, 2015.

Strang, Kenneth D, editor. *Palgrave Handbook of Research Design in Business and Management.* Palgrave Macmillan, 2015.

INDUCTIVE REASONING

Deductive Reasoning	Inductive Reasoning	Abductive Reasoning
Based on known rules and data to determine a new explanation	Based on data and previous explanations to determine a rule	Based on previous explanations and known rules to determine possible data
Conclusion is guaranteed true	Conclusion is probably true	Conclusion is a best guess

FIELDS OF STUDY

Hypothesis Testing; Research Theory

ABSTRACT

Inductive reasoning, or induction, is a process that draws general conclusions from specific examples. As such, it involves a calculation or estimation of probability: a likely outcome is always inferred from a narrower observation. As the experimental foundation of science and technology, induction is of central and multipurpose importance.

PRINCIPAL TERMS

- **abductive reasoning:** a type of reasoning in which incomplete knowledge of a specific situation is used to draw a conclusion about the simplest and most likely explanation for that situation; if the premises on which the conclusion is based are true, the conclusion is deemed to be possible or probable but not certain.
- **backward reasoning:** a reasoning process that starts with the desired conclusion or result and works backward to determine which steps will lead to that conclusion or result.
- **deductive reasoning:** a type of reasoning in which general rules are applied to a specific situation in order to draw a narrower conclusion; if the premises on which the conclusion is based are true, the conclusion is also deemed to be true.

THEORY AND PRACTICE

There are three main forms of reasoning: inductive reasoning, or induction; deductive reasoning, or deduction; and abductive reasoning, or abduction. Deduction can also be described as top-down logic, in which the truth of a specific statement is determined based on broader or more general truths. Abduction is the process of inferring a likely cause for an observed phenomenon that is incompletely understood. Induction follows the reverse process of deduction, progressing from specific truths to develop a broader or more general rule.

In a loose form, induction is as old as systematic thought itself. Since antiquity, scientists have inferred conclusions based on observation of the physical world. They saw the sun rising each morning, for example, and concluded on that basis that it would continue rising each morning. In other words, a limited number of observations was used to extrapolate a general hypothesis.

The potential for error arising from this method was as well-known as its usefulness. The fact that the harvest has been good for four years running does not mean it will be good for a fifth year. Even though something has been observed to be true in some cases, it cannot be known that it will be true in all cases. Deductively, a true premise leads axiomatically to a conclusion that must also be true, but inductively, a true conclusion does not necessarily follow from a true premise. Rather, conclusions reached by induction imply only a high probability of truth.

The first rigorous theories of induction were established in England by Francis Bacon (1561–1626) and refined by John Stuart Mill (1806–73). Bacon's *Novum Organum Scientarum* (New instrument of science, 1620) argued that a corpus of secure knowledge is reached through repeated experimentation, testing, and the systematic accumulation of data on which to build likely hypotheses. The five principles known as Mill's methods, described in Mill's book *A System of Logic* (1843), advocate induction as the most reliable foundation of a general scientific method. Mill's methods assess the probability of outcomes by selectively eliminating factors that might compromise the credibility of the hypothesis. This approach generalizes from specific circumstances. Those generalizations are not axiomatically true, as they will always rely on an inferential leap. Nonetheless, it is through induction that experimental scientific methodology operates.

INFERENTIAL EXTRAPOLATION

The chief benefits and main drawbacks of induction stem from the same source. Experimental conclusions are not confined to the scope of the experiment itself. They are extrapolated, by inference and analogy, into conjectures that have yet to be rigorously tested. This can sometimes yield far-reaching and beneficial outcomes. The curative properties of the *Penicillium* fungus were revealed by chance to Scottish scientist Alexander Fleming (1881–1955) in 1928. He noticed that the mold appeared to inhibit the spread of the *Staphylococcus* bacteria found growing in a petri dish left by an open window. Induction led Fleming to infer that the mold might then be developed medically as an antibiotic. This proved true, and penicillin has saved millions of lives since that time, although the discovery came about accidentally and not through prearranged trials. The main drawback of induction is that other conclusions, similarly promising, may be wrong. Insufficient evidence may lead to false hypotheses with potentially harmful results. Therefore, validation is always required.

PRIMARY APPLICATIONS

Real-world examples of induction at work abound in science, given that this form of reasoning is at the core of the experimental method. In some fields it is indispensable. Induction has proven vital in the development of innovative theories in cognitive and developmental psychology, where deductive approaches cannot sufficiently account for the acquisition or processing of complex information. Likewise, the Bayesian framework widely used in computational modeling relies on the use of inductive hypotheses. Bayesian probability informs statistical analysis in fields as diverse as medicine, engineering, economics, business, and management.

Another form of induction is backward reasoning, which works backward from a particular goal in order to infer the steps that are most likely to achieve that goal. Backward reasoning is prevalent in computer science (particularly artificial intelligence research), where it is also known as "backward chaining," and in game theory, where it is also known as "backward induction."

FOUNDATIONAL METHODOLOGY

Every day, everyone uses induction, usually without even knowing it. If a friend or colleague, for example, has been trustworthy in the past, it seems natural to assume that they will be in the future. As a formal research methodology, induction has near-universal application in experimental science. It generates new concepts and suggests connections between existing ones. The goal of induction is to establish a set of interrelations arising from smaller sets of discoveries. As analogies and similarities emerge, the smaller sets are assembled into a wider research framework. This process constantly adds to the sum of potentially useful hypotheses required by the progress of science. It is for this reason that science is understood as an expansive, dynamic set of disciplines.

Nicholas J. Crowe, PhD

FURTHER READING

DePoy, Elizabeth, and Laura N. Gitlin. *Introduction to Research: Understanding and Applying Multiple Strategies*. 5th ed., Elsevier, 2016.

Feeney, Aidan, and Evan Heit, editors. *Inductive Reasoning: Experimental, Developmental, and Computational Approaches*. Cambridge UP, 2007.

Glass, David J., and Ned Hall. "A Brief History of the Hypothesis." *Cell*, vol. 134, no. 3, 2008, pp. 378–81.

Hayes, Brett K., et al. "Inductive Reasoning." *Wiley Interdisciplinary Reviews: Cognitive Science*, vol. 1, no. 2, 2010, pp. 278–92, doi:10.1002/wcs.44. Accessed 16 Feb. 2017.

Heit, Evan, and Caren M. Rotello. "Relations between Inductive Reasoning and Deductive Reasoning." *Journal of Experimental Psychology*, vol. 36, no. 3, 2010, pp. 805–12. *Academic Search Complete*, search.ebscohost.com/login.aspx?direct=true&db=a9h&AN=50134344&site=ehost-live. Accessed 16 Feb. 2017.

Lassiter, Daniel, and Noah D. Goodman. "How Many Kinds of Reasoning? Inference, Probability, and Natural Language Semantics." *Cognition*, vol. 136, 2015, pp. 123–34.

Moore, Brooke Noel, and Richard Parker. *Critical Thinking*. 12th ed., McGraw-Hill Education, 2017.

Internal Validity (Causality)

> **Outside the Study Frame**
> (External Validity)
> Are the results repeatable in other studies?
>
> **Inside the Study Frame**
> (Internal Validity)
> Can the conclusions accurately be assessed from the study?

FIELDS OF STUDY

Research Theory; Experimental Design

ABSTRACT

Internal validity is one of the criteria by which quantitative research is judged and applies to research making claims about a causal relationship between variables. It measures the degree to which the results of the research prove that relationship and avoid confounding by other variables.

PRINCIPAL TERMS

- **bias:** an inclination toward a particular perspective or preconceived notion; also, in statistics, refers to a method of collecting or evaluating data that inherently and consistently produces results that inaccurately reflect the characteristics of the population or phenomenon being studied.
- **causation:** a relationship between two events, actions, or other phenomena in which one event is determined to have directly caused or otherwise affected the other.
- **confounding:** in quantitative research, describes an extraneous variable that affects both the dependent and the independent variables, the effects of which must be suppressed in order to accurately determine the relationship between the variables being studied.
- **construct validity:** the extent to which a test accurately measures what it is designed to measure.
- **external validity:** the extent to which the results of a study can be generalized to apply to other populations and settings.

Validity and Minimizing Confounding

Internal validity is one of the traditional criteria by which the success of quantitative research is judged. It measures the extent to which the results support the causal relationship between the independent and dependent variables being studied. It is only relevant to research concerned with causation. This excludes most descriptive research but includes most quantitative experimental research. While important to the consideration of the research project's success, it is distinct from external validity. Internal validity considers only factors specific to that study. External validity, on the other hand, considers the extent to which the results of the study can be generalized and applied outside of that study. Internal and external validity are both related to construct validity. This is how well an experiment designed to measure a given thing succeeds at measuring that thing. Internal and external validity are applicable to all quantitative research with causality related findings. Construct validity, however, may or may not come up depending on the work. Laboratory experiments tend to have higher internal validity than field experiments because they are conducted in controlled environments. Field experiments are conducted in the real world and usually involve more extraneous variables.

In a study, researchers manipulate an independent variable in order to observe its effects on one or more dependent variables. These variables describe the causal relationship the hypothesis has proposed and that the research is intended to support. Variables outside of that relationship are extraneous. Extraneous variables with the potential to skew the results of the research because of their own relationships to the independent and dependent variables are called confounding variables. One way of assessing internal validity is how well a study avoids the effects of confounding. Another hazard to avoid is bias, particularly when selecting or interacting with research subjects.

Threats to Internal Validity

Internal validity establishes the dependability of the evidence collected in support of the experiment's hypothesis. For this reason, researchers have long considered the issues in preserving the internal validity of their work. A 1963 research design text by Donald T. Campbell and Julian C. Stanley grouped the basic threats to internal validity into the following eight types:

- "History" refers to the results potentially being influenced by outside events that occur during the study.
- "Maturation" describes changes to the subjects' performance in a study conducted over a sufficient period that may occur because of their own experiences or other effects of age, rather than because of the manipulated variable.
- "Testing" means that subjects performing a task more than once in a study may have improved performance simply because of practice.
- "Instrumentation" is lack of consistency in the method by which performance is rated. Human scorers may be subject to bias, and measuring instruments may not be consistently calibrated.
- "Statistical regression" refers to subjects scoring in outlier ranges have a statistical tendency to score more normally when retested.
- "Selection" describes biased effects that may result when subjects in a multigroup study are not chosen randomly.
- "Experimental mortality" is the loss of subjects from an experiment before its completion potentially having an impact on its results.
- "Selection-maturation interaction" occurs when selection bias worsens the effects on internal validity of maturation and other threats above.

This remains the commonly cited model. Not all of those threats are relevant to all studies, and the list is not all-inclusive. For instance, the nature of the threats listed highlights the fact that studies with human participants especially are subject to a wider variety of internal validity concerns. However, with the exception of experimental mortality, the impact of all of these threats can be mitigated through experimental design. This can be done by using control groups, controlled variables, and randomization to minimize bias.

Internal Validity in the Lab

In a 2017 study, Sarah Ellys Harrison and her colleagues conducted an experiment to show the ability of certain types of mouse stem cells to self-assemble in vitro, producing embryos that resembled natural embryos. Historically, structures developed from embryonic stem cells (ESCs) have not resembled the products of embryogenesis. The team hypothesized that the difference in development was due to environmental differences: the lack of interactions with extra-embryonic tissues and the difference in the cells' spatial arrangement. They tested this hypothesis with an experiment that more closely replicated embryogenesis conditions. They used extra-embryonic trophoblast stem cells and a three-dimensional scaffold of extracellular matrix. Both the process and the product closely mimicked the natural formation of a mouse embryo. Extraneous variables were eliminated by conducting the experiment in controlled lab conditions. The internal validity was further assured by repeating it multiple times. The data were then analyzed for normal distribution. Also, the embryos produced by the experiment were compared against those developed from ESCs by normal lab means to confirm and quantify the differences.

The Importance of Validity

Preserving internal validity is among the most important aspects of experimental design. Other criteria deal with producing research that is generalizable to other contexts. Internal validity determines whether the change observed during the experiment was caused by the experimenter's intervention and only by that intervention. This allows the researcher to defend the causal relationship between the studied variables with high confidence.

Bill Kte'pi, MA

Further Reading

Campbell, Donald T., and Julian C. Stanley. *Experimental and Quasi-Experimental Designs for Research.* Rand-McNally, 1963.

Christensen, Larry B., et al. *Research Methods, Design, and Analysis.* 12th ed., Pearson, 2014.

Creswell, John W. *Research Design: Qualitative, Quantitative, and Mixed Methods Approaches.* 4th ed., SAGE Publications, 2014.

Goodwin, C. James, and Kerri A. Goodwin. *Research in Psychology: Methods and Design.* 8th ed., John Wiley & Sons, 2017.

Harrison, Sarah Ellys, et al. "Assembly of Embryonic and Extra-Embryonic Stem Cells to Mimic Embryogenesis in Vitro." *Science*, 2 Mar. 2017, science.sciencemag.org/content/early/2017/03/01/science.aal1810.full. Accessed 24 Mar. 2017.

Hinkelmann, Klaus, and Oscar Kempthorne. *Design and Analysis of Experiments*, vol. 1, 2nd ed., John Wiley & Sons, 2008.

Kazdin, Alan E. *Research Design in Clinical Psychology*. Pearson, 2017.

INTERPRETIVE METHODS

FIELDS OF STUDY

Research Theory

ABSTRACT

Interpretive methods are used in qualitative research that deals with the ways participants assign subjective meanings to their experiences and the phenomena they observe. This meaning-making is key to understanding human behavior and interactions. There are a number of tools employed to analyze and interpret the data collected in such studies.

PRINCIPAL TERMS

- **category analysis:** a method of examining data by sorting it or breaking it down into categories and subcategories.
- **content analysis:** the systematic study of a body of textual data in order to identify its meaning, including intended and latent messages and effects.
- **narrative analysis:** a form of qualitative research focusing on how subjects create and use stories or narratives.
- **qualitative data:** information that describes but does not numerically measure the attributes of a phenomenon.
- **subjectivity:** the quality of being true only according to the perspective of an individual subject, rather than according to any external criteria.

Interpretive Research

Interpretive methods are used in interpretive research, which involves qualitative data. While all qualitative research focuses on human behavior, interpretive methods are characterized by their handling of subjectivity and the researcher's understanding of the processes of meaning-making in which participants engage. Different schools of thought and methodologies of interpretative research find favor in different fields. There are certain common assumptions, however. Namely, people constantly create meaning out of their experiences and the world around them as they associate subjective meanings with specific phenomena. This includes not only their firsthand experiences, but phenomena they witness, secondhand experiences they learn about from others, and the events and ideas that make up their wider social and cultural world. This meaning-making occurs mainly unconsciously. It is affected both by the individual's accumulation of past personal experiences and by acculturation.

Various methods are used in interpretive research. These include category analysis, which sorts data into categories; content analysis, which identifies the patterns within that data; and narrative analysis, which examines the formation and use of narratives such as stories. Interpretive research can also overlap with critical research, which critiques its subject via a specific theoretical lens, such as feminism. Critical research is associated with social theory and social science research.

While experimental research revolves around the rigorous testing and retesting of a hypothesis, interpretive methods proceed somewhat differently. As experimental research begins with hypothesis formation, interpretive research begins with a research question. In both cases, this first step is preceded by some period of observation that inspires the hypothesis or question. Structurally, the next steps in interpretive research are similar to those of experimental research: generate, organize, and analyze data. Generating data involves conducting qualitative research on the topic or gathering preexisting research, whichever is more appropriate.

In interpretive research, data often mean participant interviews or other narratives, such as diaries, journals, letters, or recordings. Organizing the data requires separating out the signal from the noise—identifying what in the large data set is relevant to the research question. Relevant data is then analyzed with the tools appropriate to the research question.

The Challenge of Interpretive Rigor

As with all qualitative research, the strength of interpretive research is that it deals with information that quantitative research does not. While quantitative research deals strictly with numbers, interpretive research can approach its area of inquiry from entirely different angles, dealing with its subjects' experiences and stories and the meanings they make from them.

There are also clear limits to such research, however, and unique burdens on the researcher. Comparing sets of subjective data accurately is a challenge, if not impossible. For instance, consider pain reports submitted by hospital patients. Even if demographic differences are controlled for (such that the reported pain of a child is not compared to that of an adult), pain tolerances and prior experiences of pain vary among patients. Staff can attempt to classify or code different responses, such as by creating broad categories like "mild," "moderate," and "severe," to make comparisons easier. However, subjectivity remains an issue.

Some theorists have in fact challenged the notion of "rigor," a standard applied to quantitative research. They have proposed "trustworthiness" as the standard for interpretive research instead. Quantitative rigor requires research to demonstrate validity, reliability, objectivity, accuracy, and generalizability. Trustworthiness instead looks for credibility, dependability, confirmability, authenticity, and transferability.

Interpretive Methods in Entrepreneurial Research

Interpretive research is commonly associated with the social sciences but can also be found in other fields. For example, researcher Shana Ponelis examined the role of interpretive methods in entrepreneurial research. In business education, qualitative research in the form of the case study model, an in-depth study of a single business or organization, has long been the norm. Ponelis studied the use of case studies within the interpretive research paradigm, hoping to learn more about the perspectives, shared social world, and subjective experiences of the participants within the organization described in each case. Because Ponelis's research was limited to small and medium businesses lacking complex hierarchies, the owner or manager could be understood as the main decision maker and therefore the focal point of an interpretive case study. It was this person's perspectives and experiences that were given prime consideration when interpreting the business's decision-making processes and information handling.

The Importance of Interpretive Research

Interpretive research is a valuable tool to the social sciences and other disciplines that conduct qualitative research, from marketing to nursing. It provides insight into lived experiences that cannot be gained with other forms of research. Various interpretive methods help researchers understand the processes by which people make subjective meanings from their experiences, as well as the ways in which these meanings affect their decisions, feelings, and interactions.

Bill Kte'pi, MA

Further Reading

Creswell, John W. *Research Design: Qualitative, Quantitative, and Mixed Methods Approaches.* 4th ed., Sage Publications, 2013.

De Vaus, David. *Research Design in Social Research.* Sage Publications, 2013.

McBride, Dawn. *The Process of Research in Psychology.* 3rd ed., Sage Publications, 2016.

Ponelis, Shana R. "Using Interpretive Qualitative Case Studies for Exploratory Research in Doctoral Studies: A Case of Information Systems Research in Small and Medium Enterprises." *International Journal of Doctoral Studies,* vol. 10, 2015, pp. 535–50.

Schwartz-Shea, Peregrine, and Dvora Yanow. *Interpretive Research Design: Concepts and Processes.* Routledge, 2012.

Thorne, Sally. *Interpretive Description: Qualitative Research for Applied Practice.* Routledge, 2016.

Willis, Jerry W. *Foundations of Qualitative Research: Interpretive and Critical Approaches.* Sage Publications, 2007.

Yanow, Dvora, and Peregrine Schwartz-Shea. *Interpretation and Method: Empirical Research Methods and the Interpretive Turn.* 2nd ed., Routledge, 2015.

KEPLER, JOHANNES

GERMAN ASTRONOMER

Johannes Kepler was a German astronomer and mathematician whose three laws of planetary motion helped to popularize the Copernican view of a heliocentric universe and disprove the ancient idea that planetary orbits are perfect circles.

- **Born:** December 27, 1571; Weil der Stadt, Holy Roman Empire (now Germany)
- **Died:** November 15, 1630; Regensburg, Holy Roman Empire
- **Primary field:** Astronomy
- **Specialties:** Cosmology; theoretical astronomy

EARLY LIFE

Johannes Kepler was born on December 27, 1571, in Weil der Stadt, in what was then the duchy of Württemberg. Weil der Stadt at the time was an imperial free city, part of the Holy Roman Empire but enjoying extensive local rule. Heinrich, Kepler's father, was a mercenary who later fought for the Spanish army in the Netherlands; Katharina, his mother, was an herbalist and natural healer whom her son later defended against charges of witchcraft.

When Kepler was four, his family moved to nearby Leonberg; this was possibly due to the upsurge of Catholicism in Weil der Stadt, as the Keplers were Lutheran. In Leonberg, Kepler attended both the German-language elementary school and the Latin grammar school. He then attended nearby monastic schools. As a child, he saw the comet that appeared in 1577 and the lunar eclipse of 1580. However, he suffered from poor eyesight and was forced to focus on astronomical theory rather than direct observation.

After graduating from secondary school, Kepler entered the University of Tübingen in 1589 to study theology. He showed promise as a mathematician, studying under the renowned scholar Michael Maestlin, who introduced him to the works of Polish astronomer Nicolaus Copernicus. Kepler decided against entering the priesthood, and upon graduating from university, he obtained a teaching position at the Protestant university in Graz, Austria. During the six years he spent there, he married his first wife, Barbara Müller, with whom he had several children.

LIFE'S WORK

Kepler published his first major work, *Mysterium cosmographicum* (The cosmographic mystery), in 1596, setting forth a Copernican heliocentric (sun-centered) view of the universe. He lost his teaching position in 1600 when the Catholic authorities forced all Lutherans to leave Graz. Despite his Lutheranism, Kepler obtained a position at the imperial court as assistant to Tycho Brahe, Emperor Rudolf II's court mathematician and astronomer.

As Brahe's assistant, Kepler helped collect and categorize data on planetary motion based on naked-eye observations. The purpose of the project was to determine whether the Ptolemaic (geocentric, or Earth-centered) or Copernican theory of planetary

Woodcut by Jiri Daschitzsky, from: "Von einem schrecklichen und wunderbarlichen Cometen so sich den Dienstag nach Martini dieses lauffenden M. D. LXXViJ. Jahrs am Himmel erzeiget hat." (About a terrible and marvelous comet as appeared the Tuesday after St. Martin's Day (1577-11-12) on heaven. By Jiri Daschitzsky (Zentralbibliothek Zürich) [Public domain], via Wikimedia Commons

motion was the correct one. The partnership benefited both men; the keen-eyed Brahe excelled at observation but lacked the mathematical background to fit his observations into a larger system, while Kepler, unable to make astronomical observations due to his poor eyesight, excelled at systematizing and drawing conclusions from the data.

Kepler became imperial mathematician in 1601, after Brahe's death. He held the post for the next quarter of a century, serving under Rudolf II and his successors, Matthias and Ferdinand II. In this era, astronomy had not yet become clearly distinguished from astrology, and Kepler became known for his work as an astrologer in addition to his scientific work.

Perhaps Kepler's most important achievement during this period was his formulation of two of his three laws of planetary motion. He articulated the first two laws in 1609, in his work *Astronomia nova* (New astronomy). Kepler's laws did away with the complex system of "epicycles" and "deferents" created by ancient and medieval philosophers. Under both the Ptolemaic and Copernican systems, these additional orbits were required to account for the planets' otherwise unexplainable motions; Kepler's system, on the other hand, treated the planetary orbits as simple ellipses. Kepler had determined that all planets move in elliptical orbits, with the sun at one focus and the other focus empty, and that the imaginary line joining the planet to the sun sweeps over equal areas in equal time intervals (the law of areas).

In 1612, Kepler obtained a simultaneous appointment as a mathematics teacher in Linz, Austria; he held the post until 1626, when religious conflict again forced him to leave. Also in 1612, Kepler's wife died after contracting Hungarian spotted fever. Three of his children were afflicted shortly thereafter with smallpox, and one of his sons died. He was remarried a year later, to Susanna Rüttinger, with whom he had several children. In 1617, Kepler defended his mother against charges of witchcraft relating to accusations of attempted poisoning. He was ultimately successful in securing her release from prison.

Kepler published his third law of planetary motion in his 1619 work *Harmonices mundi* (The harmony of the world). Known as the law of periods, Kepler's third law states that for any planet, the square of its period of revolution is directly proportional to the cube of its mean distance from the sun. Although English scientist Isaac Newton would be the first to develop the theory of gravitation, Kepler's laws pointed in that direction as a way to explain the planets' elliptical orbits. However, *Harmonices* had a highly mystical bent, describing astronomers as priests whose duty is to praise God by revealing the "book of nature." Kepler's mysticism led him to describe the ellipses as being caused by a mysterious magnetic *anima motrix* (moving power) from the sun, which decreased in relation to distance and forced the planets to orbit.

The multivolume textbook *Epitome astronomiae Copernicanae* (Epitome of Copernican astronomy), completed in 1621, clarified the three laws and made them part of a more coherent system; the work helped to make Copernican thought even more widely influential. In 1627, Kepler published the *Tabulae Rudolphinae* (Rudolphine tables), an extensive star atlas based on Brahe's data. He left the imperial service that year, as the Holy Roman Empire became increasingly involved in political and religious conflict.

In 1628, Kepler became court mathematician and astrologer of the Silesian duchy of Sagan, working for the Swedish military leader Albrecht von Wallenstein. He died of fever on November 15, 1630, and was buried in Regensburg, Holy Roman Empire, but the grave was destroyed when Swedish troops invaded in 1632.

Impact

One of the most influential figures in the development of astronomy, Kepler provided later generations of scientists with new ways by which to understand the movements of the planets. He demonstrated, based on Brahe's experimental data, that the Copernican view of the universe is correct, largely putting an end to the view that the Earth is the center of the universe. Despite his interest in astrology, Kepler based his laws of planetary motion on data rather than mere hypotheses, thus helping to promote the scientific method of formulating theories based on empirical evidence. His three laws of planetary motion, refined by Newton to account for gravity, remain important to understanding the elliptical orbits of the planets.

In addition to his work in science, Kepler made an early contribution to the literary genre of science fiction. His novel *Somnium* (The dream), published posthumously in 1634 by his son Ludwig, uses the fictional device of a dream voyage to the moon as a means to consider how astronomical phenomena would appear in the heavens rather than on Earth. Kepler considers the practical issues relating to space flight in a manner similar to that of later writers such as Jules Verne.

—*Eric Badertscher*

FURTHER READING

Connor, James A. *Kepler's Witch: An Astronomer's Discovery of Cosmic Order amid Religious War, Political Intrigue, and the Heresy Trial of His Mother.* New York: Harper, 2005. Print. A biography of Kepler that focuses in particular on his experiences regarding religious and political conflict. Also includes translations of some of Kepler's personal writings.

Stephenson, Bruce. *Kepler's Physical Astronomy.* Princeton: Princeton UP, 1994. Print. Explains Kepler's astronomical theories and discusses the development of his laws of planetary motion.

Voelkel, James Robert. *The Composition of Kepler's Astronomia nova.* Princeton: Princeton UP, 2001. Print. Explores Kepler's process in writing *Astronomia nova* and the development of his theories about the movement of the planets.

Westman, Robert S. *The Copernican Question: Prognostication, Skepticism, and Celestial Order.* Berkeley: U of California P, 2011. Print. Chronicles the development of the Copernican model of the universe and places Kepler's work within a larger context.

JOHANNES KEPLER DEVELOPS HIS LAWS OF PLANETARY MOTION

In 1596, Johannes Kepler published his first work, Mysterium cosmographicum (The cosmographic mystery), in which he discussed what he felt was God's geometric plan of the universe. Kepler believed in the model of the universe theorized by Polish astronomer Nicolaus Copernicus, who argued that the planets orbit the sun rather than the Earth. Kepler's collaboration with Danish astronomer Tycho Brahe began when Brahe sent Kepler a letter criticizing his reliance on Copernicus's inaccurate data in Mysterium. Brahe had built an observatory that allowed him to record highly accurate observations of the positions and movements of the planets. Kepler traveled to Brahe's home in Prague in early 1600 in the hope of gaining access to this better data.

Kepler spent his first two months in Prague analyzing some of Brahe's observations of Mars. On October 24, 1601, Brahe died, leaving Kepler with years of observational data and the task of completing Brahe's work. Kepler also became the mathematical advisor to Rudolph II, emperor of the Holy Roman Empire. Using Brahe's data, Kepler continued to try to find equations that would explain the orbit of Mars and match the observations.

Hesitant to abandon his belief that planetary orbits were circular, Kepler instead began to focus on why the planets' orbital speeds varied. He imagined that there was a force connecting the sun to each planet, like the spoke of a wheel. This eventually led him to formulate what is considered his second law of planetary motion: the imaginary line connecting the sun and a planet sweeps out an equal amount of area in an equal amount of time. Therefore, the planets travel faster when closer to the sun and slower when at aphelion, the point in the orbit farthest from the sun. Kepler used this law to calculate the orbit of Mars and found that an elliptical orbit with the sun located at one focus fit both the mathematical and observational data. This conclusion is known as Kepler's first law of planetary motion.

Kepler published his results in Astronomia nova (New astronomy) in 1609. His first two laws were rejected by many because they abandoned the ideas of circular orbits and uniform velocity; however, later scientists determined that his conclusions were generally correct. Kepler published his third law of planetary motion in Harmonices mundi (The harmony of the world) in 1619, after more than seventeen years of mathematical trial and error. His third law deals with the connection between the distance of a planet from the sun and the time that it takes to orbit. Kepler eventually found that the cube of the semimajor axis of the orbit is proportional to the square of the orbital period. The semimajor axis is the average distance between the planet and the sun, while the orbital period refers to the amount of time it takes the planet to complete one trip around the sun.

Kruskal-Wallis Test

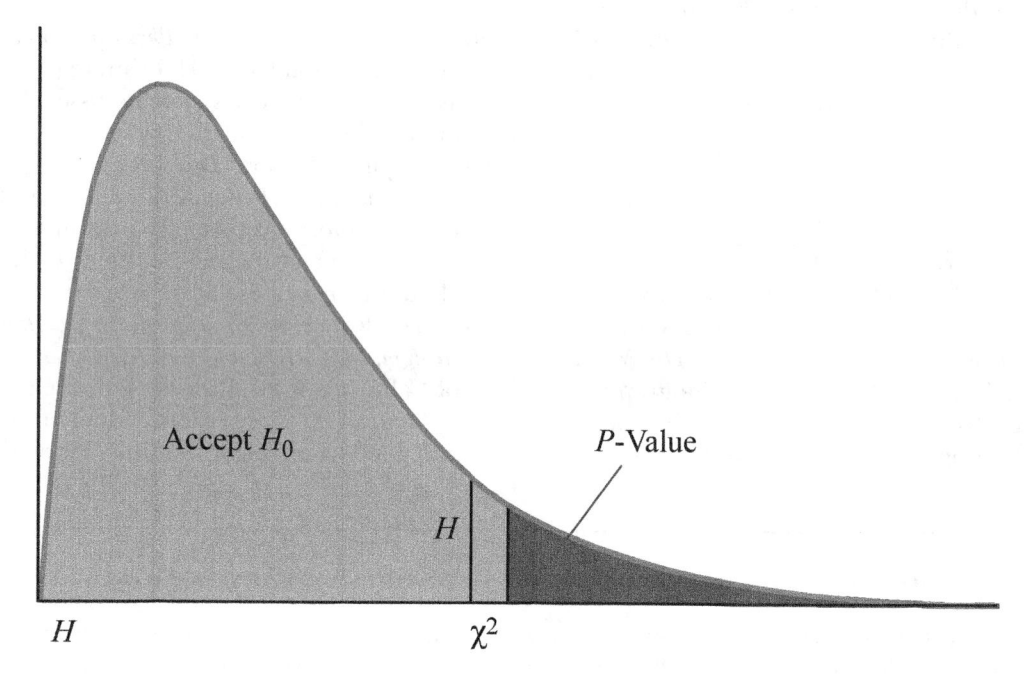

Chi-Squared Distribution

FIELDS OF STUDY

Statistical Analysis; Hypothesis Testing

ABSTRACT

The Kruskal-Wallis test is used to determine if there is a statistically significant difference among multiple categorical groups. This test is typically used when the response variable does not have a normal distribution.

PRINCIPAL TERMS

- **Mann-Whitney U test:** a nonparametric, rank-based test used to determine if a statistically significant difference exists between ordinal dependent and independent variables.
- **nominal variable:** a variable that only takes unordered values in specified categorical groups.
- **one-way analysis of variance (ANOVA):** a statistical test of the difference in means among three or more groups having normal distributions, the same variance, and statistically independent observations.
- **ordinal variable:** a variable that only takes ordered, or ranked, values in specified categorical groups.

What Is the Kruskal-Wallis Test?

The Kruskal-Wallis test is used to determine if there is a difference among ranked or continuous response values, represented by an ordinal variable, for three or more categorical groups being studied, represented by a nominal variable. For example, an experiment might be conducted to determine if a person's level of depression (ranked value) varies among groups with various exercise habits (categorical groups). Subjects would be divided into groups by exercise frequency and asked to report their level of depression. The Kruskal-Wallis test would be used to determine whether the groups originate from populations with the same distribution, or if there is a difference by category (exercise level). It does this by determining if the medians or the mean ranks of the response variable are the same among the groups.

Comparing Tests

The Kruskal-Wallis test is similar to the one-way analysis of variance (ANOVA) test, but unlike ANOVA, it is used when the response variable does not have a normal distribution. If data are normally distributed, the more powerful one-way ANOVA test is the better method. Here, "power" refers to the ability to detect a significant difference and thus reject the null hypothesis when it is false. The Mann-Whitney U test can also be used for non-normal distributions, in cases where only two groups are being studied.

The first step of the Kruskal-Wallis test is to rank the data from all the groups in ascending order. If two data points are the same, the assigned rank is the mean of the rank each would have if not tied. Then the sum of ranks for each group is found, and a significance level is determined. The Kruskal-Wallis test statistic, H, is then calculated using the following formula:

$$H = \left(\frac{12}{n(n+1)} \sum_{i=1}^{k} \frac{R_i^2}{n_i} \right) - 3(n+1)$$

where n is the total sample size, k is the total number of groups, i is the group, and R_i is the sum of ranks for the group i.

Next, the degrees of freedom (the number of data points in a sample that can vary when estimating a parameter) are computed. A chi-square distribution table is used to find the significant chi-square value and compare it to the H value. If H is greater, then the null hypothesis—the prediction that no significant differences will be found between the control and experimental groups—is rejected. These calculations are not typically done manually. In most cases, statistical software is used to perform the calculations to save time, reduce error, and allow for different views of the resulting data.

Limitations

One key limitation of the Kruskal-Wallis test is that it can only determine if two or more groups in the study differ. It cannot identify the differing groups. Another limitation is that while a normal distribution is not necessary for this test, the different groups must have the same distribution in order to compare their medians. If the groups' standard deviations differ, then they are assumed to have different distributions. If the distributions differ in shape (variability), then only their mean ranks can be compared. Lastly, too small a sample size in each group can skew the significance levels.

Most importantly, the Kruskal-Wallis is less powerful than its parametric counterpart. It is more difficult to detect a significant difference, and thereby reject the null hypothesis, with this test.

Studying Differences among Groups

Researchers in many fields, from genetics to psychology, often need to determine if there is a difference in a numerical value among categorical groups. While tests such as the one-way ANOVA and Mann-Whitney U test can be used to study such questions, their limitations may make them unsuitable for use in many common scenarios, such as when the response variable does not have a normal distribution or when multiple categorical groups are being studied. The Kruskal-Wallis test can be used to study such questions and as such provides a valuable tool that significantly expands the range of questions that can be successfully studied.

Maura Valentino, MSLIS

Further Reading

Diez, David M., et al. *OpenIntro Statistics*. 3rd ed., OpenIntro.org, 2015.

LaMorte, Wayne W. *Nonparametric Tests*. Boston U School of Public Health, 24 May 2016, sphweb.bumc.bu.edu/otlt/mph-modules/bs/bs704_nonparametric/BS704_Nonparametric-TOC.html. Accessed 21 Apr. 2017.

McClave, James T., and Terry Sincich. *Statistics*. 13th ed., Pearson, 2017.

Salkind, Neil J. *Statistics for People Who (Think They) Hate Statistics*. 6th ed., Sage Publications, 2017.

Triola, Mario F. *Essentials of Statistics*. 5th ed., Pearson Education, 2014.

Urdan, Timothy C. *Statistics in Plain English*. 4th ed., Routledge, 2017.

Kruskal-Wallis Test

PRINCIPLES OF SCIENTIFIC RESEARCH

KRUSKAL-WALLIS TEST SAMPLE PROBLEM

A researcher wants to determine whether the distance to be traveled affects pre-travel anxiety levels among people who tend toward travel anxiety. The distances to be traveled are categorized as short, medium, or long. A week before departure, twenty-one travelers are asked to rate their anxiety on a scale of 1 to 100, with 1 being no anxiety and 100 being total panic. Levels of anxiety cannot be assumed to have a normal distribution, so the Kruskal-Wallis test is used.

The researcher has ranked all the anxiety scores, as shown in the following table:

They have determined the sum of the ranks for each group (R_1, R_2, R_3) and calculated the degrees of freedom (df):

$$df = \text{number of groups} - 1 = 2$$

Based on the above data, the Kruskal-Wallis test statistic (H) is found to be 10.58.

Given the data in the chart above, the degrees of freedom, and the value of H, determine if distance to be traveled affects travel anxiety, to a significance level (α) of 0.05.

	\multicolumn{6}{c}{Rating on Anxiety Scale}					
	\multicolumn{2}{c}{Short Distance}		\multicolumn{2}{c}{Medium Distance}		\multicolumn{2}{c}{Long Distance}	
	Value	Rank	Value	Rank	Value	Rank
	15	1	25	6.5	85	21
	23	5	44	9	65	17.5
	55	16	22	3.5	46	12
	22	3.5	45	10.5	50	14.5
	18	2	65	17.5	80	20
	33	8	45	10.5	70	19
	25	6.5	50	14.5	48	13
		R_1=42		R_2=72		R_3=117
Mean	27.28571		42.28571		63.42857	
Standard Deviation	13.47484		14.739		15.58199	

Answer:

Use the degrees of freedom (2) and the desired significance level (0.05) with a chi-square distribution chart to determine the significant chi-square value (5.991). Because H is greater than the critical chi-square value (10.58 > 5.991), there is a significant difference among the groups, and the null hypothesis can be rejected. Therefore, the researcher can conclude that there is a connection between distance to be traveled and travel anxiety. Because of the limitations of the test, however, they do not know which groups vary. Two of the groups could have the same levels of anxiety regardless of the distance.

L

LABORATORY EXPERIMENT

Laboratory Experiment	Field Experiment
Low Realism	High Realism
Few Extraneous Variables	Many Extraneous Variables
High Control	Low Control
Low Cost	High Cost
Short Duration	Long Duration

FIELDS OF STUDY

Research Theory; Experimental Design; Research Design

ABSTRACT

A laboratory experiment is conducted in an artificial setting, where researchers have precise control over more environmental conditions and elements of the experiment than they would have in a real-world setting. Like other experiments, lab experiments are carefully designed in order to test a hypothesis about specific observed phenomena.

PRINCIPAL TERMS

- **control group:** the subjects in an experiment who do not receive any intervention or who receive a placebo or sham intervention.
- **dependent variable:** in experimental research, a measurable occurrence, behavior, or element with a value that varies according to the value of another variable.
- **environmental bias:** the contributions of the surroundings in which an experiment is conducted to systematic errors in the results.
- **experimental group:** the subjects in an experimental research study who receive the intervention or treatment being tested.
- **independent variable:** in experimental research, a measurable occurrence, behavior, or element that is manipulated in an effort to influence changes to other elements.
- **manipulation:** altering an independent variable in order to cause intended changes to dependent variables in experimental research.

EXPERIMENTING IN THE LABORATORY

Scientists test the validity of hypotheses (proposed explanations for phenomena based on collected data and observations) through procedures called experiments. These may be either field experiments or laboratory experiments. In the artificial setting of the lab, the conditions of the experiment are subject to much more precise control by the researcher, as opposed to the natural setting of the "field." This greater degree of control is intended to produce results that are more reliable and replicable through a standardized process.

133

Typically, a lab experiment involves two groups of subjects. The first is the control group, in which none of the variables are subject to manipulation by the experimenters. The second is the experimental group, which receives the treatment. This means that an independent variable is manipulated so that researchers can observe the changes they predicted would result in the dependent variables whose values depend on the manipulated variable. Assignment to the experimental or control group is random.

In placebo-controlled studies, which are the standard in medical studies, a control group is given an inert treatment called a placebo. In single-blind studies, participants do not know whether they have received a placebo or the actual treatment. In double-blind studies, neither the participants nor the administrators of the treatment know which participants have received the real treatment. Double-blind studies are the ideal, but are not possible with certain experiments. In the lab, researchers make an effort to control the conditions of the surroundings and keep them constant for both groups. This helps to avoid environmental bias that could skew results.

Benefits and Limitations of Lab Work

Some hypotheses can be tested through either field or lab experiments. In other cases, the nature of the phenomena being studied means that only one is a possibility. Experiments in the social sciences, for instance, often require phenomena that cannot be reproduced in the lab. Surveying people about their behavior or trying to replicate that behavior in a controlled environment does not necessarily yield the same results as observing their actual behavior in the field. Such cases include studies of voting behavior or racial discrimination. On the other hand, many experiments in physical sciences like physics and chemistry require the study of phenomena that can only be reproduced in the lab, for which there is no "field."

The primary drawbacks to lab experiments are the lack of opportunities offered by the field. These include the lack of a natural environment and the requirement to recreate relevant variables in the lab. Also, the fact that participants are aware that they are participating in a study can be undesirable in some social science experiments. Because of the controlled nature of a lab experiment, the results may not have external validity. This means they cannot be generally applied to real-life situations outside of the lab. Also, the level of control allows for a greater chance of bias on the part of the researcher to get a desired result to influence the experiment. However, lab experiments tend to have higher internal validity. The greater ability to design the study to minimize errors means that the results can more confidently be attributed to the causal relationship between the independent and dependent variables being studied.

The Large Hadron Collider

The Large Hadron Collider (LHC) is a particle accelerator built by the European Organization for Nuclear Research (CERN) to test certain particle physics hypotheses. It does this by producing, for the purposes of close and carefully controlled observation, specific phenomena through high-energy particle collisions. Built from 1998 to 2008, the LHC is one of the most complex experimental laboratories ever built. The first operational runs took place in 2009. Several experiments have been conducted simultaneously. Much of the work of the LHC has involved producing data to test the existence of various hypothetical particles. This includes the now-confirmed Higgs boson, which was first hypothesized in the 1960s.

ALICE (A Large Ion Collider Experiment) was one of those early experiments, first submitted to CERN in 1993. ALICE used the LHC to test a hypothesis from quantum chromodynamics that predicted that sufficiently high energy density would cause a phase transition from hadronic matter—the particles collided by the LHC—to quark-gluon plasma, a primordial state of matter. In other words, the experiment modeled the conditions of the early universe in the moments immediately after the big bang, as matter was being formed. The ALICE experiment was designed using systems to identify particles, track their charge and trajectory through the LHC's barrels, measure their energy, and collect data about their collision. In 2012, the ALICE results were announced: the successful production of quark-gluon plasma believed to be similar to the conditions of the universe immediately after the big bang.

Importance of Lab Research

The LHC illustrates the benefits of a lab to create a research environment to test hypotheses in which independent variables can be manipulated with incredibly precise control and with little risk of contamination by external factors. In the lab, phenomena may be created that could not be observed in any other way.

Bill Kte'pi, MA

FURTHER READING

Brown, Theodore E., et al. *Laboratory Experiments for Chemistry: The Central Science*. 13th ed., Pearson, 2015.

Cleave, Blair L., et al. "Is There Selection Bias in Laboratory Experiments? The Case of Social and Risk Preferences." *Experimental Economics*, vol. 16, no. 3, 2013, pp. 372–82. *EconLit with Full Text*, search.ebscohost.com/login.aspx?direct=true&db=eoh&AN=1392858&site=ehost-live. Accessed 9 Mar. 2017.

"Controlled Experiments." *Khan Academy*, www.khanacademy.org/science/biology/intro-to-biology/science-of-biology/a/experiments-and-observations. Accessed 9 Mar. 2017.

Johnson, Ted R., and Christine L. Case. *Laboratory Experiments in Microbiology*. Pearson, 2016.

Lemov, Rebecca. *World as Laboratory: Experiments with Mice, Mazes, and Men*. Hill and Wang, 2005.

Webster, Murray, and Jane Sell, editors. *Laboratory Experiments in the Social Sciences*. 2nd ed., Academic Press, 2014.

LATIN SQUARE DESIGNS

	Group 1	Group 2	Group 3
Round 1	A	C	B
Round 2	B	A	C
Round 3	C	B	A

3 × 3

	Group 1	Group 2	Group 3	Group 4
Round 1	A	D	C	B
Round 2	B	A	D	C
Round 3	C	B	A	D
Round 4	D	C	B	A

4 × 4

FIELDS OF STUDY

Experimental Design; Research Design; Statistical Analysis

ABSTRACT

The Latin square design is a type of blocked design that controls for the effects of two nuisance factors, or extraneous sources of variation. Latin square designs are used when an experiment involves three factors, the independent variable and the two nuisance factors. There must be a treatment for every blocking variable. Such a design is useful because it reduces the number of runs needed for an experiment.

PRINCIPAL TERMS

- **ANOVA:** short for "analysis of variance," a set of statistical models for analyzing the differences among the means of multiple data sets.
- **nuisance factor:** a factor in a research study that is of no interest to the researcher but can affect the outcome of the study; also called a nuisance variable.
- **residual:** the difference between the observed value of a variable and its estimated or expected value, represented by the mean of the group of observed values.
- **treatment:** the process or intervention applied or administered to members of an experimental group.
- **variance:** a measure of how widely data points within a group are dispersed from the group mean, expressed as the average squared distance of each data point from the mean.

BLOCK DESIGNS AND VARIABILITY

The Latin square design is a type of block experimental design. Block designs are used to control for sources of variability by repeating treatments among different experimental units, known as "blocks." Each block should be relatively homogenous, so that the variability within one block is less than the variability of the entire sample. Subjects may fall into natural category divisions (such as class years in a college), or a pretest may be used to divide them into categories with maximum internal homogeneity.

The basic block design is the randomized block design (RBD). RBD controls for one source of variability, also called a "blocking factor" or "nuisance factor." The Latin square design builds on this by allowing the researcher to control for two blocking factors. In this design, there is a single factor of interest (the treatment) and two nuisance factors. The design can also be used to study the effects of multiple factors rather than simply to reduce variability.

In a Latin square design, the number of blocks (groups) in each round must equal the number of treatments being tested. This can be represented by a square table, where the numbers of rows and columns are also equal to the numbers of blocks and treatments. Below is a square corresponding to three blocks and three treatment levels:

A	B	C
B	C	A
C	A	B

This shows which block receives which treatment in each round. Each letter represents a different treatment or treatment level. (The name "Latin square" comes from the fact that Latin letters are used to denote the treatments.) A treatment must occur only once in each row and each column. Results are then analyzed using two-way analysis of variance (ANOVA).

When the Latin square design is used to control for two nuisance factors, the composition of the blocks does not necessarily remain the same in each round. Instead, each nuisance factor is assigned to either rows or columns, and subjects in each round belong to the block that applies to them. For example, one nuisance factor may be divided into categories a_1, a_2, and a_3, and the other into categories b_1, b_2, and b_3. Nuisance factor a is assigned to columns, and nuisance factor b to rows. If both nuisance factors represent a personal trait such as age and weight and subjects can only belong to one category for each factor, then a subject who falls into the categories a_3 and b_1 will only be part of the third (far right) block in the first round of testing. However, if nuisance factor b represents a factor unrelated to the subjects (days of the week, for example), then a subject in the category a_3 will remain in the third block throughout the study, and each round will be conducted on a different day.

Latin squares were first used in agricultural experiments where soil conditions varied by row and column. Fertilizer treatments were applied to the field in strips that corresponded to the rows and columns of the square. Later, other scientific and industrial fields adopted the design.

Advantages and Disadvantages

Advantages of the Latin square design include its ability to handle more than one nuisance factor at a time, which helps reduce or even eliminate variation. It can also test multiple treatments or treatment levels with fewer experimental rounds than other designs.

One drawback is that the number of levels of every blocking factor must be equal to the number of treatments. This restricts the number of studies for which it can be used. Furthermore, the Latin square design assumes that there are no interactions between the blocking factors and the treatment factor. However, that is not always the case. It is not always possible to know beforehand if there will be any interaction between the treatment factor and one or both of the nuisance factors. Another disadvantage is that the design does not test for every possible combination of nuisance and treatment factors.

Finally, because a treatment is required for every row and column, care must be taken to prevent the study from becoming too large or too small. With too many blocks and treatments, the study may become too costly and unmanageable. With too few, the residual degrees of freedom may be too small for the results to have sufficient statistical power.

Latin Square Design in Action

A carmaker hires a researcher to test the gas mileage of five of its car models. Specifically, the manufacturer wants to know if there are any differences in the cars' everyday gas mileage during weekly commuting hours. One car of each model will be tested. Each car will be new and have the same model year.

The treatment factor being tested is the model. Nuisance factors that might affect the gas mileage, other than design, include different styles of driving and the amount of traffic on different days of the workweek. To control for these factors, the researcher decides to use a Latin square design. First, the researcher recruits five drivers who each have different driving styles, represented by the numbers

1 through 5, ranging from cautious (1) to reckless (5). This factor will be assigned to the columns of the Latin square. Next, the days of the workweek (Monday through Friday) will be assigned to the rows of the square. Finally, the researcher assigns a letter to each model (A through E) and arranges the letters in the square so that no letter is repeated in any row or column. The resulting Latin square appears as follows:

	1	2	3	4	5
M	A	B	C	D	E
T	B	C	D	E	A
W	C	D	E	A	B
Th	D	E	A	B	C
F	E	A	B	C	D

Thus, driver 1 will drive car A on Monday, car B on Tuesday, and so on.

APPLICATION

Latin squares were first used in agricultural experiments where soil conditions varied by row and column. Today, the Latin square design is used in a broad range of different fields, as it is an efficient method of testing multiple treatment levels in a relatively small number of treatment rounds.

Mandy McBroom-Zahn

FURTHER READING

De Lury, D. B. "The Analysis of Latin Squares When Some Observations Are Missing." *Journal of the American Statistical Association*, vol. 41, no. 235, 1946, pp. 370–89.

Keedwell, A. Donald, and József Dénes. *Latin Squares and Their Applications*. 2nd ed., Elsevier, 2015.

Kirk, Roger E. "Latin Square." *The SAGE Encyclopedia of Social Science Research Methods*, edited by Michael S. Lewis-Beck et al., vol. 2, Sage Publications, 2004, pp. 556–57.

Lui, Kung-Jong, and Kuang-Chao Chang. "Test Equality in Binary Data for a 4×4 Crossover Trial under a Latin-Square Design." *Statistics in Medicine*, vol. 35, no. 23, 2016, pp. 4110–23, doi:10.1002/sim.6975. Accessed 2 May 2017.

Montgomery, Douglas C. *Design and Analysis of Experiments*. 8th ed., John Wiley & Sons, 2013.

Sahai, Hardeo, and Mohammed I. Ageel. *The Analysis of Variance: Fixed, Random and Mixed Models*. Birkhäuser, 2000.

LATIN SQUARE DESIGNS SAMPLE PROBLEM

Researchers at a university want to conduct a psychological experiment to study the effects of five different learning devices. They recruit participants from each undergraduate class year as well as from the graduate school. Before the study begins, the participants will take a pretest that will assess their learning skills and then assign each one a letter grade of A, B, C, D, or E. The experiment will be a Latin square design with five rounds of testing. What is the treatment factor that will be tested, and what are the two nuisance factors that the researchers should control for?

Answer:

The treatment factor is the type of learning device. The nuisance factors are the participants' class years (first-year, sophomore, junior, senior, graduate student) and their pretest scores (A–E).

LEAKEY, MARY

Louis Seymour Bazett Leakey (1903-1972) and his wife, archeologist and anthropologist Mary Douglas Nicol Leakey (1913-1996), digging at Oduvai Gorge, Tanzania, Africa. By Smithsonian Institution from United States [No restrictions], via Wikimedia Commons

BRITISH ANTHROPOLOGIST

Mary Leakey's disciplined approach to fossil records supplied empirical support for the theory that Africa was the cradle of humankind. Her discoveries in East Africa included the approximately 1.75-million-year-old fossils of Homo habilis at Olduvai Gorge and the approximately 3.6-million-year-old footprints of three fully upright, bipedal hominids in Laetoli.

- **Born:** February 6, 1913; London, England
- **Died:** December 9, 1996; Nairobi, Kenya
- **Also known as:** Mary Nicol
- **Primary field:** Biology
- **Specialty:** Anthropology

EARLY LIFE

Mary Douglas Leakey was born Mary Nicol, the only child of British painter Erskine Edward Nicol and Cecilia Frere Nicol. Leakey was attracted to two familial interests, drawing and fossils. Very much her father's child, she began to draw when she was ten years old. In her autobiography, she jokes that perhaps her second area of interest, fossils, was also genetic: Her eighteenth-century ancestor, John Frere, had argued that flints found with bones of extinct animals were the work of a prehistoric people.

The Nicolses' life was unpredictable. They moved to wherever Erskine could paint and find a place to live. As a result, Leakey's schooling was erratic. However, she was both bright and linguistically talented. Heartbroken by her father's sudden death in 1926, she responded to her mother's attempt to introduce more formal instruction with defiant refusal. By 1930, however, Leakey had begun to attend lectures in both geology and archaeology, and she wanted to take part in excavations. She sent pleading letters offering her services for free, receiving a reply from Dorothy Liddell. Leakey and Liddell worked together for several seasons. Leakey's drawing talent became clear quickly, and her drawings of the flint tools found by Liddell's team were published. These drawings led to her introduction to anthropologist Louis S. B. Leakey.

One of the archaeologists who had used Leakey's drawings was Gertrude Caton-Thompson, who knew Louis Leakey was looking for an illustrator for his book, and she introduced them. Mary was twenty years old and talented, but naive. Louis was a thirty-year-old charismatic and dynamic visionary. He also was married with one child and another on the way. Louis would leave his wife for Mary, later marrying her, and it was not long before they were working together in Africa.

LIFE'S WORK

The Leakeys endured several difficult years of poverty, social isolation, and professional rejection of Louis's claims for the dates of fossils found at Olduvai Gorge in northern Tanzania. By mid-1937, Leakey had begun work on her own dig north of Nairobi. It was here that her best qualities began to emerge: careful and disciplined work, modern excavation techniques, cautious dating of fossils, and an ability to recognize ancient tools.

Leakey was patient, utterly devoted to her task, and able to endure privations that few would tolerate. Her skill as an artist was a tremendous asset, because photography was rare in archaeology. In 1940, after her first child was born, she did not lessen her dedication to her work, nor did she after the birth of three more children. From 1942 to 1947, she worked with her husband at Olorgesailie, a site where the abundance of tools from 1.5 million years BCE led them to hope for the hominid fossils that remained elusive. They decided to work on Rusinga Island, in the eastern part of Lake Victoria.

On Rusinga, Leakey found the pieces of a skull of *Proconsul africanus*, revealed by a single tooth sticking out of the ground. Lovingly reassembled by Leakey, *Proconsul* set the Leakeys' future on more solid ground. Her presentation of the skull to the British Museum put her name before the public. However, results, not attention, were what she craved. Unlike her husband, who was driven by his passion to prove humans evolved in Africa, Mary Leakey only wished to discover what was there. In the 1950s, the Leakeys moved to Olduvai Gorge, but even though there was ample evidence of human activity, they found no human fossils there. The gorge was remote, and the Leakeys lived once again on the edge of poverty. Louis began to suffer a series of health problems. A trip to Laetoli in 1959 proved unfruitful, so they returned to Olduvai. In July, while her ailing husband remained in his tent, Leakey went fossil hunting and found a hominid fossil the Leakeys would call *Zinjanthropus*. Dating from approximately 1.75 million years BCE, the fossil was later reclassified as *Paranthropus boisei*.

Zinjanthropus was dropped from Louis's "true man" list in 1960, after the Leakeys discovered the remains of *Homo habilis*, or "handy man." The discovery led to an enormous uproar. *Zinjanthropus*, an australopithecine, and *Homo habilis* were, roughly speaking, contemporary, which challenged many scientists' assumptions that two species of early hominids could not have coexisted.

The relationship between Leakey and her husband became increasingly difficult, and they eventually split. The split was worsened when Mary Leakey decided to accept an honorary degree from South Africa's University of the Witwatersrand, while Louis, a citizen of Kenya, angrily refused the degree offered to him, because of South Africa's policy of apartheid. The Leakeys remained married, but their partnership was over.

Beginning in September 1968, Mary Leakey was living once again in Olduvai. She had already finished excavating beds I and II at the gorge. Ahead of her lay the deposits of beds III and IV. She was also close to finishing the third volume of her study *Olduvai Gorge* (1971). Reviewing the many fossils she had found, she saw that *Homo habilis* had been well named. This ancient creature had created twenty different tools, believed to have required many years of development. A most significant find was the tools made from bone, a first for Leakey.

Leakey measured, weighed, and observed every minute curve and flake of all her finds. The simple tools she found were early versions of the hand ax, but she also found some elegant, pear-shaped hand axes. She theorized from this find that an advanced people, living side by side with the *Homo habilis* groups, had brought the hand axes to the region, a theory that countered the prevailing view of the evolution of hominids, which was assumed to have worked in straight lines of descent.

In 1972, Louis Leakey died in London. After the funeral, Mary Leakey returned to Olduvai. Within three years, she was in Laetoli, the site of remarkable archaeological discoveries. In Laetoli, Leakey and her team found hominid teeth, the nearly complete cranium of an archaic *Homo sapiens* skull, and, in July 1976, fossilized animal footprints visible on layers of volcanic ash hardened over three million years. The most dramatic discovery, however, was a set of hominid footprints, found in 1978. One adult hominid had been six feet tall; a second adult appeared to have been holding the waist of the first; and a third, quite small, had been found at the side of one of the adults. The plodding footsteps seemed to indicate that two adults and a child had fled falling volcanic ash. Leakey revealed that the footprints showed that before early humans developed large brains they had stood upright, a major discovery in the evolution of the human species.

Leakey retired in the fall of 1983 and moved back to Kenya, where her sons Phillip and Richard were living. She garnered more honors, led an active life, and wrote her autobiography. She died on December 9, 1996, in Nairobi, Kenya.

Impact

Into the early twenty-first century, the empirical evidence assembled by Leakey of the long evolutionary development of *Homo sapiens* carried an importance that Leakey herself likely could not have imagined. Her care in amassing evidence of human evolution remains an essential component of any fair-minded assessment of the principles on which modern theories of humans in the natural world are based.

—*Jane Carter Webb*

Further Reading

Bowman-Kruhm, Mary. *The Leakeys: A Biography*. Westport, CT: Greenwood, 2005. Print. Intended for general and high school-level readers and written in a style that moves between the serious and the breezy. Outline structure makes the complexities of the Leakeys' interactions with one another and of the fossil finds easy to follow. Bibliography.

Feder, Kenneth L., and Michael Alan Park. *Human Antiquity: An Introduction to Physical Anthropology and Archaeology.* 5th ed. New York: McGraw, 2007. Print. Excellent updated account of human origins and early hominid forms, written in a readable style. Photographs, illustrations, bibliography.

Leakey, Mary. *Disclosing the Past: An Autobiography.* New York: Doubleday, 1984. Print. Autobiography that examines some of the painful difficulties Leakey experienced with her marriage and with her scientific competitors. Index, bibliography.

Morell, Virginia. *Ancestral Passions: The Leakey Family and the Quest for Humankind's Beginnings.* New York: Simon, 1995. Print. Detailed work that treats all members of the complicated Leakey family; shows how the family came to dominate the field of anthropology, and discusses the animosities within the family as well as among scientists. Bibliography, index.

LINEAR AND NONLINEAR RELATIONSHIPS

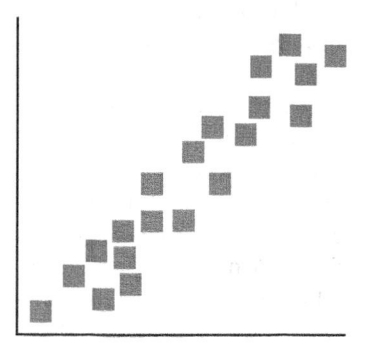

Positive Linear Relationship

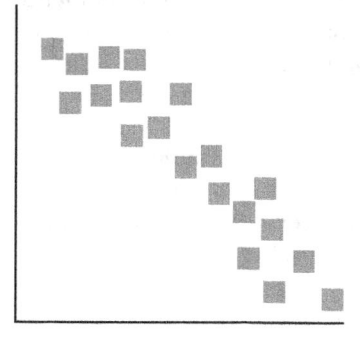

Negative Linear Relationship

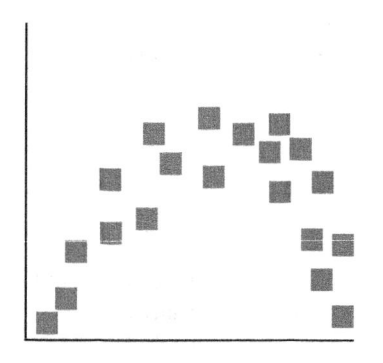

Nonlinear Relationship

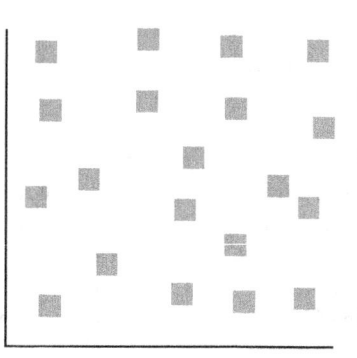

No Relationship

FIELDS OF STUDY

Statistical Analysis; Relational Analysis

ABSTRACT

Linear and nonlinear relationships are elements that represent the statistical evaluation of a variable and a constant. A linear relationship produces a straight line on a graph, whereas a nonlinear relationship will not.

PRINCIPAL TERMS

- **direct relationship:** a link between two variables in which if one variable increases, the other also increases, and if one decreases, the other also decreases.
- **inverse relationship:** a link between two variables in which if one variable increases, the other decreases, and vice versa.
- **logarithmic:** relating to the inverse of an exponent.
- **polynomial:** a mathematical statement containing one or more variables and coefficients in which the only operations are addition, subtraction, multiplication, and non-negative exponentiation.
- **variable:** in research or mathematics, an element that may take on different values under different conditions.

STRAIGHT OR CURVED LINES?

Scientists use the Cartesian coordinate system to evaluate the connection between data points from a set of variables that are plotted on a graph. René Descartes (1596–1650) and Pierre de Fermat (1601–65) figured out how to use ordered numbers with corresponding points to create algebraic equations. When graphed, the outcome resulted in a geometric shape that formed either a linear or nonlinear relationship.

In order for linear relationship to exist, three key features must be apparent: the equation can only contain two variables, the variables can only be taken to the first power, and the graph should display a straight line. The form of a linear relationship is often represented as

$$y = mx + b$$

In this equation, x and y represent the two variables, and b is a constant. (Sometimes c is used instead of b.) On a graph, this equation appears as a line on the xy plane with a slope of m and a y intercept (where y crosses the x axis) of b. The slope is the ratio between the change in x and the change in y. Slope can be either positive or negative. On a graph, a slope that increases from left to right is positive. A slope that decreases from left to right is negative.

When nonlinear relationships are plotted, they usually appear as curved lines. A linear approximation is a straight line that can be drawn tangent to part of the curve of interest. Linear approximations simplify complex functions such that a given area can be evaluated.

Determination of Linear and Nonlinear Relationships

Despite the different outcomes, similar procedures are followed to determine if a linear or nonlinear relationship exists. First, a plotting system, such as the Cartesian coordinate system, can be used to evaluate how the variables relate to data points on the graph. After the x and y variables are plotted on the graph, the set of data points will produce a geometric shape. If the shape is a straight line, then a linear relationship does exist. If the shape is curved, then a nonlinear relationship exists.

Another way to determine if a linear or nonlinear relationship exists is to model the relationship using equations and check their fit. Direct variation, or direct relationship, is a relationship where a change in one variable causes a proportional change in the other variable in the same direction. The equation for this relationship sets one variable as equal to the second variable multiplied by a constant value, as in

$$y = 3x$$

As x increases, y increases, and vice versa. Direct variation is linear.

Partial variation is a relationship that involves two variables with the addition of a fixed or constant value. One variable is equal to the value of the other variable multiplied by a constant value and changed with the addition of another constant value, which may be positive or negative. An example of a partial variation is

$$y = 3x + 2$$

Partial variation is linear as well.

Inverse variation, or inverse relationship, is a relationship where a change in one variable causes a proportional but opposite change in the other. One variable is equal to a constant divided by the second variable, as in

$$y = 30/x$$

Another way to describe this relationship is to say that one variable multiplied by the other variable is equal to a constant:

$$xy = 30$$

Thus, as x increases, y decreases, and vice versa. Inverse relationships are nonlinear.

Exponential, polynomial (including quadratic), logarithmic, and sinusoidal (wavelike trigonometric) equations all model nonlinear relationships as well.

Use in Statistical Analysis

Linear and nonlinear relationships can be used in statistical analysis of research data. In many studies, researchers are investigating whether a change in the value of an independent variable x causes a change in value in the dependent variable y. These values for x and y gathered from the study can be plotted, and a line of best fit can be drawn to try to account for the changes observed. How well that line accounts for all of the observed data points—in other words, how accurate the model is—can be determined through regression analysis. Regression calculates the R^2 (pronounced "R squared") value, or coefficient of determination, which expresses the ratio of the explained variation (that is, variation attributable to the independent variable) to the total variation. A higher R^2 value usually indicates that the model fits the data well.

A coefficient of determination can be found if the presumed relationship between x and y is linear. For nonlinear relationships, linear approximations can be used to find the coefficient. However, an R^2 value computed in this way is weaker than that of a linear relationship.

APPLICATIONS

While linear relationships are simpler than nonlinear ones, this does not mean that they have less important applications in the real world. Telecommunications, for instance, often have a linear relationship between the input and output. There is a linear relationship between the bandwidth and data rate of a wireless communications channel, for example.

Equations describing nonlinear relationships are used to model some of the most complex phenomena found in nature, such as ocean currents or air flow around the wing of an aircraft. Fluid dynamics, a subfield of physics and engineering, studies many nonlinear relationships that are necessary for a wide range of applications, such as predicting weather or designing aircraft or oil and gas pipelines.

Cindy Ferraino

FURTHER READING

"Concepts: Linear and Nonlinear." *New England Complex Systems Institute*, 2011, www.necsi.edu/guide/concepts/linearnonlinear.html. Accessed 13 Apr. 2017.

Ellenberg, Jordan. *How Not to Be Wrong: The Power of Mathematical Thinking*. Penguin Books, 2014.

Haslam, S. Alexander, and Craig McGarty. *Research Methods and Statistics in Psychology*. 2nd ed., Sage Publications, 2014.

Lomax, Richard G., and Debbie L. Hahs-Vaughn. *An Introduction to Statistical Concepts*. 3rd ed., Routledge, 2012.

Paulos, John Allen. *Beyond Numeracy: Ruminations of a Numbers Man*. Alfred A. Knopf, 1991.

Rainville, Earl D., et al. *Elementary Differential Equations*. 8th ed., Prentice Hall, 1996.

LINEAR AND NONLINEAR RELATIONSHIPS SAMPLE PROBLEM

At a local movie theater, the price of a box of candy is $2. The price of the candy is represented as the independent variable, x. The total price is represented as the dependent variable, y. The data is tabulated as follows:

Number of Boxes	Total Price
1	2
2	4
3	6
4	8
5	10

Using a linear model, the equation for the data would be

$$f(x) = 2x$$

with an R^2 value 1. Using an exponential model, the equation for the data would be

$$f(x) = 1.6e^{0.4x}$$

with an R^2 value of 0.95. Based on the R^2 values, which is the best model to use?

Answer:

Because the line with the best possible fit would have an R^2 value of 1, the linear model is the best fit for this data. However, the R^2 value of the exponential model is large enough that it could suffice if the linear model were not available.

Longitudinal Sampling

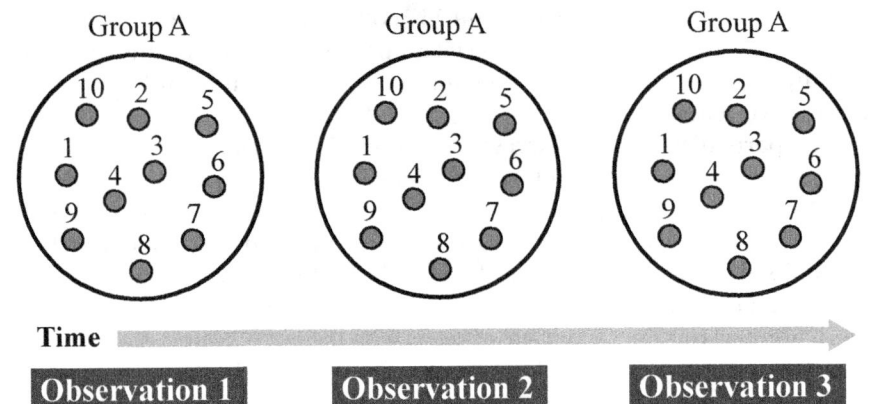

Observation 1 — Group A (subjects 1–10)
Observation 2 — Group A (subjects 1–10)
Observation 3 — Group A (subjects 1–10)

Time →

FIELDS OF STUDY

Sampling Design; Sampling Techniques; Research Design

ABSTRACT

Longitudinal sampling is a research method that allows researchers to examine and note changes among groups of individuals or within members of a designated group over time. Such studies are considered effective in accomplishing research purposes and providing information about ongoing patterns or trends within a range of fields.

PRINCIPAL TERMS

- **case study:** a research method in which selected individuals or small groups are observed and examined thoroughly.
- **cohort study:** research that takes as its subject people who share a particular characteristic such as age.
- **correlation:** an association between two variables under study.
- **observational research:** a research method in which researchers objectively examine and analyze phenomena in their natural setting, rather than under experimental conditions.
- **time-series design:** a research design in which researchers track one or more groups of subjects at designated intervals of time.

FOUNDATIONS

Longitudinal sampling is an observational research method that allows researchers to follow a subject or group of subjects over time. Longitudinal sampling is most often used in cases where researchers are interested in tracking changes over a designated period or in examining the correlations between specific events, drugs, treatments, or behaviors, and outcomes. Longitudinal sampling has been popular among researchers since the 1970s due to increased attention in tracking impacts of particular time periods. Most researchers follow a multistep process in which a research design is chosen, data is collected, statistical analysis is employed, and the study is analyzed and conclusions drawn.

Once the initial contact has been made, follow-up occurs at certain intervals. Data is collected through various instruments, such as interviews and questionnaires. When a phenomenon is rare, a researcher may choose to employ a case study, in which a certain individual or small group is observed and analyzed. The case study sample may be limited in size but allows for more in-depth responses. Unlike a cross-sectional study, which focuses on tracking multiple subjects at a specified time, longitudinal sampling uses a time-series design. This allows researchers to focus on individual changes that occur over time among the same subjects. Prospective cohort studies, a type of longitudinal study, have become increasingly common as researchers have expressed interest in studying effects on people who make up particular cohorts. For instance, baby boomers are considered a distinct group from those born before 1946 and after 1964.

The number of sampling contacts in longitudinal studies vary from two or three to thousands. In general, two or three contact periods, called "waves" or "cycles," are used. When a post-test design is used, contact between researcher and subject may be limited to the initial contact and one follow-up contact. The purpose of the post-test design is to compare changes among subgroups included in the study.

143

If an experimental design is used, group assignment may be random, resulting in groups of relatively similar subjects. This helps to cut down on the possibility of research being tainted by confounding variables that are unrelated to the study and that interfere with data collection. In nonexperimental studies, subjects are either sorted into particular groups or choose to join a specific group.

ADVANTAGES AND DISADVANTAGES
Two of the major benefits of longitudinal sampling are that it is simple to use and that it allows researchers to identify patterns and trends quickly. It also has the advantage of flexibility since the follow-up intervals are established according to the needs of a particular study. Thus, a researcher may increase the number of contacts if needed. Flexibility also results in the ability of researchers to use longitudinal sampling to either track forward (prospective) or backward (retrospective). The latter method is generally considered less reliable because it may be impacted by poor memories, hindsight, and the use of unreliable data-recording methods. Longitudinal sampling is often more expensive than other sampling methods. Such studies also have increased opportunities for sampling errors since participants may be hard to reach over time or may lose interest.

PRACTICAL APPLICATIONS
The Bill and Melinda Gates Foundation has been involved in efforts to improve contraceptive use in India, Kenya, Senegal, and Nigeria, tracking eighteen thousand women in India and Nigeria and five thousand to ten thousand in Kenya and Senegal. From 2010–11 to 2014–15, the Gates' Measurement, Learning & Evaluation (MLE) Project measured the impact of both supply-side and private-sector interventions in a longitudinal study that surveyed subjects at the initial point of contact, mid-point, and end of the contact. Longitudinal sampling allowed the foundation to determine exactly which methods of disseminating birth control information work better than others. That information is critical in countries with large populations, high fertility rates, and high infant and maternal mortality rates.

Another example of longitudinal sampling in practice is the Indonesian Children's Health Study, which looked at the impact of vitamin A deficiency on children suffering from respiratory and diarrheal infections. By following more than three thousand children over as many as six visits, researcher Alfred Sommer and his colleagues were able to show that introducing vitamin A into the diets of these children causes infections to decrease significantly.

SELECTING THE BEST METHODS
Traditionally, studies that employ longitudinal sampling have used experimental groups that are compared to a control group. This method has been considered particularly effective when tracking drug effects, treatments, attitudes, and behaviors over a period of time. For instance, a time-series design would be effective in the case of patients with osteoarthritis of the knee, allowing researchers to track the progression of the disease over time using such tools as clinical information, x-rays, and laboratory tests. Sample selections would be based on diagnosis of the condition.

Longitudinal studies have been consistently employed in twin studies to allow researchers to study the impacts of nature versus nurture. Studies of identical twins offer particular insight because such twins are twice as likely as fraternal twins to exhibit the same genes. For instance, the Western Reserve Reading and Math Project tracked twins in the first and second grades, leading researchers to conclude that first-grade language proficiency was more a response to genetics than to individual environments. By the second grade, however, environmental factors exerted a greater influence than genetics.

Using a joint research design that employs both longitudinal sampling and time-to-event methods works has been found to work better than using either model individually. It reduces possible parameter biases; allows for subjects that drop out of studies; and offers increased opportunities for analysis.

Elizabeth Rholetter Purdy, PhD

FURTHER READING
Angeles, Gustav, et al. *A Guide to Longitudinal Program Impact Evaluation*. MLE Technical Working Paper no. 1-2014, Measurement, Learning & Evaluation Project, May 2014. *Measurement, Learning & Evaluation Project for the Urban Reproductive Health Initiative*, www.urbanreproductivehealth.org/sites/mle/files/a_guide_to_longitudinal_program_impact_evaluation.pdf. Accessed 14 Apr. 2017.

Bolger, Niall, and Jean-Philippe Laurenceau. *Intensive Longitudinal Methods: An Introduction to Diary and Experience Sampling Research.* Guilford Press, 2013.

Diggle, Peter J., et al. *Analysis of Longitudinal Data.* 2nd ed., Oxford UP, 2013.

Everitt, Brian S. *A Whistle-Stop Tour of Statistics.* CRC Press, 2012.

Harlaar, Nicole, et al. "Longitudinal Effects on Early Adolescent Language: A Twin Study." *Journal of Speech, Language, and Hearing Research,* vol. 59, no. 5, 2016, pp. 1059–73. *Academic Search Complete,* search.ebscohost.com/login.aspx?direct=true&db=a9h&AN=119171594&site=ehost-live. Accessed 14 Apr. 2017.

Smith, Gary. *Essential Statistics, Regression, and Econometrics.* 2nd ed., Academic Press, 2015.

Soden, Brooke, et al. "Longitudinal Stability in Reading Comprehension Is Largely Heritable from Grades 1 to 6." *PLOS ONE,* vol. 10, 2015, doi:10.1371/journal.pone.0113807. Accessed 14 Apr. 2017.

Sudell, Maria, et al. "Joint Models for Longitudinal and Time-to-Event Data: A Review of Reporting Quality with a View to Meta-Analysis." *BMC Medical Research Methodology,* vol. 16, 2016. *Academic Search Complete,* search.ebscohost.com/login.aspx?direct=true&db=a9h&AN=120046354&site=ehost-live. Accessed 14 Apr. 2017.

Twisk, Jos W. R. *Applied Longitudinal Data Analysis for Epidemiology: A Practical Guide.* Cambridge UP, 2013.

MACH, ERNST

AUSTRIAN PHYSICIST AND PHILOSOPHER

Ernst Mach lived and worked during the nineteenth century, when the study of science was closely related to the study of philosophy, especially in German-speaking countries. Mach was influential in both disciplines but is probably best remembered for his contributions to the development of Einstein's theories of relativity. His ideas also contributed to the fields of physiology and psychology, and he is notorious for having stubbornly denied the existence of the atom.

- **Born:** February 18, 1838; Chrlice, Moravia, Austrian Empire (now part of Brno, Czech Republic)
- **Died:** February 19, 1916; Munich, German Empire
- **Primary field:** Physics
- **Specialty:** Relativity

EARLY LIFE

Ernst Waldfried Joseph Wenzel Mach was born on February 18, 1838, in what is now part of the Czech Republic. Young Mach performed so poorly as a student that after only a year, he was not allowed to return to the school near Vienna where he was studying. Since his father Johann was a teacher, Mach was able to continue his education at home, and enrolled in the University of Vienna when he was seventeen years old.

In 1860 he was awarded a doctorate in physics and went on to work as a tutor in Vienna for several years, addressing students of medicine on topics in physics. This resulted in his book "Compendium of Physics for Medical Students" (1863). In 1864, Mach joined the University of Graz as a professor of mathematics and later taught physics there.

He lectured at Graz until 1867, when he took on a professorship in experimental physics at Charles University in Prague, where he would remain for nearly thirty years. During this time Mach carried out a considerable amount of original research in areas such as optics, waves, and the study of sensory perceptions.

In 1895 Mach found himself back at the University of Vienna, as professor of inductive philosophy (inductive logic is a type of reasoning in which specific observations lead to more general conclusions). However, in 1898 he experienced a severe stroke that paralyzed the right half of his body, and he was forced to resign from his professional position. He did continue to speak, publish, and pursue independent scientific study, however. In 1901 the Austrian government honored him with an appointment to parliament.

Throughout his life Mach was particularly interested in a field called epistemology, which refers to the study of human knowledge and the limits of understanding. One of his last two books dealt with this discipline; titled *Knowledge and Error*, it appeared in 1905. Mach also wrote an autobiography, published in 1910. He died six years later on February 19, 1916, at the age of seventy-eight.

LIFE'S WORK

One of Mach's most important epistemological ideas was that all of our insights about the world are based on imperfect and fallible sensory information. According to him, what we call scientific laws are not really absolute laws of nature, but reflections of how human beings perceive the universe. Mach wrote about these ideas in his book *The Analysis of Sensations* (1897).

As a result, Mach refused to accept any scientific hypothesis that could not be proven through testing. His strict insistence on being able to verify theory was somewhat revolutionary at the time, when many complex scientific ideas were put forward and accepted without experimental proof. (This is also the reason Mach doubted that atoms were real, since during his lifetime there were no electron microscopes that could be used to view individual atoms.)

Mach's firm belief that scientific ideas needed to be provable led him to reject Isaac Newton's abstract

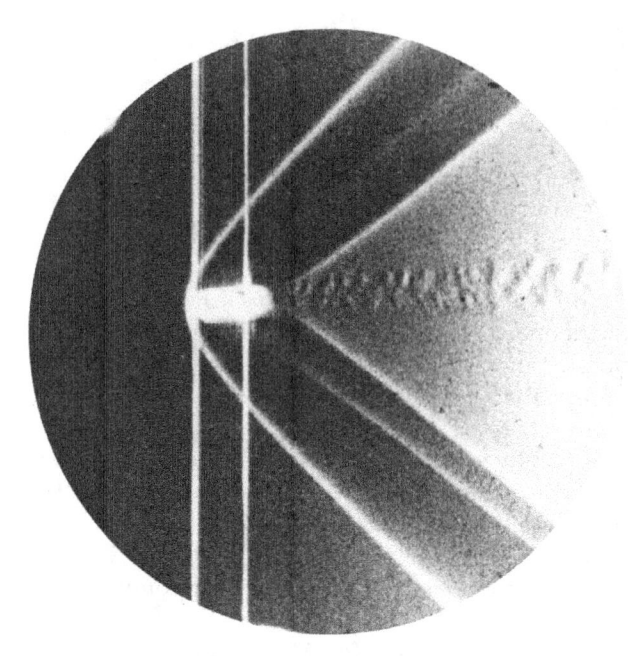

Ernst Mach's historic 1887 photograph (shadowgraph) of a bow shockwave around a supersonic bullet. (Scan from book (now lost)) [Public domain], via Wikimedia Commons.

notions of "absolute," or unchanging, space and time. Mach's book *The Science of Mechanics* (1883), generally considered his masterpiece, strongly criticized these concepts. It also introduced a sophisticated idea that Albert Einstein later came to call "Mach's principle." Part of Mach's disagreement with Newton had to do with the question of how an object's acceleration (the change in its speed) is measured through space. When physicists measure an object's change in speed, they always think about what that change is relative to. In other words, the object may be speeding up or slowing down, but speeding up or slowing down compared to what?

To solve this problem, Newton came up with the idea of absolute space: a theoretical constant that no one had ever seen, but that scientists could always use as a point of comparison. Mach, on the other hand, suggested that physicists should calculate an object's acceleration not in relation to Newton's abstract and invisible constant, but in relation to all of the other matter in the entire universe.

What Einstein dubbed "Mach's principle" is the idea that when an object has inertia (when it resists a change in its speed), the source of that inertia is the object's relationship to all the other matter in the universe. This proposal helped Einstein think about gravity and movement in new ways. Einstein always said that the concepts in *The Science of Mechanics* were enormously powerful influences on him as he was working on his general theory of relativity. Despite this fact, Mach's principle was never included in Einstein's calculations, and there is still some disagreement among physicists about whether Newton's ideas or Mach's more accurately describe the way the universe really works. One of Mach's first projects after he received his doctorate was an attempt to prove the Doppler effect, a theory proposed in 1842. The theory states that light and sound waves increase in frequency as their source draws nearer to an observer, and decrease in frequency as their source moves away from an observer. In 1860 scientists were still trying to design experiments that would show that the Doppler effect was real. Mach devised a simple, portable set of equipment that could be used to clearly demonstrate the effect.

Impact

Besides the principle Einstein named after him, Mach's name is preserved in at least two other scientific concepts. The first is the "Mach number," which describes the relationship between the speed of an object and the speed of sound in whatever medium the object happens to be traveling through. A rocket or a jet plane moving at greater than "Mach 1" is traveling at supersonic speeds, or faster than the speed of sound.

Mach was the first scientist to realize that a projectile moving faster than the speed of sound actually creates shock waves around itself. He was then able to photograph the effects of the shock waves created by a bullet moving at supersonic speeds. In order to do this, Mach invented an entirely new photographic method, taking advantage of the fact that shock waves cause light to refract (bend). The resulting image he called a shadowgraph.

A second well-known scientific phenomenon named after Mach is an optical illusion he discovered. Mach showed that when a person is looking at two regions of space that are side by side, and when those regions are of significantly different levels of lightness or darkness, the eye tends to see bands of either brighter or darker space at the border between them. The bands do not really exist, but our brains perceive them nonetheless. They are known as "Mach bands."

Mach's research into sensory awareness also included investigations of hearing, movement, and the human perception of time.

—*M. Lee*

FURTHER READING

Blackmore, John T. "Three Autobiographical Manuscripts by Ernst Mach." *Annals of Science* 35.4 (July 1978): 401–19. Print. Offers a collection of original source material as an aid to understanding Mach's role in the history of physics, psychology, and science in general.

Hoffmann, Christoph. "Representing Difference: Ernst Mach and Peter Salcher's Ballistic-Photographic Experiments." *Endeavour* 33.1 (Mar. 2009): 18–23. Print. Presents photographs depicting the bullet experiment conducted by Mach.

Karwatka, Dennis. "Ernst Mach and the Mach Number." *Tech Directions* 69.5 (Dec. 2009): 14. Print. Recounts Mach's life and career, noting his contributions to the development of several branches of physics, including optics and mechanics.

ERNST MACH AND THE SPEED OF SOUND

Ernst Mach was the first scientist to apply systematic study to objects moving faster than the speed of sound, using photography. He produced the first-ever photograph of the shockwaves produced by a bullet in supersonic motion.

Mach's photographic work was dependent on a technique called the shadowgraph. Basically, he backlit his entire experimental subject—the bullet and air it was traveling through—then photographed the shadows cast by his experiment from the other side of a sheet. His photographs were not of the shockwaves directly, which are invisible, but of the resulting distortion caused as light passes through these waves of compressed air. Mach's achievement is all the more remarkable when one considers the fine timing required to capture an image of a bullet in motion, which he achieved using trip wires in the path of his bullet. These wires are visible in the first photograph published in his paper "Photographische Fixierung der durch Projektile in der Luft eingeleiteten Vorgänge," presented to the Academy of Sciences in Vienna in 1887.

Mach's work was instrumental in solving two technology puzzles of his time. Advances in weapons technology had produced guns that could fire bullets faster than the speed of sound. Mach's experiment with this faster-than-sound firing led to two unexpected observations. The first was a second "bang" after the usual one associated with the explosion of a gun being fired. The second was the appearance of crater-like wounds and impact sites. These phenomena are explained by the shockwave of compressed air created by a supersonic bullet.

Eventually, Mach's name would be attached to the Mach number—the speed of an object moving through a fluid divided by the speed of sound—in common usage, it is a measure of speed for objects (such as jets) traveling at or near the speed of sound. The formula for the Mach number is $M = V/a$, where M is Mach number, V is the velocity of the object in question, and a is the speed of sound for the medium through which the object is traveling (somewhere around 770 miles per hour through dry air). An object traveling at subsonic speeds is traveling at a Mach number of less than 1. An object traveling at exactly the speed of sound is traveling at Mach 1. Anything above Mach 1 but below Mach 5 is supersonic flight, Mach 5–10 is hypersonic, and Mach 10+ is high-hypersonic.

Mann-Whitney *U* Test

FIELDS OF STUDY

Statistical Analysis

ABSTRACT

The Mann-Whitney *U* test, also called the Wilcoxon rank-sum test, is a distribution-free test that compares two populations in order to determine whether their distributions have the same median. It can also show differences in distribution spread and shape.

PRINCIPAL TERMS

- **independent variable:** in experimental research, a measurable occurrence, behavior, or element that is manipulated in an effort to influence changes to other elements.
- **Kruskal-Wallis *H* test:** a nonparametric, rank-based test used to determine if a statistically significant difference exists between an ordinal dependent variable and two or more categorical groups of an independent variable.
- **nonparametric:** characterized by having a non-normal distribution.
- **normal distribution:** a probability or frequency distribution in which plotting the values contained in a data set, according to either the probability of their occurring or the frequency with which they occur, results in the appearance of a symmetrical bell-shaped curve, with the majority of values clustered around the middle; also called a bell curve.
- **ordinal variable:** an independent variable that only takes ordered, or ranked, values in specified categorical groups.

Testing Nonparametric Data

In medical and public health research, data often follow a normal distribution, in which all values in a data set fall symmetrically about the arithmetic mean. However, sometimes data do not follow a parametric distribution. In such cases, parametric methods of analysis do not apply, since those methods are based on the assumption of a normal distribution, and measures of central tendency cannot be used. Ordinal variables, which are ordered, categorical independent variables (or "predictor variables"), can present in a non-normal distribution. Parametric statistics can sometimes be applied to ordinal data, but often a nonparametric approach must be used.

A common method when working with ordinal data is to measure the difference between medians. The null hypothesis is that the two populations have equal medians, and the two-tailed alternative hypothesis is that their medians are unequal. ("Two-tailed" means that the question is investigated in both directions, higher and lower.) The Mann-Whitney *U* test is used to compare two independent samples with continuous or ordinal outcomes. It can be seen as an alternative to a two-sample *t* test, which looks for the difference of means. Differences in medians detected in a Mann-Whitney *U* test can also point to differences in the spread or overall shape of the distributions being tested.

Comparing Statistical Tests

The Mann-Whitney *U* test assumes an ordinal measurement scale, a continuous response variable, independent samples, and similar distributions. Another nonparametric statistical test, the Kruskal-Wallis *H* test, assumes an ordinal or continuous response variable, a predictor variable with two or more categorical groups, independent observations (that is, no relationship between observations within or between groups), and the same distribution variability (shape). The drawback to this test is that one can find whether a statistical difference exists, but not where.

For both tests, to set up the problem, values in each population are ranked and the ranks ordered from smallest to largest. Next, a decision rule is chosen to determine when the null hypothesis is rejected. The specific test statistic is computed, and the result is interpreted. For the Mann-Whitney *U* test, the sample data are combined before ranking.

Rank of Newest James Bond Actor by Generation

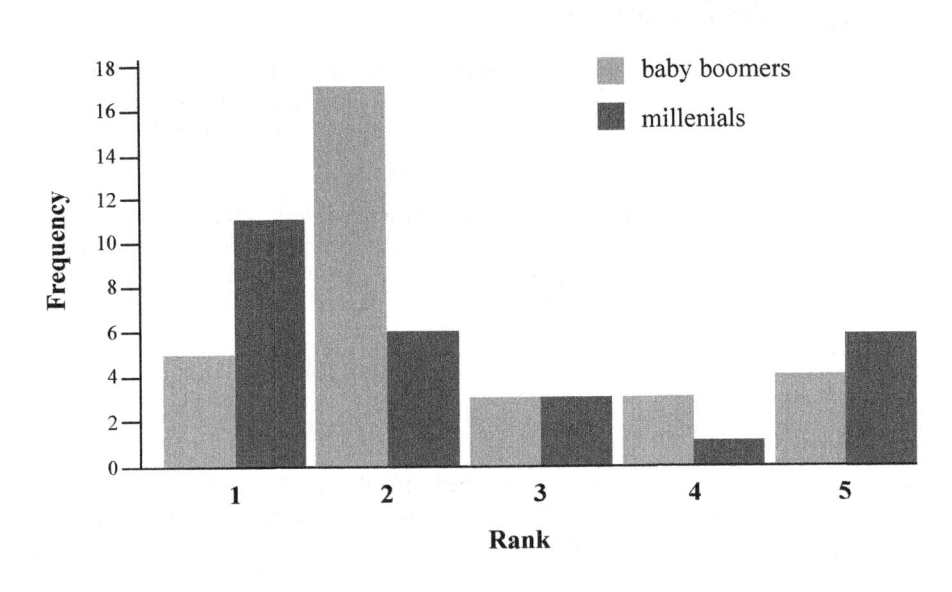

The Mann-Whitney U test is a particularly valuable tool in medicine, psychology and the social sciences, business, and marketing. Often the test is used in circumstances like those in which Student's t test is preferred, except that the data are not normally distributed. In medical research, for instance, the Mann-Whitney U test is used to compare whether two treatments have equal efficacy in addressing an ailment. In marketing and the social sciences, the test is commonly employed in studies of preferences, such as whether consumer preferences within a product category vary by region or how demographics affect political attitudes and voting behaviors.

Its test statistic is the smaller of either U_1 or U_2, as defined by the equations

$$U_1 = n_1 n_2 + \frac{n_1(n_1+1)}{2} - R_1$$

$$U_2 = n_1 n_2 + \frac{n_2(n_2+1)}{2} - R_2$$

where n is the sample size of each population and R is the sum of the ranks for each.

The Mann-Whitney U test has lower power than parametric alternatives, such as the t test and ANOVA. Having lower power means that if there is a real statistical difference and the alternative hypothesis is true, the Mann-Whitney U test is less likely to detect it. For a small sample, a frequency histogram should be drawn, or some other method used, to determine whether the distribution of outcome values is likely normal or not before applying this test.

APPLICATIONS

The Mann-Whitney U test is a particularly valuable tool in medicine, psychology and the social sciences,

Mandy McBroom-Zahn

FURTHER READING

Daniel, Wayne W., and Chad Lee Cross. *Biostatistics: A Foundation for Analysis in the Health Sciences.* 10th ed., John Wiley & Sons, 2013.

Gibbons, Jean Dickinson. *Nonparametric Methods for Quantitative Analysis.* 2nd ed., American Sciences Press, 1985.

Kruskal, William H., and W. Allen Wallis. "Use of Ranks in One-Criterion Variance Analysis." *Journal of the American Statistical Association*, vol. 47, no. 260, 1952, pp. 583–621. *JSTOR*, www.jstor.org/stable/2280779. Accessed 21 Apr. 2017.

LaMorte, Wayne W. *Nonparametric Tests.* Boston U School of Public Health, 24 May 2016, sphweb.bumc.bu.edu/otlt/mph-modules/bs/bs704_nonparametric/BS704_Nonparametric-TOC.html. Accessed 21 Apr. 2017.

Mood, Alexander M., et al. *Introduction to the Theory of Statistics.* 3rd ed., McGraw-Hill, 1974.

Urdan, Timothy C. *Statistics in Plain English.* 4th ed., Routledge, 2017.

SAMPLE PROBLEM

A high school science teacher wants to know if sophomores who took a test prep course performed better on a standardized exam than those who did not take the course. The teacher sampled twenty students, ten of whom took the prep course and ten of whom did not. Below is a table showing the scores for both samples and the corresponding rankings.

		Total Sample Ascending Order		Ranks	
Test Prep	Control	Test Prep	Control	Test Prep	Control
345	280	345	280	4	1
582	449	468	295	7	2
670	500	500	300	8.5	3
720	358	550	358	12	5
500	295	582	449	13	6
650	601	600	500	14	8.5
780	700	650	520	16	10
468	520	670	530	17	11
600	300	720	601	19	15
550	530	780	700	20	18
				$R_1 = 130.5$	$R_2 = 79.5$

The teacher next calculated the U value, using the Mann-Whitney U test:

$$U_1 = n_1 n_2 + \frac{n_1(n_1+1)}{2} - R_1$$

$$U_2 = n_1 n_2 + \frac{n_2(n_2+1)}{2} - R_2$$

$$U_1 = n_1 n_2 + \frac{n_1(n_1+1)}{2} - R_1$$

$$= (10 \times 10) + \left(\frac{10(10+1)}{2}\right) - 130.5$$

$$= 100 + 55 - 30.5$$

$$= 24.5$$

$$U_2 = n_1 n_2 + \frac{n_2(n_2+1)}{2} - R_2$$

$$= (10 \times 10) + \left(\frac{10(10+1)}{2}\right) - 79.5$$

$$= 100 + 55 - 79.5$$

$$= 75.5$$

Explain whether the teacher used a one- or two-tailed test and why. Based on a 95 percent confidence level, determine whether the test prep course improved the students' scores.

Answer:

Because the teacher only wants to test for the median of the test prep course group's scores being greater than that of the control group, a one-tailed hypothesis must be used. The null and alternative hypotheses are

$$H_0 = M_{TP} \leq M_C$$

$$H_A = M_{TP} > M_C$$

where M_{TP} is the median of the population of students who took the prep course and M_C is the median of the control group population.

For a confidence level of 95 percent, let $\alpha = 0.05$. Consulting a table of critical values shows that the critical value of U at $p < 0.05$ is 27. This is larger than the smaller U value of 24.5. Therefore, the result is significant at the $\alpha = 0.05$ level, and the null hypothesis can be rejected.

Meitner, Lise

AUSTRIAN PHYSICIST

Physicist Lise Meitner's joint research with chemist Otto Hahn, and later Fritz Strassmann, yielded the discovery of new radioactive elements and their properties and paved the way for the discovery of uranium fission.

- **Born:** November 7, 1878; Vienna, Austria-Hungary
- **Died:** October 27, 1968; Cambridge, England
- **Primary field:** Physics
- **Specialties:** Nuclear physics; atomic and molecular physics

EARLY LIFE

Lise Meitner was the third of eight children born to Hedwig Skovran and Philipp Meitner. She was interested in mathematics and physics from a young age. After receiving a matriculation certificate from the gymnasium (secondary school) in Vienna, she went on to the University of Vienna, where, from 1901 to 1905, she studied mathematics, physics, and philosophy. In 1902, she began her study of theoretical physics under the tutelage of Ludwig Boltzmann, who was an advocate of atomic theory—the idea that all matter is composed of tiny, invisible, and (at the time) indivisible components. This was by no means generally accepted by physicists of the day, but in Boltzmann's view, the discovery of radioactivity supplied the experimental proof that tiny particles, or atoms, form the building blocks of all things.

In 1905, Meitner finished her doctoral thesis on heat conduction in nonhomogeneous bodies, becoming only the second woman to receive a doctorate in science from the University of Vienna. She soon became familiar with the new field of radioactivity and was ready to enter the realm of atomic physics at the beginning of a promising new period in that branch of science.

LIFE'S WORK

After graduation, Meitner went to Berlin to attend the lectures of theoretical physicist Max Planck, and what was intended to be a short stay became a thirty-one-year period of research on the frontiers of atomic physics and radioactivity. Planck was one of the world's most notable scientists, having formulated the theory of thermal radiation in 1900—a major advance in the developing field of quantum mechanics—and having been one of the first to recognize and stress the importance of Albert Einstein's special theory of relativity.

As important as Meitner's association with Planck was, however, it was her friendship and collaboration with Otto Hahn that would change the course of atomic science. Meitner and Hahn met in 1907, the same year Meitner came to Berlin, and she decided she wanted to work with Hahn and keep to the study of radioactivity. After some persuasion, the two received permission from Emil Fischer, the director of the Chemical Institute of Berlin, where Hahn was working, to become a research team, with the provision that Meitner promise not to go into the chemistry department, where the male students did their research. For the first few years, Meitner and Hahn's research was confined to a small room originally planned as a carpenter's shop. When women's education was officially sanctioned and regulated in Germany in 1909, Fischer gave Meitner permission to finally enter the chemistry department.

In 1912, the Kaiser Wilhelm Institute for Chemistry was opened as a part of the University of Berlin. Hahn became a member, and Meitner became an assistant to Planck at the university's Institute for Theoretical Physics, allowing the Meitner-Hahn partnership to continue with greater facilities and an enlarged staff. Hahn, a future Nobel laureate, brought to the team his knowledge of organic chemistry; Meitner brought her expertise in theoretical physics and mathematics. Together they would be responsible for some important advances, including their 1917 discovery of the rare radioactive element protactinium (atomic number 91).

From 1917 to 1926, Meitner continued to conduct her own research on the nature of beta rays. The interpretation of the physical properties of radioactive substances continued to be an area of personal interest, and she was the first to maintain that in the process of disintegration of radioactive materials, the emission of radiation follows rather than precedes the emission of the particles. During this time, she won considerable acclaim, and in 1926 she was named professor extraordinary at the University of Berlin. Adolf Hitler's anti-Semitic decrees would ultimately force her to leave this post;

although Meitner had been raised Protestant, both of her parents were of Jewish background. For a few years after Hitler came to power in 1933, however, the change in government did not affect Meitner's collaboration with Hahn.

In the early 1930s, nuclear physics made profound and dramatic advances, with James Chadwick discovering the neutron and Frédéric and Irène Joliot-Curie discovering artificial radioactivity. In 1934, Meitner and Hahn began to follow up the work of a group of scientists in Italy, headed by Enrico Fermi, who had bombarded uranium with neutrons and found several radioactive products thought to be transuranic elements (elements with atomic numbers higher than that of uranium, 92). Meitner and Hahn soon found a new group of radioactive substances that could not be identified as any of the elements just below uranium in the periodic table. Only one assumption seemed possible: that they were higher. Still, unanswered questions and puzzling results remained, even though another scientist, Fritz Strassmann, had joined Meitner and Hahn.

As the spring of 1938 arrived and Austria was occupied by the Nazis, Meitner was forced to leave Germany. She went first to the Netherlands; then to Copenhagen, Denmark, as a guest of Niels Bohr and his wife; and finally to Stockholm, Sweden, to work in the new Nobel Institute, where a cyclotron was being constructed.

In Berlin, Hahn and Strassmann had continued their work after Meitner left. She wrote to Hahn for data on the properties of the substances produced by their experiments. Hahn and Strassmann conducted more tests to prove the existence of radium but could only identify products resembling barium isotopes. Meitner discussed this new information with her nephew, the physicist Otto Frisch. Meitner and Frisch concluded that the uranium nuclei had split into two fragments and that a large amount of energy had been released. Immediately, they prepared a communication for the British science journal *Nature* in which they introduced the term *nuclear fission* to elucidate scientific principles previously thought to be impossible. For a short time after 1939, Meitner continued to investigate the nature of fission. In 1950, independently of others doing similar research, she advanced ideas concerning the asymmetry of fission fragments and worked on various aspects of the shell model of the nucleus.

Lise Meitner around 1906 in Vienna. [Public domain or Public domain], via Wikimedia Commons

In 1947, after spending half a year as a visiting professor in the United States, Meitner became a citizen of Sweden, retired from the Nobel Institute, and went to work in a small laboratory that the Swedish Atomic Energy Commission had established for her at the Royal Institute of Technology. In 1960, she left Sweden and retired to Cambridge, England. She died in 1968, a few days prior to her ninetieth birthday.

IMPACT

Meitner helped revolutionize the science of physics and its concepts. Her active participation in nuclear research resulted in the discovery of new elements and paved the way for the discovery of atomic fission, a term she helped coin and a process she helped interpret correctly. She entered her field at a time when women in science were not just a rarity but an oddity, and she overcame the prejudices and preconceptions she faced to make some revolutionary contributions to nuclear physics.

Honors came to Meitner from all quarters throughout her long life. She earned a distinguished reputation in the 1920s, receiving the 1924 Leibniz Medal of the Berlin Academy of Sciences and the 1925 Lieben Prize of the Austrian Academy of Sciences. In 1947, she was awarded the Prize of the City of Vienna, and in 1949, she won the Max Planck Medal. She was elected a foreign member of the Royal Society of London in 1955 and of the American Academy of Arts and Sciences in 1960. In 1966, she shared the United States Atomic Energy Commission's Enrico Fermi Award with Hahn and Strassmann. In addition, four American educational institutions (Syracuse University, Rutgers University, Smith College, and Adelphi College) awarded her honorary doctorates in science.

—*Andrew C. Skinner*

FURTHER READING

Graetzer, Hans G., and David L. Anderson. *The Discovery of Nuclear Fission: A Documentary History*. New York: Reinhold, 1970. Print. Reprints the original papers and reports by scientists who first uncovered the problem and meaning of nuclear fission.

Meitner, Lise. "Looking Back." *Bulletin of the Atomic Scientists* 20.9 (1964): 2–7. Print. An autobiographical account of Meitner's life, covering her youth through the discovery of atomic fission.

Rife, Patricia. *Lise Meitner and the Dawn of the Nuclear Age*. Boston: Birkhäuser, 1999. Print. A biography

---

LISE MEITNER DISCOVERS NUCLEAR FISSION

In 1934, a physicist named Ida Noddack theorized that it was possible to split atomic nuclei. Noddack suggested that physicist Enrico Fermi's experiments resulted not in the production of new elements, as was widely accepted, but rather in the disintegration of the uranium atom's nucleus into isotopes of known elements. Noddack was unable to prove her theory, and it was largely dismissed.

It was not until 1938 that Lise Meitner was able to prove Noddack's hypothesis, based on the experiments conducted by her colleagues Otto Hahn and Fritz Strassmann. Initially, Meitner was baffled by the results of Hahn's experiments. According to Hahn's data, the uranium produced the much lighter elements barium and radium. While radium seemed a somewhat plausible result, Hahn and Meitner both knew that for uranium to produce barium, the single neutron aimed at it would have had to knock off almost one hundred particles. This, they agreed, was impossible.

In Sweden, Meitner and physicist Otto Frisch eventually revised their way of thinking in accordance with a theory perpetuated by Niels Bohr. They began to realize that if they thought of the nucleus as a drop of liquid being stretched until it broke in two parts, rather than thinking of it as being chipped in half, then the conclusion could be that fission was taking place. In fact, during transformation, the nucleus changed shape and vibrated before splitting apart. The physicists, who shared their discovery with Bohr, quickly realized that the amount of energy released by the split nucleus would be far greater than any known process.

Meitner's greatest contribution to the discovery of nuclear fission was her reasoning as to why it created such a large amount of energy. Due to what is known as the packing fraction of the nucleus, she knew that the mass of the nucleus is less than the sum of its parts because the conversion of mass into energy is necessary to hold the nucleus together. Thus, using Albert Einstein's mass-energy equivalence equations to do the calculations, Meitner concluded that the nuclei created as a result of fission have different packing fractions, and it is this difference in mass that repels the atoms from one another, giving off energy in the process. Meitner calculated the amount of energy released by one atom splitting and theorized that the splitting could happen in a chain reaction. Further calculations revealed that a long chain reaction could potentially be responsible for an incredible and historically unparalleled release of energy.

Meitner's discovery eventually led to nuclear power and, more frighteningly, the atomic bomb. The scientists at work on the Manhattan Project, the United States' effort to harness atomic energy and develop a bomb, invited Meitner to join them. She refused to participate in the project, however, and was dismayed that her work led to the atomic bombings of Japan in 1945.

interpreting Meitner's life and describing her work leading to the discovery of fission.

Sime, Ruth Lewin. *Lise Meitner: A Life in Physics.* Berkeley: U of California P, 1996. Print. Provides a comprehensive chronicle of Meitner's life, career, and contributions to atomic and nuclear physics.

Sparberg, Esther B. "A Study of the Discovery of Fission." *American Journal of Physics* 32.1 (1964): 2–8. Print. Reviews the history of the discovery of fission and discusses Meitner's place in that history.

MENDEL, GREGOR

AUSTRIAN BOTANIST AND GENETICIST

Nineteenth-century monk and teacher Gregor Mendel's pea plant experiments demonstrated principles of heredity that would eventually evolve into the new disciplines of genetics and molecular biology. His work not only helped solve the mysteries of diseases such as cystic fibrosis, Tay-Sachs disease, sickle cell anemia, and hemophilia, it also laid the foundation for the Human Genome Project.

- **Born:** July 22, 1822; Heinzendorf, Austrian Empire (now Hyncice, Czech Republic)
- **Died:** January 6, 1884; Brünn, Moravia (now Brno, Czech Republic)
- **Primary field:** Biology
- **Specialties:** Genetics; botany

EARLY LIFE

Gregor Mendel was born Johann Mendel to a peasant family in Heinzendorf, Austria, on July 22, 1822. He did not inherit the financial means to pursue a rewarding profession outside of farming, and most children of commoners did not attend school beyond the elementary years. However, young Mendel was such a brilliant student, usually the top in his class, that his family sacrificed to pay his tuition at the gymnasium (high school), followed by two years at the Olmütz Philosophical Institute, a preparatory school for students planning to attend college.

Mendel worked very hard as a student. To help pay his expenses, which included food, he worked as a tutor in his free time. Sometimes he was forced to go hungry, and he suffered from stress-related illnesses. Whether due to depression or a mysterious illness, Mendel spent an entire year at home in bed when he was seventeen.

Because of the financial strain, Mendel's future as a scholar looked bleak. Fortunately, one of his teachers, an Augustinian monk, came to his rescue, suggesting that he join the Abbey of St. Thomas. The monastery, located in Brünn (Brno), the capital of Moravia (in the present-day Czech Republic), was known for teaching and other intellectual pursuits. In 1843, Mendel entered the monastery as a novice and received the name Gregor. This sanctuary had much to offer, including a botanical garden, a rich library, and brethren engaged in many interesting scholarly pursuits. Mendel studied at the Brünn Theological College and was ordained as a priest while he was still in his twenties.

Among the vows that Mendel accepted in exchange for the monastic lifestyle was that of poverty. He kept to his vows throughout his entire life, owning just a few precious items such as a microscope, a collection of scientific books, some caged birds, and mice used for breeding experiments.

In keeping with the expectations of the Order of St. Augustine, Mendel began teaching science courses part time at the Oberrealschule, the technical high school in Brünn. A superior at the monastery then recommended him for a long-term substitute position at the gymnasium in Znaim (Znojmo), located in southern Moravia, where he taught mathematics and Greek.

LIFE'S WORK

Teaching seemed to be Mendel's calling. His students raved about him, and he was devoted to them. He decided to become a permanent teacher in natural sciences, his avocation. When he took the qualifying exams, however, he failed.

Disappointed, he returned to the St. Thomas monastery. His fellow monks insisted he failed because he lacked the necessary college background, and so in 1851 the abbot sent Mendel to the University of Vienna to study science. During his two years there, Mendel studied with distinguished professors and learned important scientific research skills. When he returned from Vienna in 1853, he resumed teaching in Brünn. He eventually took the certifying exams again, and again he failed. Mendel apparently suffered from anxiety and barely completed the first exam question.

He returned to the monastery devastated and concerned about his future. Fortunately, the Augustinians encouraged scientific pursuits as well as teaching, and so Mendel immersed himself in various scientific activities, including beekeeping and meteorology, while continuing to teach part time. In 1856, he began his famous study of pea plants in a brand new greenhouse built especially for his work.

The nineteenth century was bubbling with scientific research, but no one had yet grasped the intricacies of fertilization. One popular theory called "blending" had evolved from the ancient Greek philosopher Hippocrates. It claimed that each parent passes along traits in the form of tiny particles, which then mix together, resulting in either separate, or blended, traits in the offspring. For instance, people believed that medium-height offspring resulted from the blending of tall parent's blood with a shorter parent's blood.

As theories about evolution developed, from the French botanist Jean-Baptiste Lamarck to Erasmus Darwin and his grandson Charles Darwin, the scientific community desperately needed to find the answers to the puzzle of heredity, which left a gap in the theories. European universities and scientific organizations put pressure on scientists to find the solution.

In 1828, A. F. Wiegmann had won a prize sponsored by the Berlin Academy of Sciences for an experiment involving the crossing of pea plant varieties that disproved the blending concept, but he was unable to provide any explanations. Charles Darwin obtained similar results in his experiments with snapdragons, but he could not defend his outcome, either.

Mendel had previously experimented with crossbreeding ornamental plants, hoping for new color varieties. However, the hybrids always looked like one parent or the other, and he was determined to find out why they were not blending. It would require the observation of multiple generations, which meant studying thousands of plants over several years.

Mendel began the experiments by testing thirty-four varieties of pea plants for purity. The first round of crossbreeding involved the selection of fourteen pairs of pea plants, two pairs each with one of seven different traits. He collected data from the hybrids and then again from each successive generation. He also crossed plants with two traits, called dihybrids. After two years, Mendel had completed the entire research study and discovered that the traits passed on by parents were independent of each other and either dominant or recessive, thereby explaining the lack of blending.

In 1865, Mendel presented his research, "Experiments in Plant Hybridization," to an audience at the Brünn Society for Natural Science. It was published in the society's journal and Mendel personally sent copies to various colleagues, including Charles Darwin, although he had never met the British naturalist. Mendel waited months for some response, but the scientific community largely ignored his research. Contemporary scientists believe that few people understood the magnitude of Mendel's experiments involving genetic inheritance, and it was largely forgotten for more than thirty years.

Although disappointed with the lack of response, Mendel kept busy. In 1868, he was elected abbot, a position that brought many new responsibilities, including certain civic obligations. It left him little time for either science or teaching. In his later years, he gradually withdrew from public life and lived a quiet, isolated life at the monastery. He died on January 6, 1884, from kidney disease compounded by heart problems.

Gregor Mendel. [Public domain], via Wikimedia Commons

Impact

Gregor Mendel was the first botanist to incorporate basic mathematics and algebra in his experiments. This approach allowed Mendel to solve the puzzle of heredity that had eluded Charles Darwin and other pioneering scientists. His published research lay dormant until the turn of the century, when four scientists from different countries stumbled upon his experiment simultaneously. From then on, Mendel was recognized as the father of genetics and earned the fame that he was denied during his lifetime.

Fame included some controversy though, when Mendel's laws of heredity were misused to support the Nazi movement. In the 1950s, the Communist government of Czechoslovakia shut down the Mendel Museum housed at the Abbey of St. Thomas, as well as the monastery, in its efforts to suppress science. It has since reopened and displays Mendel's possessions, including his copy of Darwin's *The Origin of Species*, with Mendel's own notes in the margins.

Although Mendel's laws of heredity cannot address sex-linked or codominant traits, Mendelian genetics have been used successfully in medical research to help determine the cause of such hereditary illnesses as cystic fibrosis, Tay-Sachs disease, sickle cell anemia, and hemophilia. Thanks in part to Mendel's research, in 2003 the Human Genome Project was able to sequence all of the genes of the human body, an endeavor that holds great promise for future medical care.

—*Sally Driscoll*

MENDEL AND HIS PEAS: GENETICS OF HEREDITY

Mendel was a Central European monk with a passion for horticulture and experimentation, but he was not a formal scientist. He spent much of his early life in the mid-nineteenth century in a monastery in the present-day Czech Republic breeding common peas (Pisum sativum) to satisfy his curiosity about heredity.

Common garden peas are excellent for long-term research into inheritance because they are easy to grow in bulk, they can produce more than one generation per year, and their reproduction is easy to control. Their flowers have male and female reproductive organs, and Mendel could easily self-pollinate individual plants or cross-pollinate multiple plants.

So what would happen if he crossed a purebred white-flowered plant with a purebred purple-flowered plant? Leading inheritance theory at the time said that the two traits should blend together, perhaps creating a shade of lilac. As it turned out, all of the offspring of a purple/white cross either had white or purple flowers, but none showed evidence of blending.

Mendel eventually identified seven different traits in his peas that refused to blend. These traits were easy to identify and were either binary (one of two forms) or along a short continuum: flower position (axial or terminal); stems (short or long); seed texture (round or wrinkled); seed color (yellow to green); seed coat color (whitish grey to brown); pod shape (inflated or constricted); and pod color (yellow to green). Clearly, the prevailing theory of blending across generations was not quite right.

By crossing pairs of peas purebred for each of these traits and carefully tracking their offspring, Mendel soon discovered that in all seven of his traits, one form appeared dominant over the other. For instance, crossing yellow-seed plants with green-seed plants creates a first generation that always has only yellow seeds. Mendel took it a step further, though, self-pollinating this new batch of all yellow-seeded plants to see what would happen. (In his purebred yellow-seed plants, only yellow-seeded offspring would occur.) What he found was that this second generation had a three-to-one ratio of yellow to green. The green-seed trait had reappeared.

By carrying out many such crosses and carefully tracking the ratios of different trait types in the resultant peas, Mendel came to three major conclusions. First, inheritance is determined by "factors" that are passed on unchanged. Second, each plant has one such factor from each parent for each trait (for instance, one seed-color factor from its "father" and one seed-color factor from its "mother"). Third, traits can be masked, but the "factor" for this trait may still be present in an individual and may still be passed on to the next generation.

Mendel's factors are now known to be genes, and the different versions of a gene (such as green-seed versus yellow-seed) for a given trait are called alleles. There are a variety of situations for which Mendel's rules do not account, such as codominance, incomplete dominance, or linked traits, but his work remains fundamental to the current understanding of inheritance and, by extension, evolution.

Further Reading

Griffiths, Anthony J. F., et al. *An Introduction to Genetic Analysis*. 10th ed. New York: Freeman, 2012. Print. A chapter on Mendelian genetics provides a clear explanation of Mendel's experiments, using contemporary terminology. Includes problems with answers, a glossary, and an index.

Henig, Robin Marantz. *The Monk in the Garden: The Lost and Found Genius of Gregor Mendel, the Father of Genetics*. Boston: Houghton Mifflin, 2001. Print. A biography of Mendel, recounting his life, his experiments, and the importance of his genetic discoveries.

Mendel, Gregor. *Experiments in Plant-Hybridisation*. Trans. William Bateson. Charleston: Nabu, 2011. Print. Reprints Bateson's 1909 translation of Mendel's lectures on his groundbreaking experiments.

Mill's Methods of Causal Reasoning

FIELDS OF STUDY

Research Theory

ABSTRACT

Philosopher John Stuart Mill (1806–73) developed five methods for determining the causes of events through examination and observation. Since the methods were designed for reaching conclusions through comparison and contrast, they are most helpful when causes are already known or suspected.

PRINCIPAL TERMS

- **joint method:** method that looks at both similarities and differences of observed cases to determine the cause of a particular phenomenon.
- **method of agreement:** method that looks for a single common cause to explain the common effect in all observed instances.
- **method of concomitant variation:** method that attempts to determine the degree of impact of a particular cause by examining all aspects of observed phenomena.
- **method of difference:** method that looks for the one factor that is present in cases when the phenomenon occurs but is absent in cases when the phenomenon does not occur.
- **method of residues:** method that explains cause by looking at all previously observed causes and effects to determine the chief cause of the effect under investigation.

DETERMINING CAUSE

John Stuart Mill was a nineteenth-century philosopher, political theorist, and economist. He argued that all research should use inductive reasoning instead of deductive reasoning as a starting point.

Mill established five methods of causal reasoning to determine important relationships between causes and effects. The first method is the method of agreement. This method requires researchers to develop hypotheses by looking for the common cause that is suspected of creating a certain common effect. For example, John, Sue, Jane, and Ted attend a reception. Wine is served along with shrimp, chicken, or sushi appetizers. In the first instance, John, Sue, Jane, and Ted all eat shrimp. John, Ted, and Sue eat chicken but Jane does not. All of the students except Sue eat sushi. All four students become ill after the reception. This effect leads to the conclusion that the shrimp was tainted because it was the only food that they all ate.

According to Mill's method of difference, the researcher uses observation to examine all differences in outcome to identify the sole element that explains the cause of a particular outcome. In another case involving the students, John, Sue, and Jane all eat chicken, but Ted does not. John and Ted eat shrimp, but Sue and Jane do not. All of the students except John eat sushi. John, Sue, and Jane all become ill after the reception, but Ted does not. This leads to the conclusion that the chicken was tainted as it was the only food that Ted did not eat.

Mill's joint method looks at both similarities and differences to explain the cause of a particular effect.

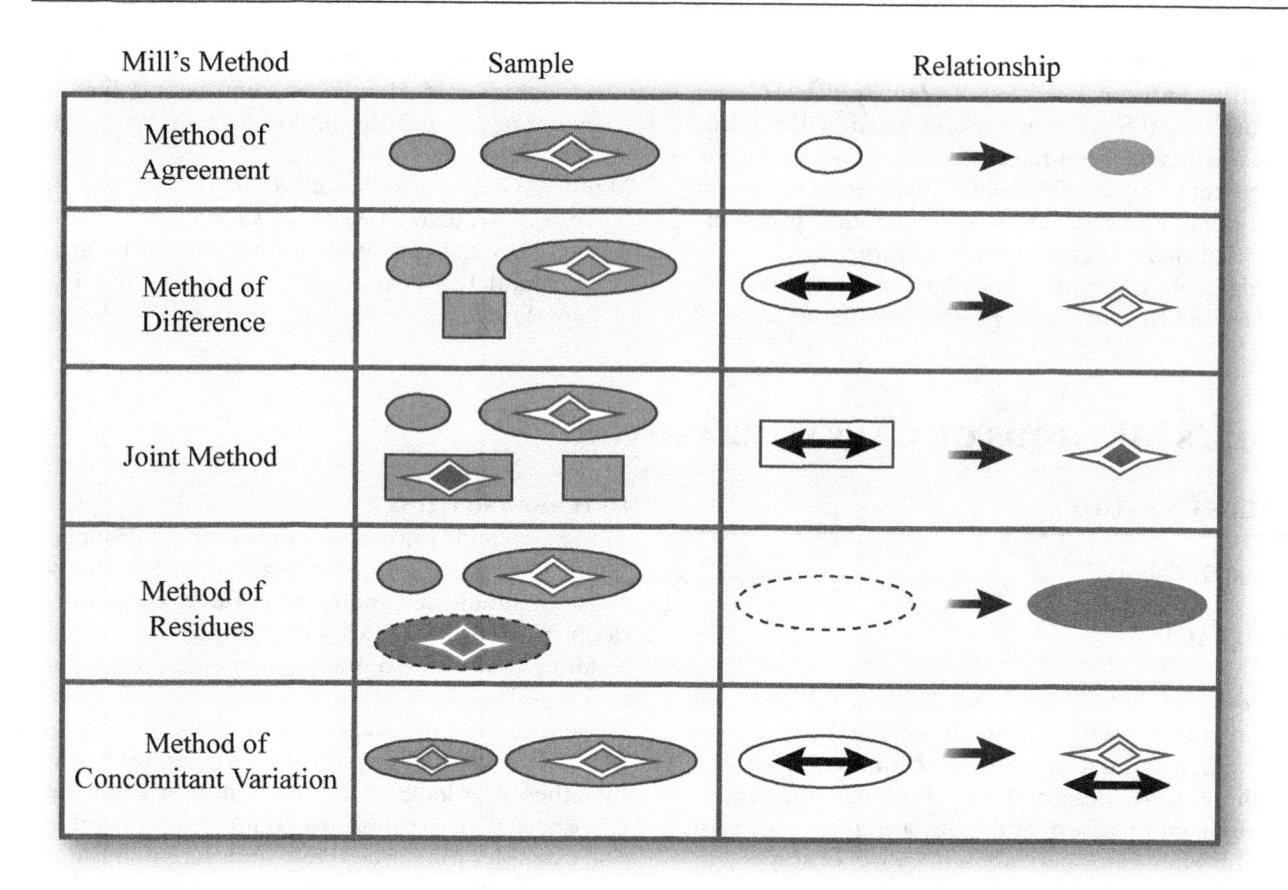

Mill's Method	Sample	Relationship
Method of Agreement		
Method of Difference		
Joint Method		
Method of Residues		
Method of Concomitant Variation		

In this scenario, all four students drank wine; John and Sue ate shrimp; all except John ate chicken; and all except Jane ate sushi. If Jane is the only one who did not become ill, the food most likely to be tainted was the sushi, which she did not eat. Mill's method of concomitant variation requires researchers to look at all possible variations in a suspected cause to pinpoint the degree to which the cause resulted in an observed outcome. Among the four students attending the reception, suppose that the shrimp was the tainted food. John ate one shrimp and became nauseated; Sue ate two and experienced stomach pain; Jane ate three and began throwing up; Ted ate seven and was taken to the hospital before the reception ended. The conclusion is that the amount of shrimp consumed was directly related to the level of illness experienced by each student.

Finally, Mill's method of residues calls for an examination of all causes already observed and the effects produced by those causes to identify the one cause that explains the problem that remains. In another scenario, it is already known that eating the shrimp and chicken at the reception caused certain effects. A student who ate the sushi as well experienced those effects as well as another, different effect. That new effect can then likely be attributed to having been caused by the sushi.

Positives and Negatives

Mill's methods of causal reasoning have helped to answer numerous questions in a range of fields in the physical, biological, and social sciences. The methods have been particularly useful in dealing with such issues as finding correlations between disease outbreaks and their causes.

The most serious drawback to Mill's methods is the fact that they are generally limited to situations where the cause is either known or highly suspected. It may be that a situation is not completely explained by known variables but is the result of causes unknown at a particular time. Mill's methods are not always able to explain effects caused by multiple variables rather than a single cause. Some scholars are highly critical of the fact that mistaken conclusions may be drawn because of examination of a limited number of possible causes.

Causal Reasoning in Practice

One of the most useful examples of understanding the importance of Mill's methods is to examine the link between historical outbreaks of cholera and the development of the germ theory of disease. Even before the causes of cholera were fully understood, physicians understood that outbreaks were often linked to poverty, overcrowding, and poor sanitation. Some suggested the disease was caused by a miasma in the air. Others, however, contended that it was spread through contaminated water. In 1854, British physician John Snow was able to link a London cholera outbreak to sewage-tainted water. He compared and contrasted the sources of water supply and the frequency of outbreaks. In a more contemporary example, American scientists used Mill's methods to link lower incidences of high blood pressure and fatal heart disease to states with soil high in selenium.

An example of the weakness of Mill's methods occurred long before he developed them. In the mid-seventeenth century, it was learned that British tobacco sellers seemed to be exempt from contracting the plague. Officials at Eton College ordered all students to begin smoking to protect them from plague. However, the disease was likely carried by fleas from infected rodents. Therefore, students continued to become ill, negating the suspected link between tobacco and plague.

Place in Research

Mill's methods have survived because they continue to assist researchers in finding answers to situations when it is possible to use existing knowledge and draw conclusions on available evidence. They are particularly helpful in the case of controlled experiments. Researchers use them to examine groups that are similar except for a specific cause suspected of creating an observed effect.

Elizabeth Rholetter Purdy, PhD

Further Reading

Ducheyne, Steffen. "J. S. Mill's Canons of Induction: From True Causes to Provisional Ones." *History and Philosophy of Logic*, vol. 29, no. 4, 2008, pp. 361–76. *Academic Search Complete*, search.ebscohost.com/login.aspx?direct=true&db=a9h&AN=34899641&site=ehost-live. Accessed 14 Apr. 2017.

Loizides, Antis, editor. *Mill's* A System of Logic: *Critical Appraisals*. Routledge, 2014.

Rosen, Frederick. *Mill*. Oxford UP, 2013.

Tulodziecki, Dana. "Principles of Reasoning in Historical Epidemiology." *Journal of Evaluation in Clinical Practice*, vol. 18, no. 5, 2012, pp. 968–73. *Academic Search Complete*, search.ebscohost.com/login.aspx?direct=true&db=a9h&AN=80204058&site=ehost-live. Accessed 14 Apr. 2017.

Van Heuveln, Bram. "A Preferred Treatment of Mill's Methods: Some Misinterpretations by Modern Textbooks." *Informal Logic*, vol. 20, no. 1, 2000, pp. 19–42.

White, Peter A. "Causal Attribution and Mill's Methods of Experimental Inquiry: Past, Present and Prospect." *British Journal of Social Psychology*, Sept. 2000, pp. 429–47.

Multiple Case Study

FIELDS OF STUDY

Research Theory; Research Design; Experimental Design

ABSTRACT

A multiple case study is a research method that analyzes the findings of more than one case study. A case study is an in-depth examination of a person, group, institution, or program over a specific time period. Multiple case studies generate data in order to elucidate why or how a phenomenon is occurring. Researchers conducting multiple case studies select cases with specific attributes so that comparisons can be made between them and conclusions can be generalized.

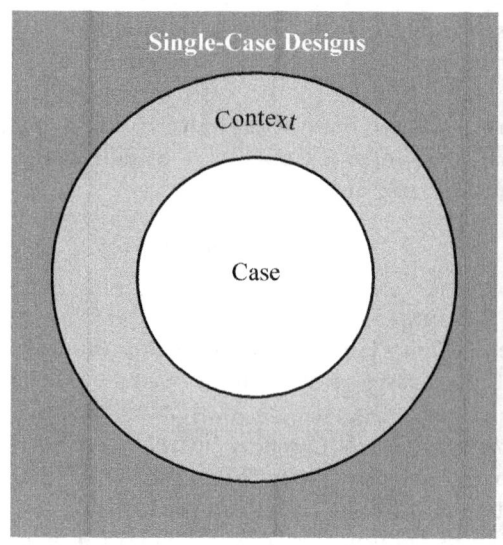

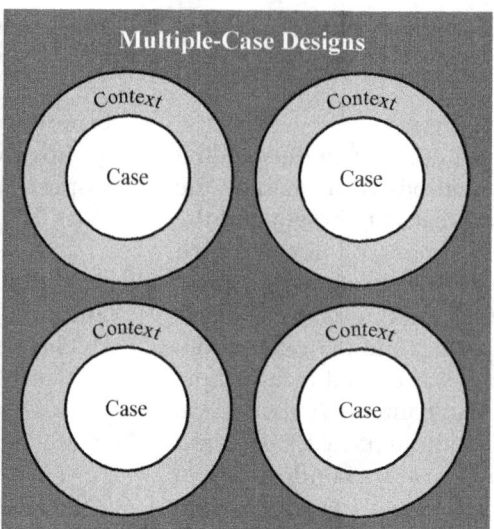

PRINCIPAL TERMS

- **non-probabilistic sampling:** one of several sampling methods in which the subjects of study are not chosen at random.
- **pilot case:** the original case study, which is used to determine if a multiple case study is warranted.
- **robustness:** the ability to reproduce similar results across studies despite slight variations in test conditions.
- **sample size:** the number of subjects included in a survey or experiment.

What Is a Multiple Case Study?

A case study is an in-depth observation and analysis of a particular phenomenon in its natural environment. Case studies are useful for illustrating particular trends and for understanding how or why certain phenomena are occurring. However, the results of an individual case study are not easy to generalize. A multiple case study analyzes more than one case. A case may be an individual person who is representative of a large group. Or a case may be a particular institution, organization, ecosystem, or program. The sample size of multiple case studies is greater than individual case studies. Therefore, multiple case studies yield greater robustness. Studying more than one case provides stronger evidence that the study's conclusions can be generalized. This is similar to replication of a controlled experiment. Just as researchers compare the results of replicated experiments, researchers compare cases in multiple case studies to each other in order to see whether the findings hold true across various situations.

Researchers can aggregate the results of previous single case studies to create a multiple case study. Multiple case studies can also be designed from the outset. From each case, multiple sources of data can be collected, including interviews, direct observations, surveys, and existing records. A multiple case study is not to be confused with multiple subjects in a study. In a study of a manufacturing company, the case is that company. However, there are many employees that may be subjects in the study. A multiple case study would examine many manufacturing companies and compare the findings from each case. Sometimes a pilot case is conducted to hone the processes of the study or to predict if the researchers' hypothesis holds true in a single case. If the hypothesis holds true, a multiple case study may be appropriate to support the findings of the pilot case.

Multiple case studies, as in single case studies, use non-probabilistic sampling. Researchers seek to study specific subjects who meet predefined criteria. If the research question involves studying towns where more than 50 percent of the population works in one factory, only towns that meet that criteria may be used in the study. Furthermore, case studies are time intensive for both the researchers and the participants. Therefore, researchers may choose to study particular factories because they are willing to allow access to their facilities, records, and staff. This means the subjects of case studies are not chosen probabilistically or randomly. In studies that use non-probabilistic samples, it is not always clear whether the subjects are truly representative of the larger population.

Considerations of Multiple Case Studies

Researchers select cases that share an element or elements related to the research question. However, the cases also have significant differences, such as the location and the individual participants. In this way, the researchers can better understand how the element in question operates across various situations. The number of cases included depends upon the complexity of the research question, the availability of cases, the degree of certainty desired, and the number of rival theories. For each strong rival theory, a specific case may be used in an attempt to debunk or confirm it.

The most basic multiple case study involves analyzing at least two similar cases. But including a greater number of cases increases the likelihood that the results can be generalized. For example, examining several institutions that have implemented a specific program offers more robust results than a study that examines only two institutions. A multiple case study can also study dissimilar phenomena. For example, studying both the successful and unsuccessful implementations of a specific program can yield useful insights. Best practice dictates least two cases for each phenomenon being studied in order for a study to be successful. However, such cases may not exist or be available for study.

Disadvantages of multiple case studies include the time intensive nature of conducting in-depth analyses of several cases. Furthermore, even with multiple cases, the findings might be difficult to generalize to other populations outside of the groups studied. If there are a large number of cases available to study, the researcher must choose which the cases to study. This can introduce bias and raises questions of objectivity. Interview questions and style can also introduce bias. Although case study researchers seek to analyze cases in a natural environment, direct observations can influence the behaviors of the research subjects, thereby skewing the results. The large amount of information gathered for a multiple case study can make the data difficult or impossible to analyze.

Multiple Case Studies in Practice

Multiple case studies are used for comparison. Analyzing the results from multiple cases helps researchers to better understand how the research subject operates across different situations. When designing a multiple case study, the researchers must first define their research question. Conducting a literature review can help researchers determine whether previous case studies on a similar topic have been conducted. Once researchers have established the focus of their study, they can then select their data sources and decide upon their methods of data collection. Observation and interviews can be conducted with each case, and statistical information from surveys or existing records can be collected. Researchers must then decide how to categorize and store their data and field notes. Then they analyze their data and draw conclusions based on those results. Upon reviewing their data, researchers may decide to revise the focus of their study, depending on their findings.

Multiple case studies are used across many fields. A case can be an individual person, which lends itself to studies in medicine, psychology, and social sciences. A case can be a group of people, which lends itself to studies in anthropology and sociology. A case can also be a town, a school, a factory, or another type of institution, which adds city planners, managers, educators, and policy makers to the mix of researchers. Any field or endeavor that can benefit from in-depth comparisons of various people, programs, or events can benefit from multiple case study research. Multiple case studies are often more illustrative of the causes of particular phenomenon than pure statistical data are.

Maura Valentino, MSLIS

Further Reading

Bartlett, Lesley, and Frances Vavrus. *Rethinking Case Study Research: A Comparative Approach.* Routledge, 2017.

Hancock, Dawson R., and Bob Algozzine. *Doing Case Study Research: A Practical Guide for Beginning Researchers.* 3rd ed., Teachers College Press, 2017.

Lapan, Stephen D., MaryLynn T. Quartaroli, and Frances J. Riemer, editors. *Qualitative Research: An Introduction to Methods and Designs.* John Wiley & Sons, 2012.

Stake, Robert E. *Multiple Case Study Analysis.* Guilford Press, 2006.

Taylor, Steven J., Robert Bogdan, and Marjorie L. DeVault. *Introduction to Qualitative Research Methods: A Guidebook and Resource.* 4th ed., John Wiley & Sons, 2016.

Tight, Malcolm. *Understanding Case Study Research: Small-Scale Research with Meaning.* Sage Publications, 2017.

Yin, Robert K. *Applications of Case Study Research.* 3rd ed., Sage Publications, 2012.

Yin, Robert K. *Case Study Research: Design and Methods.* 5th ed., Sage Publications, 2014.

Multistage Sampling

FIELDS OF STUDY

Sampling Design; Sampling Techniques; Experimental Design

ABSTRACT

Multistage sampling is a sampling method that is commonly used when it would be too costly or time-consuming to study the entire population. Researchers begin by dividing the population into smaller groups, called sampling units. Multistage sampling builds upon cluster sampling, in which the researcher only makes one division of the population into subgroups called clusters before drawing a simple random sample of a certain number of clusters and then studying all members of the selected clusters. If those clusters are too large then the researcher will make further divisions into secondary units or beyond. This process of multiple divisions is known as multistage sampling.

PRINCIPAL TERMS

- **cluster sampling:** a probabilistic sampling method in which the population being studied is divided into subsets called "clusters" based on a shared characteristic unrelated to the research question, such as geographic location, and then multiple clusters are selected at random for comprehensive examination.
- **primary units:** the first and largest sampling units drawn from a population.
- **random sampling:** one of several probabilistic sampling methods in which individuals within the population being studied are selected for examination at random, so that each member of the population has a chance (often an equal chance) of being included.
- **secondary units:** the second sampling units drawn from a primary unit.

OVERVIEW

Multistage sampling is a method in which a population is divided into smaller groups, known as sampling units. This sampling method is useful when the population under study is very large and complete lists of all the members of a population do not exist. For example, a state legislature might use multistage sampling to survey the residents of the state. In the first step, researchers divide the population into primary units. In the example above, the state population could be divided into primary units based on county lines. These units contain all members of the population and do not overlap with one another. If the researcher then used a random sampling method to select a certain number of these units for comprehensive examination, the research method would be cluster sampling.

However, the population size of each primary unit may still be very large. For these reasons, a researcher can use a sampling method, such as random or stratified sampling, to select a certain number of primary units to make another division into secondary units. In the state residents example, the researcher might make a second division by dividing some of the primary units by town lines. The researchers may continue to divide a secondary unit into tertiary units and so forth. Or the researchers could draw samples from one or more secondary units. The researchers can use a variety of sampling methods, such as stratified, random, or cluster sampling, to select subjects. The sampling methods used can differ between the sampling units.

ADVANTAGES AND DISADVANTAGES

By using multistage sampling, researchers are able to efficiently survey a large population. Multistage sampling is especially useful when a complete list of the subjects in a population does not exist. However, despite its advantages, multistage sampling can be less precise than stratified random sampling or simple random sampling. The subjects within each unit are more likely to share characteristics with one another than with subjects in other units. Therefore, each stage in a multistage sampling design introduces a greater risk of sampling error. However, this loss of precision can be an acceptable trade-off when the cost of data collection would otherwise be too high.

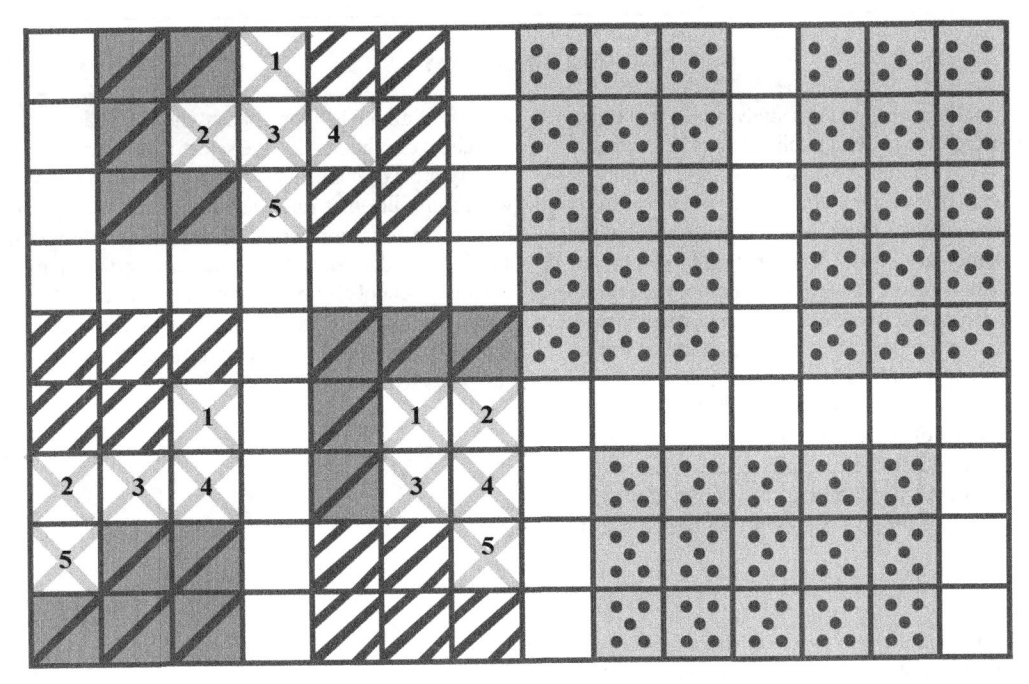

Use in Many Fields

Multistage sampling is used in fields and professions that deal with large data sets. For example, the performance management company Nielsen measures the ways that audiences respond to television programs. They cannot reasonably survey every television viewer. Instead, they use multistage sampling to ensure that they have a representative sample. Similarly, the US National Center for Health Statistics uses multistage sampling to conduct the National Health Interview Survey by dividing the country into counties and then dividing those units into households.

Allison Hahn, PhD

Further Reading

Henry, Colette, Lene Foss, and Helene Ahl. "Gender and Entrepreneurship Research: A Review of Methodological Approaches." *International Small Business Journal*, vol. 34, no. 3, 2016, pp. 217–41.

Levy, Paul S., and Stanley Lemeshow. *Sampling of Populations: Methods and Applications*. John Wiley & Sons, 2013.

Lynn, Peter, and Olena Kaminska. "Survey-Based Cross-Country Comparisons Where Countries Vary in Sample Design: Issues and Solutions." *Journal of Official Statistics*, 2016.

Robinson, Oliver C. "Sampling in Interview-Based Qualitative Research: A Theoretical and Practical Guide." *Qualitative Research in Psychology*, vol. 11, no. 1, 2014, pp. 25–41.

Rubin, Allen, and Earl R. Babbie. *Empowerment Series: Research Methods for Social Work*. Cengage Learning, 2016.

Tillé, Yves, and Alina Matei. "21 Basics of Sampling for Survey Research." *The SAGE Handbook of Survey Methodology*. Sage, 2016.

Watters, John K., and Patrick Biernacki. "Targeted Sampling: Options for the Study of Hidden Populations." *Social Problems*, vol. 36, no. 4, 1989, pp. 416–30.

SAMPLE PROBLEM

Imagine that a small research team is studying obesity rates in American children who attend public elementary school. There are millions of elementary students who attend public schools in the United States, and the team needs to conduct this study within one year. There is not enough time to interview each elementary school student. Furthermore, the research team does not have enough time or money to travel to each school. However, the researchers also want to ensure that they have a proportionate representation of rural, suburban, and urban schools. The team has access to a list of all the public elementary schools in each state. However, they do not have a list of all the elementary school students.

How would you advise the research team to use multistage sampling? Begin by identifying the primary and secondary units. Then, indicate how the researchers could select units for the study.

Answer:

The research team should begin by using states as the primary units. The researchers can use a simple random sampling method to select a certain number of states as their primary units. Because the researchers already have access to a list of all public elementary schools in each state, the researchers can then divide the selected primary units into secondary units of public elementary schools. If the researchers want to ensure that rural, suburban, and urban schools are proportionately represented, the researchers could then use stratified random sampling to select a certain number of elementary schools from each category for study. From the selected secondary units, the researchers could draw random samples of students or they could survey all of the students at each selected school.

Nested Analysis of Variance

FIELDS OF STUDY

Variance Analysis; Statistical Analysis; Experimental Design

ABSTRACT

Nested analysis of variance (ANOVA) is a robust statistical test of difference (variation) among groups. It builds on simpler ANOVA designs in that it allows researchers to test multiple levels of nominal or attribute variables "nested" in subgroups under the main effect variable. It is most commonly used to test additional treatment effects within a main treatment effect in life sciences, psychology, and sociology, and less commonly in medicine.

PRINCIPAL TERMS:

- **F-statistic:** the ratio of the mean square between groups and mean square within groups in an analysis of variance (ANOVA) test.
- **mean square:** the ratio of the sum of squares and degrees of freedom in an ANOVA test.
- **nested variable:** a variable located within another variable; also called "hierarchical variables."
- **nominal variable:** a variable that only takes unordered values in specified categorical groups.
- **variance:** a measure of how widely data points within a group are dispersed from the group mean, expressed as the average squared distance of each data point from the mean.

A Test for Nested Variables

Nested analysis of variance (ANOVA) is a statistical test commonly used in the life, behavioral, and social sciences. It is employed when a study has one measurement variable and two or more nominal variables (attribute variables) that are also nested variables. The term "nested" indicates that each nominal variable in a subgroup occurs with only one group, or higher-level nominal variable. The subgroup variables are considered to have random effects because they are randomly sampled from the larger set of potential subgroups. The highest-level attribute variable can be either model I (fixed effects) or model II (random effects), but all lower-level attribute variables must be model II. However, the overall design can be fixed, random, or mixed effect.

The nested ANOVA design is distinguished from the crossed model, a way of testing variance in fixed factors. The crossed model takes two or more factors and studies their effects and interactions in all combinations. A two-way or higher nested ANOVA, in contrast, forms a hierarchy of factors. While this means some combinations of factors are not studied, noise due to individual subjects can be measured. It also allows greater generalization of the effects of the highest-level factor. Hybrid designs do exist, in which crossed designs are nested or nested designs are crossed.

Nested ANOVA also differs from single-classification ANOVA. In a single-classification ANOVA, there is one F-statistic. This is the ratio of the mean square between groups and the mean square within groups. It indicates whether variables are statistically significant as a group. A two-level nested ANOVA has two F-statistics: one for the subgroup variables and one for higher-level groups. The F-statistic is calculated the same way but follows the hierarchy of groups.

Benefits and Limitations

Researchers use the nested design to test for differences between experimental and control outcomes and the variability among the hierarchy of levels. Using this study design can lower study costs. However, statistical conclusions can only be obtained for those subgroups that are replicated, due to issues of statistical power.

	Testers	A				B				C			
	Subjects	A	B	C	D	A	B	C	D	A	B	C	D
Measurements		y_{1A1}	y_{1B1}	y_{1C1}	y_{1D1}	y_{2A1}	y_{2B1}	y_{2C1}	y_{2D1}	y_{3A1}	y_{3B1}	y_{3C1}	y_{3D1}
		y_{1A2}	y_{1B2}	y_{1C2}	y_{1D2}	y_{2A2}	y_{2B2}	y_{2C2}	y_{2D2}	y_{3A2}	y_{3B2}	y_{3C2}	y_{3D2}
		y_{1A3}	y_{1B3}	y_{1C3}	y_{1D3}	y_{2A3}	y_{2B3}	y_{2C3}	y_{2D3}	y_{3A3}	y_{3B3}	y_{3C3}	y_{3D3}

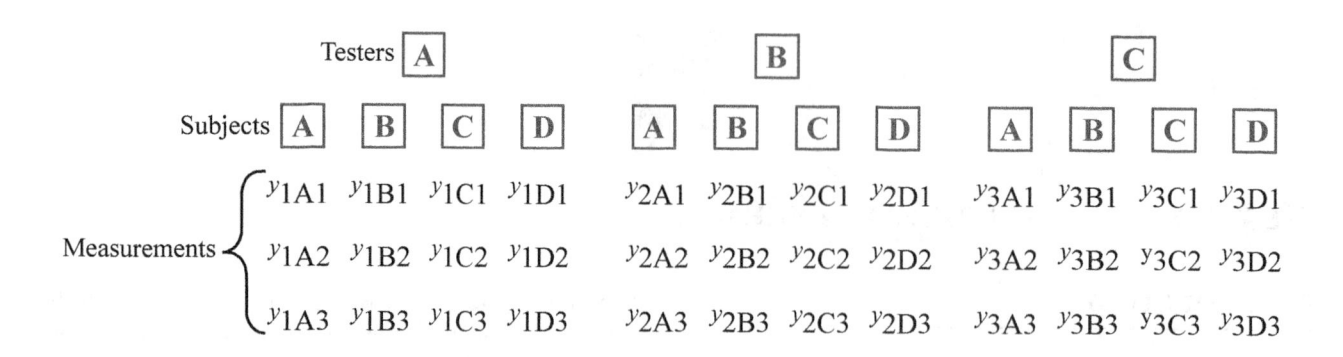

Limitations for using the nested ANOVA design are mainly in understanding the correct conditions under which to employ the model. Many times, a crossed two-way model can be used instead. Identifying the need to evaluate subgroups before the experiment begins is paramount to ensuring the statistical power is sufficient for the outcome of interest. Several assumptions should be verified before starting the ANOVA procedure. For nested ANOVA, errors should be randomly distributed with a symmetric mean and variance, and should be uncorrelated. There should be a restriction for the main factor A (fixed versus random). The distribution should approximate a normal distribution and be independent. When properly used, nested ANOVA is a powerful and cost-effective means to test several hypotheses at once.

Nested ANOVA in Practice

Nested ANOVA is a common technique in the life sciences, sociology, and psychology. Experiments in these fields often involve a single measurement variable and multiple nominal variables. For example, in biology, the natural variation of individual animals, tissue types, and cells takes a hierarchical form. This introduces noise at every level. Nested ANOVA is a strong fit to assess this noise.

Applications

Nested ANOVA is an important statistical technique in biomedical, psychological, and sociological sciences. It is commonly used in agriculture to test the effects of various treatments on plants and animals, such as feed quality and veterinary uses. Medical science more rarely uses nested ANOVA, mainly because it is not well understood in the context of clinical trials. However, it is a very cost-effective means of collecting and analyzing a large amount of data in one experiment. Understanding how it works and when to use it is important in the broader context of variance analysis and statistics.

Mandy McBroom-Zahn

Further Reading

Cardinal, Rudolf N., and Michael R. F. Aitken. *ANOVA for the Behavioural Sciences Researcher.* Lawrence Erlbaum Associates, 2006.

Damon, Richard A., and Walter R. Harvey. *Experimental Design, ANOVA, and Regression.* Harper and Row, 1987.

Krzywinski, Martin, et al. "Points of Significance: Nested Designs." *Nature Methods*, vol. 11, no. 10, 2014, pp. 977–78, doi:10.1038/nmeth.3137.

Sahai, Hardeo, and Mohammed I. Ageel. *Analysis of Variance: Fixed, Random and Mixed Models.* Birkhäuser, 2012.

Welham, S. J., et al. *Statistical Methods in Biology: Design and Analysis of Experiments and Regression.* Chapman & Hall, 2015.

Wright, Daniel B., and Kamala London. "Multilevel Modelling: Beyond the Basic Applications." *British Journal of Mathematical and Statistical Psychology*, vol. 62, no. 2, 2009, pp. 439–56, doi:10.1348/000711008X327632.

SAMPLE PROBLEM

In the following problem using data from Richard Damon and Walter Harvey, the weights in grams (from ten to twenty weeks) of chickens placed on four treatments (low calcium, low lysine; low calcium, high lysine; high calcium, low lysine; high calcium, high lysine) were recorded over a period of time. Each of the four treatments was given to two different pens, each containing six chickens. The data are as follows:

	\multicolumn{8}{c	}{Treatment (TRT)}
	LoCa	LoL	LoCa	HiL	HiCa	LoL	HiCa	HiL
Pens	1	2	1	2	1	2	1	2
	573	1041	618	943	731	416	518	416
	636	814	926	640	845	729	782	729
	883	498	717	373	866	590	938	590
	550	890	677	907	729	552	755	552
	613	636	659	734	770	776	672	776
	901	685	817	1050	787	657	576	657

Source: Adapted from Richard A. Damon and Walter R. Harvey, *Experimental Design, ANOVA, and Regression* (Harper and Row, 1987, p. 26).

The main factor is treatment, which has four levels. The pens variable is nested within the treatment level. A total of eight pens were used for the experiment. Perform a nested ANOVA to test the two-tailed hypothesis that there is a statistical difference between treatments overall and at the pen-level. Let $\alpha = 0.05$.

H_0(null): There is no significant difference between treatments to produce weight gain.

H_A(alternative): There is a statistically significant difference between treatments to produce weight gain.

Answer:

The resulting ANOVA table to test the overall hypothesis is as follows:

Source	Degrees of Freedom (df)	Sum of Squares (SS)	Mean Square (MS)	F	P value
Treatment (TRT)	3	53943.417	17981.14	0.732	0.539
Pen (TRT)	4	125688.167	31422.04	1.279	0.294
Error	40	982654.333	24566.36		
Total	47	1162285.917			

According to the ANOVA, neither treatment nor pen was significant at the 0.05 level, as the *P* value is higher than 0.05.

Nested Designs

FIELDS OF STUDY

Research Design; Experimental Design

ABSTRACT

The nested design is a type of research design that is used to evaluate multiple factors, or independent variables, with differing levels. It is most often compared to the crossed, or factorial, design, which is used to screen multilevel factors.

PRINCIPAL TERMS

- **ANOVA:** short for "analysis of variance," a set of statistical models for analyzing the differences among the means of multiple data sets.
- **fixed effect:** in an experiment, the levels of a factor that a researcher selects and controls.
- **random effect:** in an experiment, a random sample of levels from all the possible levels of a factor.

NESTING FACTORS

The choice of research design is essential to how a researcher evaluates an experiment. The nested design, or hierarchical design, is used to evaluate studies involving multiple factors. A factor is an independent variable or a confounding variable that affects the outcome, or dependent, variable. Each factor can have two or more levels, which may be quantitative but are most often categorical.

In order to use the nested design, the researcher must determine whether the factors are crossed or nested. Factors are considered crossed when every level of each factor occurs in combination with each level of every other factor. Consider the case of a café where three different bakers are using two ovens. If the researcher wishes to study the rate of burned food being produced, both oven and baker are factors. Both oven A and oven B would need to be paired with each baker for the factors to be crossed.

However, it is not always possible to ensure that each factor will be crossed. Consider the case of a café chain where one oven and three bakers (1, 2, and 3) are located in Ohio and another oven and three different bakers (4, 5, and 6) are in Michigan. In this case, it is not possible to combine all ovens with all bakers. Rather, the factors would be nested in the following way: oven A would be paired with bakers 1, 2, and 3, and oven B would be paired with bakers 4, 5, and 6. Thus, the baker factor would be nested within the oven factor because the levels of the baker factor each exist in only one level of the oven factor. Research designs can included crossed, nested, or both crossed and nested factors. Knowing which factor is of which type helps the researcher select the right method of analysis.

For analysis purposes, a factor can be considered a fixed effect or random effect. In general, fixed effects are the levels that the researcher chooses to manipulate, whereas random effects come from random samples of the level from a larger population of levels. When a study includes both types of factors, it is known as a "mixed model." When using a mixed model, nested ANOVA is used to estimate variance at different levels. Random effects increase the uncertainty in these calculations. A nested design assumes that the random effects of factors will be normally distributed.

LIMITATIONS, BENEFITS, AND COMPARISON

Each experimental design involves limitations, benefits, and assumptions. Commonly, the nested design follows the pre-experimental design because it does not include a control group and often does not randomize subjects to experimental conditions. In many cases, such as in genetics studies, randomization is impossible. The lack of control group and randomization threatens the internal validity of the experiment. Nested designs can provide estimates of precision when experimental conditions are repeated. They allow cannot detect interaction variance as well as crossed designs can.

When the nested design is unbalanced (that is, when there are different numbers of observations in each hierarchical level), a staggered nested design is used. This can provide better estimates for some variance components and their sums.

Crossed designs, or factorial designs, are true experimental designs. Crossed designs also examine the main effects and interactions of two or more

Nested Design

Factor 1	Factor 2					
	a	b	c	d	e	f
A	■	■				
B			■	■		
C					■	■

Partially Crossed Design

Factor 1	Factor 2					
	a	b	c	d	e	f
A	■	■				
B	■	■				
C			■	■		
D			■	■		
E					■	■
F					■	■

multilevel factors. However, subjects are randomized to experimental conditions to study the levels of those factors. The full factorial design analyzes all combinations of factors, making it thorough but costly and time-consuming. The fractional factorial design examines only a fraction of the factors, which is useful when there are many factors and limited resources. However, some degree of confounding occurs.

NESTED EXPERIMENTS IN PRACTICE

Research scenarios often lend themselves to the use of nested factors. For example, a researcher may be studying the effects of the application of two different pesticides (pesticides A and B) on four different crops (crops 1, 2, 3, and 4) for an agribusiness located in California and New York. However, due to climatic conditions, crops 1 and 2 can only be grown in California and crops 3 and 4 can only be grown in New York. In addition, pesticide 2 is banned in California. In this case, it is not possible to pair all crops with all pesticides in order to used crossed variables. Therefore, pesticide 1 will be nested with crops 1 and 2 and pesticide 2 with crops 3 and 4.

APPLICATIONS

Because nested designs help examine the factors in complex research scenarios, they are important across scientific fields. They are common in fields like genetics, ecology, and education where it is often difficult, if not impossible, to assign subjects randomly to groups. Thus, the nested design can be used where full factorial designs would not be possible.

Pamelyn Witteman, PhD

FURTHER READING

Anderson, Mark J., and Patrick J. Whitcomb. "Two-Level Factorial Design." *DOE Simplified: Practical Tools for Effective Experimentation.* 3rd ed., CRC Press, 2015.

Creswell, John. *Research Design: Qualitative, Quantitative, and Mixed Approaches.* 4th ed., SAGE Publications, 2014.

Field, Andy. *Discovering Statistics Using IBM SPSS Statistics.* 4th ed., SAGE Publications, 2013.

Krzywinski, Martin, et al. "Points of Significance: Nested Designs." *Nature Methods*, vol. 11, 2014, pp. 977–78. doi:10.1038/nmeth.3137. *Nature*, www.nature.com/nmeth/journal/v11/n10/full/nmeth.3137.html. Accessed 11 May 2017.

Vogt, Paul. *Quantitative Research Methods for Professionals.* Allan & Bacon, 2007.

Yoshikazu Ojima. "Generalized Staggered Nested Designs for Variance Components Estimation." *Journal of Applied Statistics*, vol. 27, no. 5, 2000, pp. 541–53. *Business Source Complete*, search.ebscohost.com/login.aspx?direct=true&db=bth&AN=3420576&site=eds-live. Accessed 10 May 2017.

> **SAMPLE PROBLEM**
>
> Researchers for a cab company wish to study the rate of accidents that drivers have when driving a hybrid or a gas-powered car. The company operates in Boston and Philadelphia. Drivers work in only one city. Only hybrid cars are driven in Boston. Only gas-powered cars are driven in Philadelphia. Given this scenario, which factor is nested within the other? What would need to be changed in order for the study to become a crossed design?
>
> gas-powered car level, and the Boston drivers nested within the hybrid car level.
>
> In order for the study to become a crossed design, all drivers would have to access to both vehicle types. This might be achieved by deploying gas-powered cars to Boston and hybrid-powered cars to Philadelphia.
>
> **Answer:**
>
> In this scenario, the driver factor would be nested within the car factor. The Philadelphia drivers would be nested within

Nonequivalent Dependent Variables Design

FIELDS OF STUDY

Research Design

ABSTRACT

Nonequivalent dependent variables design is a research method used when one wishes to administer a treatment to an entire group of participants, rather than dividing the group into an experimental group and a control group, with the control group not receiving the treatment. In effect, the nonequivalent dependent variable acts as the control in the experiment.

PRINCIPAL TERMS

- **control group:** the subjects in an experiment who do not receive any intervention or who receive a placebo or sham intervention.
- **dependent variable:** in experimental research, a measurable occurrence, behavior, or element with a value that varies according to the value of another variable.
- **experimental group:** the subjects in an experimental research study who receive the intervention or treatment being tested.
- **independent variable:** in experimental research, a measurable occurrence, behavior, or element that is manipulated in an effort to influence changes to other elements.
- **nonrandom assignment:** the process of purposefully dividing the subjects in an experiment into different experimental or control groups based on specific traits that are relevant to the treatment being studied.

OVERVIEW

The nonequivalent dependent variables design (NEDV) is a type of quasi-experimental design that has interesting features making it desirable in some research situations. Most research studies require an experimental group and a control group. Participants in the experiment are often randomly assigned to one of the two groups. In the experimental group, an independent variable is changed, and a dependent variable is measured to see how it responds to the change. The control group serves to show what would happen if the independent variable had not been changed. NEDV uses a much different approach.

Instead of randomly assigning participants to the two different groups, all participants are grouped together in the experimental group. This makes NEDV akin to other research designs based on nonrandom assignment. NEDV remains unusual, however, in its use of a single group. Aspects of NEDV can also be employed to combat the bias typically inherent in nonrandom assignment. These include the use of pretests and posttests. Rather than establishing a separate control group, the researcher selects a characteristic of the group that is similar to the one being studied, yet distinct enough that its expected lack of change can serve as the experiment's control. The advantage to using an NEDV design is that the experimental treatment is applied to all participants. This can be desirable when there is a high level of confidence that the treatment will be advantageous. Therefore, the control group need not be deprived of the treatment.

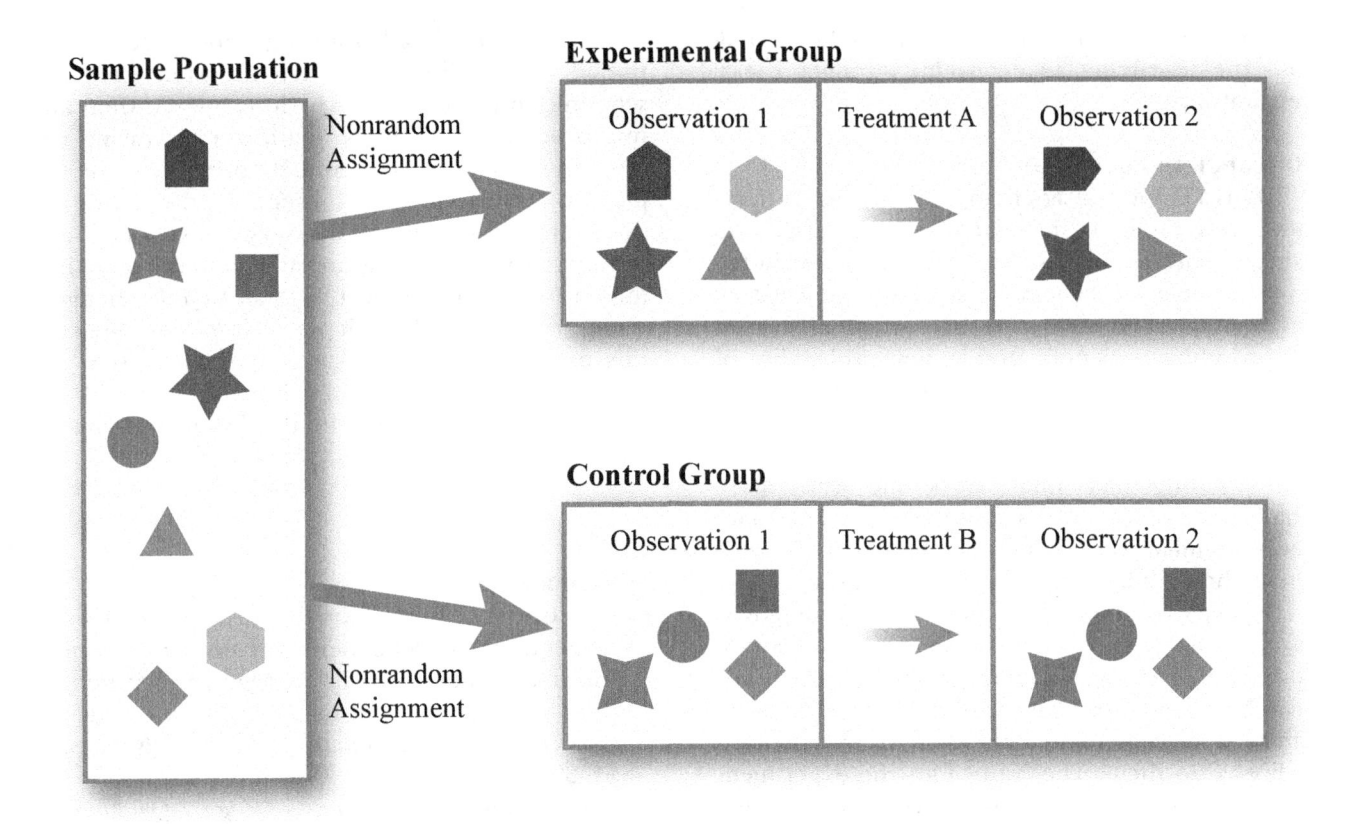

Challenges and Ensuring Effectiveness

A challenging feature of the NEDV design is the need to find a nonequivalent variable for use as the control group. This is because the nonequivalent variable that is selected must be similar enough to the variable being studied that both variables would be subject to the same general influences. At the same time, it must differ enough that it will not be affected by the treatment or intervention that is being evaluated by the experiment. This is a fine line to walk, and in many scenarios a suitable variable cannot be identified. This means that the NEDV design cannot be used and an alternative approach must be devised.

Some implementations of NEDV attempt to turn this difficulty from a weakness into a strength. These NEDV studies use a technique known as pattern matching. Rather than using a single, similar variable as a control in an experiment, pattern matching NEDV uses a large number of similar variables alongside the main variable of interest. Doing this makes it possible to assess the impact of the treatment or intervention on many different characteristics. This could potentially provide a broader understanding of its full impact and of its mechanisms of operation. This type of NEDV research can be invaluable. It creates opportunities for researchers to uncover interactions and effects that were not anticipated, which in turn may lead to whole new areas of study being opened up. What is more, pattern matching NEDV tends to be more reliable. The use of multiple variables increases a study's internal validity and allows the researcher to determine stronger causal relationships.

RESEARCH IN EDUCATION

NEDV is a helpful design in many fields, but perhaps none rely on it to the extent that research into educational effectiveness does. This can be attributed to a number of factors. First, many educational interventions subject to study have a high likelihood of providing a benefit and a low probability of harm. Therefore, providing the intervention to the entire study group is seen as desirable or ethically acceptable. In addition, students in traditional school settings meet what can be one of the more challenging requirements of using NEDV. They have several similar yet distinct variables that describe them: their grades in their individual classes.

A simple example of how the NEDV design is used in educational research would be a study into the effectiveness of a math tutoring program. The researcher would investigate a group of students' mathematical aptitude before and after the program was delivered to them. The control for the experiment would be the students' performance in another subject (history, science, English, etc.) during the same period as the math study. By using this approach, all of the students can receive the intervention being studied. The student performance in the subject not being studied can act as the control for the experimental group, which is defined by the math grade.

AN IMPORTANT DESIGN OPTION

When randomized selection is not suitable, NEDV is a valid alternative method for research. The greatest benefit offered by the use of NEDV may be its utility in avoiding occasions when the variable being studied is influenced during the study by a factor that is separate from the intervention or treatment that is the focus of the research. When this occurs, the changes in the variable are likely to be incorrectly attributed to the intervention or treatment, when in fact they are unrelated to it and would have occurred even if it had not been administered.

If, during the math tutoring experiment mentioned above, the school had adjusted its daily schedule to make all class periods longer, then an improvement to students' math grades during or after the intervention would be difficult to interpret. In order to tell which event was the source of the change, it would be necessary to look at an additional variable that was not subject to the math intervention but was subject to the schedule change. Using a nonequivalent dependent variable makes this possible. NEDV design can thus function as a way of reducing internal validity threats and allowing more confidence in the inferences made based on results.

Scott Zimmer, JD

FURTHER READING

Coryn, Chris L. S., and Kristin A. Hobson. "Using Nonequivalent Dependent Variables to Reduce Internal Validity Threats in Quasi-Experiments: Rationale, History, and Examples from Practice." *New Directions for Evaluation*, vol. 131, 2011, pp. 31–39.

Creswell, John W. *Research Design: Qualitative, Quantitative, and Mixed Methods Approaches.* 4th ed., Sage, 2014.

Holgado-Tello, Fco. P., et al. "A Simulation of Threats to Validity in Quasi-Experimental Designs: Interrelationship between Design, Measurement, and Analysis." *Frontiers in Psychology*, 16 June 2016, pp. 1–9.

Martin, William E., and Krista D. Bridgmon. *Quantitative and Statistical Research Methods: From Hypothesis to Results.* Jossey-Bass, 2012.

Salkind, Neil J. *Encyclopedia of Research Design.* Sage, 2010.

Seawright, Jason. "The Case for Selecting Cases That Are Deviant or Extreme on the Independent Variable." *Sociological Methods & Research*, vol. 45, no. 3, 2016, pp. 493–525.

Thyer, Bruce A. *Quasi-Experimental Research Designs.* Oxford UP, 2012.

Non-probabilistic Sampling

FIELDS OF STUDY

Sampling Techniques; Sampling Design

ABSTRACT

Non-probabilistic sampling refers to a group of techniques for selecting research subjects from a larger population. These techniques are typically simpler and more cost-effective than probabilistic sampling techniques, but they eliminate randomness in the selection process. Because randomness is widely held to be critical in generating reliable sampling data, non-probabilistic sampling has limited generalizability and is prone to bias.

PRINCIPAL TERMS

- **convenience sampling:** a non-probabilistic sampling method in which members of the sample group are drawn from a population based on their availability and ease of access; also called accidental sampling or haphazard sampling.
- **diversity sampling:** a non-probabilistic sampling method in which members of the sample group are specifically selected according to the researcher's best judgment to establish a diverse sample, which may or may not be representative of the population.
- **quota sampling:** a non-probabilistic sampling method in which members of the sample group are selected to fulfill preestablished quotas for certain demographic characteristics, often in proportion to the demographics of the larger population.
- **random sampling:** one of several probabilistic sampling methods in which individuals within the population being studied are selected for examination at random, so that each member of the population has a chance (often an equal chance) of being included.

Definition of Sampling Methodologies

When researchers study large groups, or populations, they typically do not study every member of the population. In most cases, doing so would be too time consuming and expensive. Instead, researchers use a process called sampling in order to select a representative portion of the population to study. They then use the data they gather from this sample to estimate characteristics of the larger population. The larger the sample, the more reliable the results. Also, as a general rule, the more random the sample, the more reliable the data. This is because theoretically every element of the larger population has an equal chance, or at least a nonzero chance, of being selected. There are various different random sampling techniques, known collectively as probabilistic sampling, in which random, chance-based selection plays a primary role in the sampling process.

In some cases, though, random sampling may be unfeasible or otherwise impractical. Instead, researchers may use a number of strategies for controlling the range and scope of the subjects selected for their study using predetermined criteria. These strategies, known as non-probabilistic sampling, provide data that can provide useful results that are both cost-effective and timely. However, because the sampling pool is not random, researchers cannot effectively estimate error range. For this reason, resulting data must be used with care.

Non-probabilistic Sampling Techniques

There are a number of different non-probabilistic sampling techniques. Each has advantages and significant limitations. The majority of these techniques are purposive, meaning that subjects are specifically chosen for the sample according to certain standards established by the researchers. The exception to this is convenience sampling, also called accidental or haphazard sampling. In this method, subjects are selected for study simply based on availability and ease of access. It is the simplest, easiest, and cheapest way to gather data. Pollsters, for example, may simply go to a public place and ask any passers-by to answer their questions. The sample group consists of whoever happens to be there at the time and agrees to respond to the poll. Other convenience sampling methods include mail-out questionnaires, cold calling, and customer surveys available at checkout stations.

This methodology secures data quickly, but extrapolating reliable results can be dicey. There is

```
                    Nonprobabilistic Sampling
          ┌──────────────────┴──────────────────┐
       Purposive                             Accidental
   ┌────┬────┼────┬────┐                   ┌────┴────┐
Snowball Quota Diversity Target Expert  Voluntary Convenience
sampling sampling sampling audience sampling  sampling  sampling
```

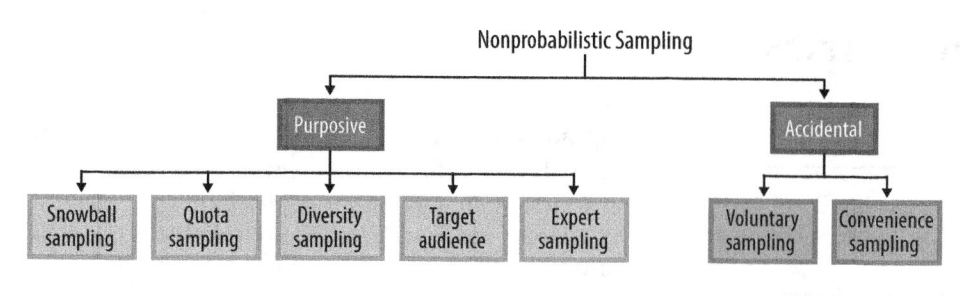

no quantitative way to measure the representative nature of the subjects that are tested. No attempt is made to measure who gets asked, and no criteria are used except availability. While convenience sampling is more chance based than other non-probabilistic techniques, it is still not as reliable or as generalizable as probabilistic sampling. Members of the sample group have already self-selected simply by making themselves available for sampling.

More sophisticated is a purposive non-probabilistic technique known as quota sampling. In this method, subjects are chosen for the sample based on certain predetermined criteria. Criteria are typically gauged to match the general distribution of demographic characteristics within the broader population. One challenge inherent in this method is determining which particular characteristics the quotas should be based on. Quota sampling is best used when researching a question about a particular subgroup or about how different subgroups within a population interact. Diversity sampling is similar to quota sampling in that the researcher attempts to capture all demographics within the broader population. However, the goal of diversity sampling is simply to sample from the broadest range of characteristics, and the researcher is not concerned with meeting predetermined criteria regarding distribution of those demographics. Often the inclusion of samples is based on the researcher's judgment of the individual and the whole sample.

Non-probabilistic Sampling in Action

A political polling organization wants to know how a city's residents feel about a particular political candidate. Also of interest is whether and how those attitudes correspond to gender and race or ethnicity. The researchers decide to use quota sampling in order to obtain a sample that is representative of the city's demographics. To do so, the researchers first obtain a sampling frame, such as the results of a city census. They then establish quotas based on the city's racial or ethnic makeup.

If the city's population is 15 percent Hispanic, for example, then 15 percent of the sample group—perhaps 150 out of a total of 1,000 poll respondents—should be Hispanic. In addition, because the researchers are interested in gender as well as race or ethnicity, they control for gender representation as well. So, for example, if the Hispanic population of the city is approximately 52 percent female and 48 percent male, then 52 percent of those 150 respondents, or 78 in total, should be women. The other 48 percent, or 72 total, should be men.

With these quotas in place, the researchers then go about enlisting members of the sample group. Various methods may be used to contact potential respondents, such as mailed questionnaires, direct phone calls, or simply stopping people in a public place. The first questions would ask the respondent's race or ethnicity and gender. Then, for example, if the researchers used mailed questionnaires, the first seventy-eight completed forms that they receive from Hispanic women will be included in the study. Any further responses from Hispanic women will be discarded, as that quota has been met. Responses are collected until all quotas are filled.

Using these controls, once the study is completed, the researchers will have a good idea of how city residents across gender and race or ethnicity lines feel about the political candidate. However, while this information may serve as a guide for the candidate's campaign, the results may not be truly representative. The sampling process was not sufficiently random, and cannot be generalized to the larger population with any degree of statistical confidence.

Implications of Non-probabilistic Sampling

While there are many good reasons to use non-probabilistic sampling, any researcher who does so compromises the randomness of their study. Researchers simply cannot know the probability of any one element of the pool being used as a subject. Non-probabilistic sampling is best used as a preliminary stage of data gathering, subject to later, more precise measuring. Companies, politicians, or other organizations who desire to be responsive to their customer base or constituents can use nonprobabilistic sampling to test the waters as a first step to developing strategies for long-term changes and improvements.

Joseph Dewey

FURTHER READING

Andres, Lesley. *Designing & Doing Survey Research.* Sage Publications, 2012.

Blair, Edward, and Johnny Blair. *Applied Survey Sampling.* Sage Publications, 2015.

Daniel, Johnnie. *Sampling Essentials: Practical Guidelines for Making Sampling Choices.* Sage Publications, 2012.

Thompson, Steven K. *Sampling.* 3rd ed., John Wiley & Sons, 2012.

Trochim, William M. K. "Nonprobability Sampling." *Research Methods Knowledge Base,* 2006, www.socialresearchmethods.net/kb/sampnon.php. Accessed 21 Feb. 2017.

Weisberg, Herbert F., et al. *An Introduction to Survey Research, Polling, and Data Analysis.* 3rd ed., Sage Publications, 1996.

Nonresponse Error

FIELDS OF STUDY

Sampling Design; Sampling Techniques

ABSTRACT

In a research study, nonresponse is when sample units selected for a study fail to participate, are unable to respond, or refuse to answer survey questions. Depending on the cause of this lack of participation and the dimension of the missing values, nonresponse may result in nonresponse error, which may create significant problems in the reliability of any inferences derived from that data pool.

PRINCIPAL TERMS

- **bias:** an inclination toward a particular perspective or preconceived notion; also, in statistics, refers to a method of collecting or evaluating data that inherently and consistently produces results that inaccurately reflect the characteristics of the population or phenomenon being studied.
- **overcoverage:** a type of bias in which certain members of the target population are overrepresented in the sample group surveyed or in the group of respondents.
- **responsiveness:** the willingness of selected survey subjects to provide answers.
- **undercoverage:** a type of bias in which some members of the target population are not sufficiently represented in the sample group surveyed or in the group of respondents.

RESPONSIVENESS AND RESULTS

To save time, money, and resources, researchers interested in obtaining information about a larger population typically choose to survey a sample group rather than attempting to interview the entire population. This sample group should be chosen carefully to ensure that the study results accurately reflect the target population. Randomization of sample selection is a method that is often used to accomplish this. Every member of the larger population has an equal chance to be surveyed.

However, even if the sample frame is carefully chosen, researchers must still keep possible bias in mind when analyzing the level of responsiveness. Even in randomized studies, survey results can still be biased due to nonresponse. Nonresponse is when some members of the sample group fail to respond to the survey. When analyzing the results of the survey, researchers must consider the people who did not respond. Depending on which and how many people did not respond, their nonresponse may skew the results of the survey so that they are not truly representative of the sample and, as a result, of the target population. This is a nonresponse error. These differences could mean that there is overcoverage or undercoverage of some members of the sample group that skews the results. For example, if the majority of older people contacted for an online survey do not respond because they do not have access to the Internet, they will not be represented in the survey results.

CAUSES OF NONRESPONSE ERROR
There are numerous causes for nonresponse error. In any pool of randomly selected subjects, there will be a percentage who are simply unavailable. Other times, if the survey is sent by mail, they may throw it away; if the subjects are cold called, they may not pick up; if they are e-mailed, the survey may be sent to a spam folder. In addition, some subjects may attempt to participate but become frustrated by the format. Others may complete a mail-in survey but forget to mail it back, or they may intend to respond but simply put it off or forget until after the deadline. Some people may also decline to participate because of some overriding factor. This could include politics, religion, or general interests. Additionally, some may prefer not to reveal what they see as personal data about themselves despite guarantees by the researchers of confidentiality.

Researchers can take some measures to try to avoid nonresponse error. In the design stage, they can make sure that the survey is conducted in a format suitable for the sample group and free of any possible technological issues. They can also keep the survey to an appropriate length and ensure that there is adequate time for people to respond. In addition to clearly stating that the responses are confidential, they can send reminders to complete the survey. If the researchers expect a certain nonresponse rate prior to sending out the survey, they can survey additional people so that any shortfall in responses will be covered. For example, if a researcher needs one thousand responses to a survey and expects a nonresponse rate of 50 percent, then they can send out two thousand surveys. Then, if 50 percent of people do not respond, the researcher will still have enough responses to complete their study.

If a significant amount of nonresponse still occurs, researchers can adjust the results of the survey to compensate. The two main methods of doing this are imputation and weighting. Imputation is when a researcher uses information provided by a respondent with similar characteristics to the nonrespondent in the place of data that the nonrespondent failed to provide. This is usually done in the case of item nonresponse—that is, when a subject has returned the survey with some information filled in, but has failed to respond to one or more items on the survey. In these cases, the information that the nonrespondent did supply assists in identifying a similar respondent. Weighting is when a researcher assigns different weights to responses from different subgroups, to compensate for different response rates among those groups.

NONRESPONSE ERROR IN TELEPHONE SURVEYS
While, despite the rise of the Internet, conducting surveys by telephone remains a viable method, concerns about the influence of nonresponse error also remain. If the calls are only made at certain times of day, for instance, an economic difference may exist in the response results. People with lower incomes often work more than one job and may be less available to answer a phone survey than those with higher incomes. If this characteristic is important to the study, the results may then be biased.

IMPORTANCE OF CONSIDERING NONRESPONSE ERROR
Surveys are important for many fields of research, including government, industry, and social sciences. As nonresponse is almost unavoidable, it is crucial for researchers to keep nonresponse error in mind to ensure the reliability of the data culled from the responses.

Joseph Dewey

FURTHER READING
Biemer, Paul P., et al., editors. *Total Survey Error in Practice.* John Wiley & Sons, 2016.
Blair, Edward, and Johnny Blair. *Applied Survey Sampling.* Sage Publications, 2015.
Christensen, Larry B., et al. *Research Methods, Design, and Analysis.* 12th ed., Pearson, 2014.
Dooley, Larry M., and James R. Lindner. "The Handling of Nonresponse Error." *Human Resources*

Development Quarterly, vol. 14, no. 1, 2003, pp. 99–110. *Business Source Complete*, search.ebscohost.com/login.aspx?direct=true&db=bth&AN=11578762&site=ehost-live. Accessed 31 May 2017.

Groves, Robert M. "Nonresponse Rates and Nonresponse Bias in Household Surveys." *Public Opinion Quarterly*, vol. 70, no. 5, 2006, pp. 646–75. *Academic Search Complete*, search.ebscohost.com/login.aspx?direct=true&db=a9h&AN=23631890&site=ehost-live. Accessed 31 May 2017.

Lindner, James R., et al. "Handling Nonresponse in Social Sciences Research." *Journal of Agricultural Education*, vol. 42, no. 4, 2001, pp. 43–53.

Pike, Gary R. "Adjusting for Nonresponse in Surveys." *Higher Education: Handbook of Theory and Research*, edited by John C. Smart, vol. 22, Springer, 2007, pp. 411–49.

Sakshaug, Joseph W., et al. "Nonresponse Error, Measurement Error, and Mode of Data Collection: Tradeoffs in a Multi-mode Survey of Sensitive and Non-sensitive Items." *Public Opinion Quarterly*, vol. 74, no. 5, 2010, pp. 907–33. *Academic Search Complete*, search.ebscohost.com/login.aspx?direct=true&db=a9h&AN=58147065&site=ehost-live. Accessed 31 May 2017.

SAMPLE PROBLEM

A state university system wants to conduct a survey of students who matriculated at its schools within a certain time period. During this period, a total of 37,560 students matriculated within the state college system. Canvassing that entire population would be cost prohibitive and too time consuming, so researchers determined that a random sample of 5,018 students would produce sufficiently representative results. However, the researchers only expect a response rate of about 40 percent. How many surveys should they send out in order to achieve the necessary 5,018 responses?

Answer:

The researchers will need to send out enough surveys so that 40 percent of that number will be equal to 5,018. This can be solved using a simple ratio equation:

$$\frac{5018}{x} = \frac{40}{100}$$

$$40x = 5018 \times 100 = 501800$$

$$x = \frac{501800}{40}$$

$$x = 12545$$

The researchers would have to send out 12,545 surveys in order to receive the desired number of responses.

Objectivity

FIELDS OF STUDY

Research Theory; Research Design

ABSTRACT

In science, objectivity, or the quality of being uninfluenced by individual perspectives or biases, is considered a major goal. Yet researchers are unavoidably individuals and therefore always subjective to some degree. For this reason, both natural and social sciences devote considerable effort to ensuring steps are taken to make research as objective as possible. Part of this is a conscious decision by researchers, who generally strive to be objective in following the scientific method. Yet biases remain, often unconsciously. It is important to recognize inherent biases in order to minimize them and their impacts on research findings.

PRINCIPAL TERMS

- **cognitive bias:** an ingrained irrational or illogical judgment based on subjective perspective.
- **cultural bias:** the interpretation of a phenomenon from the perspective of the interpreter's own culture.
- **double-blind:** the condition of a study in which neither the researchers nor the subjects know which experimental unit is a control group or a test group.
- **qualitative data:** information that describes but does not numerically measure the attributes of a phenomenon.
- **quantitative data:** information that expresses attributes of a phenomenon in terms of numerical measurements, such as amount or quantity.
- **subjectivity:** the quality of being true only according to the perspective of an individual subject, rather than according to any external criteria.

Understanding Objectivity

Regardless of their factual knowledge, every researcher approaches a project as an individual with unique perspectives that have evolved throughout their life experiences. Some of this subjectivity is trivial, such as a preference for coffee over tea. Such trivial subjectivity either does not impact the validity of research or can be consciously dismissed if relevant to a study. Education, training, and experience all push researchers to be as detached as possible from their individual preferences and perspectives. This quality, known as objectivity, is generally seen as critical to scientific research. The scientific method, with its insistence on testing and reproduction of results, is often considered a way to strive for objectivity.

However, maintaining objectivity is far from simple. In addition to basic subjective preferences, all researchers and all human subjects of research are subject to subtler biases, including sampling bias, cognitive bias, and cultural bias. These influences, which are typically unconscious and are thought to be deeply ingrained in human psychology, can detract from objectivity in powerful ways. For example, a white researcher living in a predominantly white town may recruit only white people for a study. If the study is focused on a local phenomenon, this may not be problematic. However, if the researcher attempts to generalize the results of the study to the whole US population, both sampling and cultural bias will likely compromise objectivity and therefore the validity of the findings. Systematic bias can affect the development and structure of experiments, the accuracy of measurements, and the interpretation of results.

Objectivity is particularly at issue when working with qualitative data, which is by nature more subjective than quantitative data. Some have argued that qualitative research can never be truly objective, as it necessarily involves interpretation of subjective data. Other philosophers of science claim that even quantitative research cannot be fully free from bias. Some research ethicists suggest that arguments over objectivity in research are simply the result of scholars attempting to define the various paradigms of their disciplines. Regardless, scientists agree that objectivity should be maintained whenever possible.

MAINTAINING OBJECTIVITY
Scientists strive to maintain objectivity at every step in a study, from initial conception to evaluation of results. Experts on research integrity agree that researchers should never begin with the idea that they know exactly what they will find. The purpose of research is to arrive at answers that reflect actual findings. Thus, findings may support initial hypotheses, but they may also turn a researcher in a new direction. In the initial stages of research, when hypotheses are being formed and research questions are being stated, it is essential that language be as unbiased as possible so as not to distort results. The size and makeup of samples in a study are also important. Randomization and other methods are used to make research subjects as representative of the general population as possible.

One of the most effective methods for maintaining objectivity in data collection is to set up blind research projects. Single-blind studies eliminate bias among research subjects by keeping them unaware of whether they are members of a control group or the test group. A double-blind study attempts to remove bias among researchers as well, and is considered the gold standard for much research. However, blind trials are not always possible and require significant time and investment.

Once data is collected, researchers face additional objectivity challenges in analyzing and reporting their findings. Possible issues include failure to record data correctly, errors in statistical analysis, and cultural biases in interpretation. Objectivity may also be lost if a researcher opts for a method of analysis that does not reflect what was actually learned. Replication of research and peer review of material submitted for publication are two common methods used to encourage the highest possible degree of objectivity. Even so, studies that take steps to maintain objectivity may still later be found to be biased.

REAL-WORLD RESEARCH
An ongoing concern is whether or not it is possible for research conducted or sponsored by private, for-profit companies to be as objective as that undertaken by government or nonprofit organizations. Scholars have identified a "funding effect" bias, noting that results of industry-funded studies consistently present the industry's products more favorably than do results of research funded by outside parties. Examples include research funded by the tobacco industry that downplays the health effects of smoking and research funded by fossil-fuel companies that questions climate change. Critics claim that this apparent lack of objectivity threatens the integrity of scientific research.

Because of major problems with conflicts of interest in research, significant interest has been directed toward the promotion of objectivity. Various academic, national, and international organizations have formed to uphold good research practices. Much research itself remains devoted to studying the causes and effects of objectivity, subjectivity, and bias. By better understanding the issues and their implications, scientists can better pursue truth.

RESEARCH ETHICS
Objectivity remains vital to science even as some researchers question its limits. The belief that natural science lends itself to objectivity while social science is more likely to be subjective remains common, even to the point of contention between the two broad fields. Yet it is apparent that objectivity is an important goal, if perhaps one impossible to fully achieve, across the sciences and beyond. It has important applications to other disciplines, from philosophy to international relations, and even to everyday life. As a concept, objectivity stands as a critical idea that illustrates the great complexity of human consciousness.

Elizabeth Rholetter Purdy, PhD

FURTHER READING
Agazzi, Evandro. *Scientific Objectivity and Its Contexts.* Springer, 2014.

Grinnell, Frederick. "Research Integrity and Everyday Practice of Science." *Science and Engineering Ethics*, vol. 19, no. 3, 2013, pp. 685–701. *Academic Search Complete*, search.ebscohost.com/login.aspx?direct=true&db=a9h&AN=89583462&site=ehost-live. Accessed 19 Apr. 2017.

Harding, Sandra. *Objectivity and Diversity: Another Logic of Scientific Research.* U of Chicago P, 2015.

Krimsky, Sheldon. "Do Financial Conflicts of Interest Bias Research? An Inquiry into the 'Funding Effect' Hypothesis." *Science, Technology, & Human Values*, vol. 38, no. 4, 2013, pp. 566–87.

Letherby, Gayle, et al. *Objectivity and Subjectivity in Social Research*. Sage Publications, 2013.

Steck, Andreas J., and Barbara Steck. *Brain and Mind: Subjective Experience and Scientific Objectivity*. Springer, 2016.

Yetiv, Steve A. *National Security through a Cockeyed Lens: How Cognitive Bias Impacts US Foreign Policy*. Johns Hopkins UP, 2013.

One-Tailed and Two-Tailed *t*-Tests

FIELDS OF STUDY

Statistical Analysis; Variance Analysis

ABSTRACT

A *t*-test compares the means of two different sets of data to determine whether the difference between the means is statistically significant. One-tailed and two-tailed *t*-tests are intended to determine whether the likelihood of a particular experimental result is of sufficient size to justify rejecting the null hypothesis.

PRINCIPAL TERMS

- **alternative hypothesis:** a prediction that a significant difference will be found between the control and experimental groups.
- **confidence interval:** a range of values within which the value of an unknown population parameter has a given probability of falling.
- **null hypothesis:** a prediction that no significant differences will be found between the control and experimental groups.
- **sample distribution:** a probability or frequency distribution for the values obtained from a random population sample, plotted according to either the probability of their occurring or the frequency with which they occur.
- **standard deviation:** a measurement of the degree of dispersion present in a distribution; greater dispersion results in a larger standard deviation.

Understanding One-Tailed and Two-Tailed *t*-Tests

Data sets are often compared using the *standard deviation* of each. However, sometimes these cannot be known or are too small to be useful. In those cases, the standard deviation is estimated. A *t*-test can then compare the means of two data sets to determine whether the difference between the means is statistically significant, thereby proving or disproving the *null hypothesis*. These tests answer the basic question, is it more probable that a particular piece of data in a *sample distribution* was produced by random chance or by a process that confirms the *alternative hypothesis*?

The term "tail" refers to the way that a distribution is displayed on a graph. In a normal distribution, the graph has the shape of a bell curve, with a high point in the middle of the graph and downward slopes on either side as the values approach zero. The upper and lower ends of the graph are referred to as "tails" because their appearance is reminiscent of an animal's tail.

A graph of a distribution can have one tail or two tails. The distinction depends upon the type of data being collected and whether extreme values within the data (making up the tail) can be logically expected to occur in one or two directions. For example, a teacher giving a math exam could expect that while most students will receive grades in the middle of the grading scale, a few will score much higher than this and a few will score much lower. Because the extreme values can reasonably be expected to occur both above and below the mean, this type of situation would call for a two-tailed *t*-test.

In other situations, researchers may only be concerned with testing in one direction, leading them to select a one-tailed test. For instance, if a company wanted to prove that its new type of weight-loss shake is as effective as the leading brand, the goal of the test is only to show that the new shake is no less effective. A one-tailed test could be used.

Using Confidence Intervals

Confidence intervals (CIs) allow prediction of the range of values within which an unknown population

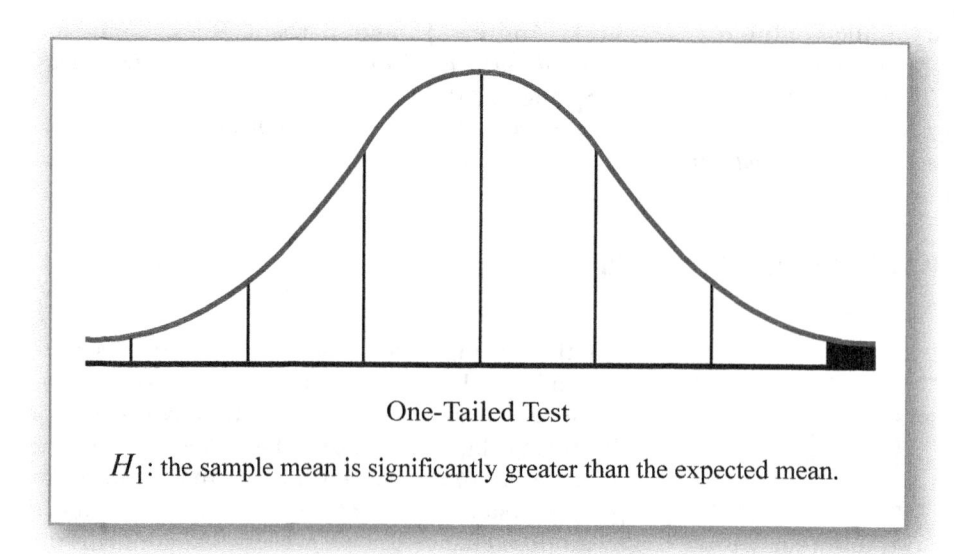

One-Tailed Test

H_1: the sample mean is significantly greater than the expected mean.

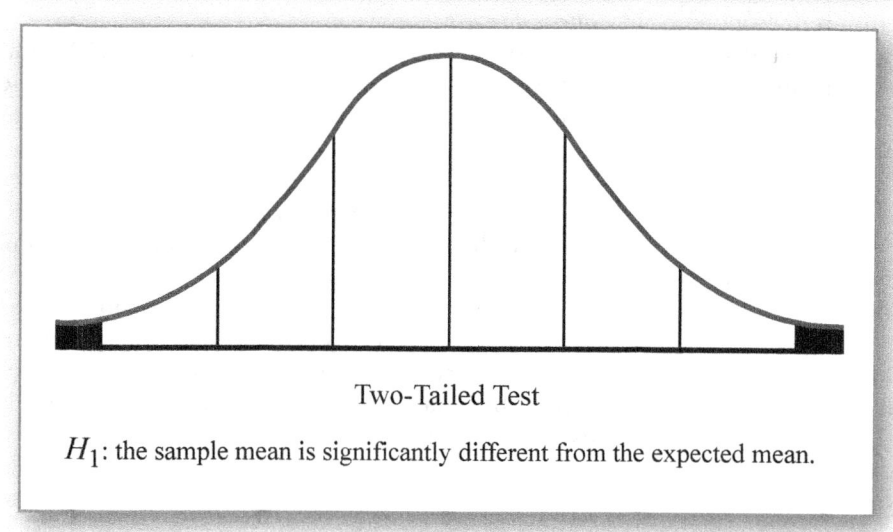

Two-Tailed Test

H_1: the sample mean is significantly different from the expected mean.

parameter would be expected to fall, assuming multiple random samples are drawn from that population. The confidence level describes the probability that the parameter falls within the CI. In hypothesis testing, if the null value is contained within the CI, the null hypothesis cannot be rejected.

Note that CIs differ between one-tailed and two-tailed tests. When using a two-tailed test at a confidence level of 90 percent, the CI is centered on the sample mean. Thus, 5 percent of the possible values would be in one tail of the distribution on the positive side and 5 percent in the second tail on the negative side. However, when using a one-tailed test with a confidence level of 90 percent, the CI begins at one end of the distribution curve with 10 percent of the possible values in the tail being tested. For a one-tailed test, the area of significance is equal to 1 minus the confidence level. For a two-tailed test, the area of significance is that result divided by 2, representing the two tail regions of the plot. This difference between the tests is important when converting results between t tables.

LIMITATIONS AND CONSIDERATIONS

While useful in many scenarios, *t*-tests do have some limitations. For example, the *t*-test can test differences between only two datasets. The use of *t*-tests also assumes that the populations being studied are normally distributed, the variances of the populations are equal, and independent (unrelated), random samples were used. Thus, one-tailed and two-tailed tests are not always possible or appropriate.

Furthermore, the two-tailed *t*-test has lower power than the one-tailed test. The penalty for lower power means that if there is a real statistical difference and the alternative hypothesis is true, the two-tailed *t*-test is less likely to detect it. However, the null hypothesis is larger when using a one-tailed *t*-test because it consists not only of no effect being observed but any effect that is not the direction of interest. One-tailed *t*-tests must therefore be used and reviewed with care in order to avoid bias or even unethical practice, particularly in areas such as medicine.

One-Tailed and Two-Tailed *t*-Tests in Action

In many research scenarios *t*-tests are useful. For example, a drug company may wish to test the efficacy of a new blood pressure drug compared to an existing one. The test could be conducted using a two-tailed *t*-test. To conduct this study, a group of patients would be given the existing drug, and blood pressure data would be collected from the test subjects. A second group would be given the new drug, and their blood pressure data would be collected. A two-tailed test could then be conducted to determine if the new drug significantly lowered or raised patients' blood pressure when compared to the existing drug.

The Importance of *t*-Tests

Researchers use *t*-tests when they need to compare the means of two different sets of data to determine whether the difference between the means is statistically significant. Such comparisons are common in many fields ranging from economics to education to the sciences, which has led to widespread use of one- and two-tailed *t*-tests.

Scott Zimmer, JD

Further Reading

Frieman, Jerome, et al. *Principles & Methods of Statistical Analysis.* SAGE Publications, 2018.

Ruxton, Graeme D., and Markus Neuhäuser. "When Should We Use One-Tailed Hypothesis Testing?" *Methods in Ecology and Evolution*, vol. 1, no. 2, 2010, pp. 114–17. British Ecological Society, onlinelibrary.wiley.com/doi/10.1111/j.2041-210X.2010.00014.x/full. Accessed 9 May 2017.

Spiegel, Murray R., and Larry J. Stephens. *Schaum's Outlines: Statistics.* 5th ed., McGraw-Hill, 2014.

Triola, Mario F. *Elementary Statistics.* 12th ed., Pearson Education, 2014.

Vik, Peter. *Regression, ANOVA, and the General Linear Model: A Statistics Primer.* SAGE Publications, 2014.

Yanow, Dvora, and Peregrine Schwartz-Shea, editors. *Interpretation and Method: Empirical Research Methods and the Interpretive Turn.* 2nd ed., Routledge, 2014.

SAMPLE PROBLEM

A researcher has developed a new method of studying for exams. They wish to test whether the method demonstrably improves students' grades. What would be the hypotheses? Should a one- or a two-tailed test be used to test the hypothesis?

Answer:

The alternate hypothesis is that the new method improves student grades by a statistically significant amount. The null hypothesis is that the study method has no significant effect. Because the researcher is most interested in showing that the grades of students using the method are higher than those of students who do not use the method, a one-tailed test would be appropriate.

Oppenheimer, J. Robert

American Physicist

J. Robert Oppenheimer was an American physicist noted for his work during World War II as scientific director of the Manhattan Project, which developed the first atomic bomb. After the war, he became director of the Institute for Advanced Study in Princeton, New Jersey. Before joining the Manhattan Project, Oppenheimer worked as a physics professor, with a joint appointment to the University of California, Berkeley, and the California Institute of Technology.

- **Born:** April 22, 1904; New York, New York
- **Died:** February 18, 1967; Princeton, New Jersey
- **Primary field:** Physics
- **Specialties:** Atomic and molecular physics; nuclear physics

Early Life

Julius Robert Oppenheimer was born in New York City on April 22, 1904, into a wealthy German Jewish family. His father, Julius S. Oppenheimer, was a first-generation immigrant who had become a successful textile merchant. His mother, Ella Friedman, a noted artist, came from a wealthy Jewish family from Baltimore.

Official portrait of J. Robert Oppenheimer, first director of Los Alamos National Laboratory. By Department of Energy, Office of Public Affairs [Public domain or Public domain], via Wikimedia Commons

Oppenheimer's parents were not religious but belonged to a secular humanist offshoot of Reform Judaism known as the Society for Ethical Culture. Oppenheimer and his younger brother, Frank, attended the Ethical Culture Society School. Robert was a solitary child, partly because of his own intellectual interests and partly because of his parents' overprotective nature. His lack of social interaction made it difficult for him to relate to other children, except for his brother Frank, who shared Robert's scientific interests.

Robert graduated from the society's school in 1921. He waited a year before entering Harvard University in order to recover from an attack of dysentery and colitis contracted during a trip to Germany. To help him recover, his parents sent him to a ranch in New Mexico.

Oppenheimer entered Harvard in 1922, majoring in chemistry. His interest began to shift toward physics, however, partly through the influence of physics professor Percy Williams Bridgman. Oppenheimer graduated from Harvard in 1925 with a chemistry degree, but he decided to go to Europe to do doctoral work in theoretical physics.

In England, Oppenheimer studied at Cambridge University's Cavendish Laboratory. His work was supervised by Sir Joseph John Thomson, the physicist who discovered the electron. Oppenheimer found experimental work difficult, and was encouraged to focus on theoretical physics. Therefore, he traveled to Germany in 1926 to study under Max Born at the University of Göttingen.

Born, a founder of quantum mechanics, helped the young prodigy develop what became known as the "Born-Oppenheimer approximation" of molecular motion. The approximation contributed to quantum theory by considering nuclear motion and electronic motion separately.

Oppenheimer graduated with his doctorate in physics in 1927, leaving Göttingen to do postdoctoral research under Paul Ehrenfest at the University of Leiden, in the Netherlands. At Leiden, he acquired the nickname "Opje," which his American students later changed to "Oppie." Leaving Leiden, Oppenheimer also studied briefly under Austrian physicist Wolfgang Pauli in Zurich, Switzerland.

Life's Work

Returning to the United States, Oppenheimer took up studies at the California Institute of Technology (Caltech). In 1928, he accepted a dual teaching position in physics, commuting between Caltech and the University of California, Berkeley. He held the joint appointment until joining the Manhattan Project in 1941. At Berkeley, he befriended the experimental physicist Ernest O. Lawrence, developer of the cyclotron, a type of particle accelerator.

Oppenheimer became one of the most popular professors at Berkeley, as well as a leading figure in theoretical physics. He wrote extensively, publishing papers on many branches of physics, including quantum theory. Writing on gravitational collapse, he was the first to publish papers on what became known as black holes. However, despite Oppenheimer's accomplishments and great popularity, many of his colleagues did not consider him a first-rank physicist. Their feelings were based partly on his lack of focus and perhaps because of his interest in Eastern mysticism.

In 1937, upon his father's death, Oppenheimer spent much of his inheritance on political causes, supporting the antifascist side in the Spanish Civil War. Unlike his brother, Frank, however, he never joined the Communist Party.

In August 1939, Oppenheimer met Katherine "Kitty" Puening Harrison, a Berkeley student and political activist. Her estranged first husband, Joe Dallet, had been a labor organizer who was killed in the Spanish Civil War. Robert and Kitty married in November 1940, and their first child, Peter, was born in May 1941. Their daughter, Katherine, was born in 1944.

Oppenheimer became involved in the race to develop an atomic bomb almost as soon as the United States entered World War II. The Lawrence Radiation Laboratory at Berkeley was already involved in work on an atomic weapon. Brigadier General Leslie Richard Groves, the army officer responsible for overseeing construction of the Pentagon, became the military head of the government's secret bomb development program, called the Manhattan Project. In June 1942, Groves appointed Oppenheimer the Manhattan Project's scientific director.

Oppenheimer was in charge of establishing a facility at Los Alamos, New Mexico, to consolidate the various US atomic-bomb projects. He was closely involved in the day-to-day operations of the project, as well as the project's overall strategy. He also demonstrated great diplomatic skill in handling relations between the military and civilian personnel; many of the latter were his former students.

On July 16, 1945, the Manhattan Project conducted its first successful test at Alamogordo, New Mexico. Oppenheimer named the test "Trinity," which some scholars have considered a reference to a poem by the English author John Donne. Later, Oppenheimer recalled that as he watched the explosion, he was reminded of verses from the Bhagavad Gita, the Hindu scriptures, in which the god Vishnu says, "Now I am become Death, the destroyer of worlds."

US President Harry Truman decided to use the atomic bomb to end World War II quickly and avoid a proposed invasion of Japan. On August 6, 1945, the B-29 bomber *Enola Gay* dropped an atomic bomb on Hiroshima, Japan, and on August 9 on Nagasaki. The ultimate death toll from both bombings has been estimated to be between 100,000 and 200,000 people, mostly civilians. Many more suffered severe injuries from the direct explosion and from radiation.

After the Manhattan Project, Oppenheimer continued his work with nuclear energy. From 1947 to 1952, he served as chairman of the General Advisory Committee to the US Atomic Energy Commission (AEC). In 1947, Oppenheimer also gave up his professorship at Berkeley and became director of the Institute for Advanced Study in Princeton, New Jersey. Oppenheimer served as an advisor to both the United States and the United Nations on nuclear energy. He strongly supported nuclear arms control and opposed the development of a hydrogen bomb.

Impact

Oppenheimer used his position to support international control of atomic energy and weapons, a stance that upset many military and political figures, particularly AEC commissioner Lewis Strauss and Manhattan Project colleague Edward Teller, who wanted the United States to develop a hydrogen bomb. Oppenheimer eventually came to support the H-bomb, after Teller and Stanislaw Ulam produced a feasible design in 1951. The first H-bomb was successfully tested by the United States in 1952.

During the Red Scare of the early 1950s, Oppenheimer's critics pointed out his record of support for left-wing causes, as well as inconsistencies in his wartime comments to governmental investigators. In 1954, Oppenheimer's security clearance was suspended, a move that many people around the world saw as an attack on liberalism and freedom of thought.

Oppenheimer largely withdrew from public view following the loss of his clearance, though he continued as head of the Institute for Advanced Study. In 1963, President Lyndon B. Johnson showed official favor to Oppenheimer by presenting him with the Enrico Fermi Award. Oppenheimer died of throat cancer on February 18, 1967. His funeral was attended by political, military, and scientific dignitaries.

In the decades since his death, Oppenheimer has remained a symbol of the tension between science and the state and between scientists' ethical and political responsibilities. His story has been dramatized numerous times, notably in the stage play *In the Matter* of *J. Robert Oppenheimer* (1969), the film *Fat Man and Little Boy* (1989), and the opera *Doctor Atomic* (2005).

—Eric Badertscher

FURTHER READING

Bernstein, Jeremy. *Oppenheimer: Portrait of an Enigma.* Chicago: Dee, 2005. Print. Biography of Oppenheimer by a physicist who studied under him and was a staff writer at the *New Yorker*.

Bird, Kai, and Martin J. Sherwin. *American Prometheus: The Triumph and Tragedy of J. Robert Oppenheimer.* New York: Vintage, 2006. Print. A Pulitzer Prize–winning biography copiously footnoted and containing information drawn from a multitude of interviews.

Pais, Abraham, and Robert Crease. *J. Robert Oppenheimer: A Life.* New York: Oxford UP, 2007. Print. Biography cowritten by a physicist who knew Oppenheimer and a historian at the Brookhaven National Laboratory.

THE FIRST ATOMIC BOMB IS DETONATED

While visiting the United States in 1939, physicist Niels Bohr derived a theory of fission with John Wheeler of Princeton University, leading Bohr to predict that the common isotope uranium-238 would require fast neutrons for fission, but that the rarer uranium-235 would fission with neutrons of any energy. Thus, uranium-235 would be far more suitable for use in a bomb. Uranium bombardment in a cyclotron led to the discovery of plutonium in 1940, as well as the discovery that plutonium-239 was fissionable and thus potentially good bomb material. Uranium-238 was used to create plutonium-239, which was separated from the uranium by chemical methods.

Studies of fast-neutron reactions for an atomic bomb were brought together in Chicago in June 1942 under the leadership of J. Robert Oppenheimer. He became a personal adviser to American brigadier general Leslie Richard Groves, who built Oppenheimer a laboratory for the design and construction of an atomic bomb at Los Alamos, New Mexico.

In 1943, Oppenheimer began overseeing the Manhattan Project, during which two bomb designs were developed. A gun-type bomb called "Little Boy" used 35 pounds of uranium-235 in a 9,900-pound cylinder about 6 feet long and 1.5 feet in diameter, in which a uranium bullet could be fired into three uranium target rings to form a critical mass. An implosion-type bomb called "Fat Man" had a 12-pound spherical core of plutonium about the size of an orange, which could be squeezed inside a 5,000-pound sphere about 5 feet in diameter.

A flat scrub region about 125 miles southeast of Alamogordo was chosen for the test site, and observer bunkers were built about 6 miles from a 100-foot steel tower. On July 13, 1945, one of the plutonium bombs was assembled at the site; the next morning, it was raised to the top of the tower. The bomb was detonated on July 16. The resulting implosion initiated a chain reaction of nearly 60 fission generations in about a microsecond. It produced an intense flash of light and a fireball that expanded to a diameter of about 2,000 feet in two seconds, rose to a height of some 8 miles, and formed an ominous mushroom shape. Forty seconds later, an air blast hit the observer bunkers, followed by a sustained and awesome roar. Measurements confirmed that the explosion had the power of 18.6 kilotons of trinitrotoluene (TNT), nearly four times the predicted value.

On March 9, 1945, 325 American B-29 bombers dropped 2,000 tons of incendiary bombs on Tokyo, resulting in 100,000 deaths from the fire storms that swept the city. Still, the Japanese military refused to surrender. On August 6, 1945, the B-29 Enola Gay dropped Little Boy bomb on Hiroshima. On August 9, Fat Man was dropped on Nagasaki. Between 100,000 and 200,000 people died as a result of the bombs. Japan officially surrendered on August 15.

PEIRCE, CHARLES SANDERS

AMERICAN PHILOSOPHER

Although he was largely unrecognized by his contemporaries—apart from his contribution to pragmatism—Peirce developed a system of philosophy that attempted to reconcile the nineteenth century's faith in empirical science with its love of the metaphysical absolute. His difficult and often confusing ideas anticipated problems central to twentieth century philosophy.

- **Born:** September 10, 1839
- **Birthplace:** Cambridge, Massachusetts
- **Died:** April 19, 1914
- **Place of death:** Near Milford, Pennsylvania

EARLY LIFE

Charles Sanders Peirce was the son of Benjamin Peirce, one of the foremost American mathematicians. During his childhood, his mother, Sarah Hunt (Mills) Peirce, took second place to his dynamic father, who personally supervised the boy's education and provided a role model that inspired but also proved impossible to emulate. Convinced of his son's genius, Benjamin Peirce encouraged his precocious development. Charles began the study of chemistry at the age of eight, started an intense scrutiny of logic at twelve, and faced rigorous training in mathematics throughout his childhood. In the latter case, he was seldom given general principles or theorems. Instead, he was expected to work them out on his own.

At sixteen, Peirce entered Harvard, where his father was professor of mathematics. Contrary to expectations, Peirce proved a less than brilliant student, and he was graduated, in 1859, seventy-first out of a class of ninety-one. Probably too young and certainly too much the nonconformist to fit into the rigid educational system of nineteenth century Harvard, Peirce's inauspicious beginning in institutional academics was prophetic. Though he would continue his education, receiving a master's degree from Harvard in 1862 and a bachelor's degree in chemistry the following year, his future did not lead to a distinguished career in academics or, indeed, in any conventional pursuit. His lot in life, in spite of so much promise, was frustration and apparent failure.

Peirce's difficulty in adjusting to the world of ordinary men was related to his unusual and often trying personality. Always his father's favorite, Peirce became convinced of his own genius and impatient with those who failed to recognize the obvious. Shielded and overindulged as a child, Peirce never developed the social skills required for practical affairs nor the self-discipline necessary to make his own grandiose vision a reality. Such problems were exaggerated by his passion for perfection and his abstract turn of mind. Peirce found real happiness only in the rarefied world of his own philosophical speculation.

As a youth, Peirce both attracted and repelled. Always prone to the dramatic gesture and, when he was inclined, a brilliant conversationalist, he could be an entertaining companion, but he could also use his rapier wit as a weapon. Of medium height, dark, swarthy, and fastidious in matters of dress, the handsome young Peirce reveled in his reputation as a lady's man and spent much energy in seeking the "good life." He actually paid an expert to train his palate so that he could become a connoisseur of fine wines. In 1862, Peirce married Harriet Melusina Fay, three years his senior and infinitely more mature and self-possessed. A feminist and intellectual in her own right, "Zina" worshipped her captive "genius" and labored for years to keep him out of serious trouble while restraining his extravagance. However, she could also be jealous and possessive, and, though Peirce would experience some stability under Zina's influence, the marriage was doomed.

LIFE'S WORK

Upon his graduation from Harvard, Peirce went to work for the United States Coast and Geodetic Survey, a position acquired through his father's influence. Benjamin Peirce served as a consulting geometer for the organization and became its superintendent

Charles Sanders Peirce (1839 – 1914). [Public domain], via Wikimedia Commons

in 1867. Charles Peirce remained with the survey in various capacities until 1891, when he was asked to resign. This bureaucratic career, while terminated in less-than-desirable circumstances, was not without accomplishments. His deep commitment to the experimental method helped put the survey on a firm scientific basis, and Peirce himself became internationally known for his work on gravity research.

Peirce also continued an association with Harvard, once again through his father's influence, holding temporary lectureships in logic in 1865-1866 and 1869-1870 and from 1872 to 1875 serving as assistant at the Harvard Observatory. His observatory work on the measurement of light provided data for the only book he published during his lifetime, *Photometric Researches*

(1878). Peirce hoped for a permanent appointment at Harvard, but his lack of a doctoral degree, his erratic lifestyle, and a typically personal quarrel with Harvard president Charles W. Eliot made the dream impossible.

More important than Peirce's actual work, the atmosphere and personal contacts at Harvard helped mold his philosophical outlook. Never idle, Peirce spent his spare time studying the work of Immanuel Kant, the ideas of the medieval scholastics, and various theories in logic and mathematics. The most useful forum for his developing ideas was the so-called Metaphysical Club.

In the meetings of this unusual group, which included William James, Oliver Wendell Holmes, Jr., Francis E. Abbot, and Chauncey Wright, among others, Peirce had the opportunity to test his theories before a critical audience. It was there that he used the term "pragmatism" to describe the relationship between a conception and its effects that allows one to understand the actual meaning of the original conception by knowing its effects. Although Peirce intended his idea as a theory of meaning, William James, more than twenty years later, would popularize the term and expand it far beyond the original intention. In fact, objecting to his friend's interpretation, Peirce, in 1905, coined the term "pragmaticism" to distinguish his thought from James's version.

In his Harvard years, Peirce began to write articles for *The Journal of Speculative Philosophy* and other scholarly publications, as well as more popular magazines such as *Popular Science Monthly*. Such articles, along with numerous book reviews, provided his major public outlet for the remainder of his life. Ignored by much of the philosophical community, these writings contained important contributions to logic, mathematics, and metaphysics.

Peirce finally got his chance to teach when he was hired as a part-time lecturer at Johns Hopkins University in 1879. Apparently an effective teacher, he produced some of his best work in logic and scientific methodology at Johns Hopkins. However, his erratic behavior, coupled with his divorce from his first wife and remarriage to a twenty-six-year-old French woman, the mysterious Mme Juliette Pourtalai, made it difficult for the authorities to accept him, no matter how brilliant, as part of the faculty. In 1884, Peirce was dismissed from his position because of unsuitable activities of a moral nature, probably connected with his divorce and remarriage.

Peirce's second marriage began a phase of his life that would be philosophically productive but personally frustrating, ending in self-imposed exile. In 1887, his academic career hopelessly in shambles and his labors for the survey drawing to a conclusion, Peirce moved to Milford, Pennsylvania, a resort area on the Delaware river. With a small inheritance, he was able to purchase land and begin construction of an elaborately planned home he called "Arisbe." Though Peirce was able to live in his retreat for the remainder of his life, the mansion was never really completed.

Typically, Peirce had overextended himself. When he lost his government salary in 1891 and suffered severe losses in the depression of 1893, he began a long slide into poverty. His closest and always tolerant friend, William James, tried to help as much as possible, arranging for a series of lectures in Boston in 1898 and finally persuading Harvard to allow the notorious philosopher to give a series of lectures at the university in 1903. No effort, however, even by the most famous American philosopher, would make Peirce acceptable to established society in the nineteenth century. Finally, James began collecting donations for a Peirce fund from interested and unnamed friends. From 1907 until his death in 1914, Peirce was largely supported by this fund, which amounted to about thirteen hundred dollars a year. Peirce, who had often been jealous of James and attacked his version of pragmatism with undisguised contempt, paid his friend a typical compliment by adopting Santiago (St. James) as part of his name in 1909.

Even in his last years, which were marred by illness, Peirce was productive. He continued to work in isolation, leaving behind a massive collection of papers. Ironically, Harvard, the institution that had so often rejected him, recognized his worth and purchased the manuscripts from his widow. Between 1931 and 1935, the six volumes of the *Collected Papers of Charles Sanders Peirce*, edited by Charles Hartshorne and Paul Weiss, were published by Harvard. This collection began what amounted to a revolution in American academic philosophy, making the ideas of Peirce a touchstone for twentieth century philosophical inquiry.

The exact nature of Peirce's contribution to understanding is by no means clear. Numerous scholars have spent careers examining his writings, never reaching a consensus. The confusion is rooted in the nature of Peirce's work itself. Not satisfied with a contribution in a single area of inquiry, Peirce envisioned a vast architectonic system ending in a complete explanation of all human knowledge. In short, Peirce strove to be a modern Aristotle. Although admirable, this goal ran up against a central dilemma in human thought, providing a source of tension within Peirce's system as well as within the world in which he lived.

Science, in the last years of the nineteenth century, revealed a limited vision of reality, of what could be known. The world, according to this view, consisted of matter and could be fully explained through the scientific method. Many thinkers, unable to accept this so-called positivistic version of reality, countered with an explanation based on the mind itself as the source of everything. Best represented in the Idealism of Georg Wilhelm Friedrich Hegel, this view had spawned many variations. Peirce could not fully accept either position. Positivism seemed to deny the possibility of metaphysics or, perhaps better, a universe with meaning that could be understood by human beings. Idealism seemed hopelessly subjective, denying the possibility of actually knowing the physical universe.

Peirce set out to reconcile the irreconcilable by carefully examining immediate experience. Characteristically, this examination would be grounded on clear and precise thinking such as his famous "pragmatic maxim." He also rejected nominalism and accepted the position of the medieval scholastic Duns Scotus on the reality of Universals. Peirce insisted that cognition itself is reality, and everything that is real is knowable. The structure of experience is revealed in what he called "phaneroscopy." This term is typical of Peirce's obsession with the invention of new words to explain concepts, which is one of the reasons his ideas are so difficult.

Phaneroscopy is roughly analogous to the modern concept of phenomenology. From his phenomenological basis, Peirce deduced three categories or qualities of experience that he termed Firstness, Secondness, and Thirdness. This division of experience allowed him to move from an essentially psychological analysis to logic itself through what he called the "semiotic," or the doctrine of signs. By signs, Peirce essentially meant those things in the mind that stand for the real things of the world. A word, for example, would be a sign but only one kind of sign. Peirce's analysis of signs and their relationships was a vast and complicated explanation of how human beings think and provides the logical basis for his whole system.

A complete discussion of this difficult and obscure argument is not possible in this context, but most modern philosophers would agree that it constitutes Peirce's most important contribution to philosophy, particularly logic. Its obscurity, however, has led to many different interpretations. Phenomenologists, for example, find considerable comfort in his explanation of experience, while the logical positivists, who seldom agree with phenomenologists, also see their ideas reflected in Peirce's theory of signs. In fact, most philosophical systems in the twentieth century found some part of Peirce's ideas important in either a positive or a negative way.

Peirce's logic, however, was only the foundation of a broad system that included a complete theory of knowledge as well as cosmological speculations. This system, while not as widely accepted as his semiotic, includes a number of important concepts. For example, Peirce develops what he calls "tychism," or the doctrine of chance, which explains irregularities within nature. This idea should be balanced with "synechism," which is the doctrine that continuity is a basic feature of the world. Here again, Peirce reconciles the irreconcilable, and the result provides a reasonable picture of the actual condition of scientific inquiry.

Synechism represents scientific law, which Peirce calls habit, without which one could not understand the operation of the natural world. Tychism, however, explains how change is possible and prevents a deterministic version of reality, which is the logical result of scientific law. Science then, while based on research that, if pursued to infinity, will result in "truth," must in the practical world be based on probability. Even in logic itself, one cannot be sure that all statements are correct. Although not denying absolute truth, this concept, which Peirce called "fallibilism," provides a healthy corrective to those who are convinced that they have found the ultimate answer to reality.

SIGNIFICANCE

Few can profess to understand all of Peirce's philosophy, and his work will probably never appeal to the average person unschooled in the mysteries of philosophical discourse. Nevertheless, his attack on the central dilemma of modern thought, created by scientific advance and its almost inevitable clash with human values, is the necessary starting point for many twentieth century philosophers and, through their work, has a profound influence on the way the world is viewed.

It may be true that Peirce ultimately failed in his attempt to reconcile the "hard" world of science with cherished human values represented by the "soft" world of Idealism, but, unlike his tragic personal life, his philosophy was certainly a glorious failure. Moreover, Peirce remained a true optimist who believed in the inevitability of human progress through reason. His system of thought, while far from perfect, did provide a view of reality that would make such progress possible. His first rule of reason demanded that the road to new knowledge always be left open. The greatest sin against reasoning, he believed, consisted in adopting a set of beliefs that would erect a barrier in the path of the search for truth.

—*David Warren Bowen*

FURTHER READING

Almeder, Robert F. *The Philosophy of Charles S. Peirce: A Critical Introduction.* Totowa, N.J.: Rowman & Littlefield, 1980. An analysis of Peirce's philosophy, stressing his epistemological realism, which contains a perceptive and detailed discussion of his theory of knowledge.

Conkin, Paul K. *Puritans and Pragmatists: Eight Eminent American Thinkers.* Bloomington: Indiana University Press, 1968. One of the finest overviews of American intellectual history. Places Peirce within the context of the development of American thought between Jonathan Edwards and George Santayana.

De Waal, Cornelis. *On Peirce.* Belmont, Calif.: Wadsworth/Thomson Learning, 2001. Brief (91-page) overview of Peirce's philosophy, designed to introduce his ideas to students.

_____. *On Pragmatism.* Belmont, Calif.: Wadsworth/Thomson Learning, 2005. Describes the main figures and the central issues of pragmatism, with individual chapters devoted to Peirce, Dewey, James, and other philosophers.

Goudge, Thomas A. *The Thought of C. S. Peirce.* Toronto: University of Toronto Press, 1950. One of the most perceptive studies of Peirce's thought. Sees Peirce's philosophy as resting on a conflict within his personality that produced tendencies toward both naturalism and Transcendentalism.

Misak, Cheryl, ed. *The Cambridge Companion to Peirce.* New York: Cambridge University Press, 2004. Collection of essays discussing Peirce's philosophy and his place within the pragmatist tradition. The essays include examinations of Peirce and medieval thought, his account of perception, and his theory of signs.

Moore, Edward C. *American Pragmatism: Peirce, James, and Dewey.* New York: Columbia University Press, 1961. An analysis of American pragmatism based on its three primary figures. Provides an excellent comparison of their different positions.

Potter, Vincent G. *Charles S. Peirce: On Norms and Ideals.* Amherst: University of Massachusetts Press, 1967. An analysis of Peirce's attempt to establish aesthetics, ethics, and logic as the three normative sciences. The author places particular emphasis on the role of "habit" in the universe.

Reilly, Francis E. *Charles Peirce's Theory of Scientific Method.* New York: Fordham University Press, 1970. A discussion of Peirce's ideas concerning the method and the philosophy of science.

Skagestad, Peter. *The Road of Inquiry: Charles Peirce's Realism.* New York: Columbia University Press, 1981. Focuses on Peirce's theory of scientific method but also contains an introduction with considerable biographical information.

PLACKETT-BURMAN DESIGN

FIELDS OF STUDY

Research Design; Experimental Design

ABSTRACT

The Plackett-Burman design is a type of experimental research design used to screen factors and assess their relative impact on the dependent variable. This design is useful for identifying the main effects when there are a large number of factors. As a result, the experiment and subsequent analysis become much less complex, as the most important factors have been determined via screening.

PRINCIPAL TERMS

- **active factor:** the independent variable that has the most profound effect on the dependent variable or outcome.
- **full factorial design:** an experimental research design that analyzes all possible combinations of factors in an experiment.
- **two-way interaction:** in statistics, an effect by which the impact of one independent variable on the dependent variable is altered, in a nonadditive way, by the influence of a second independent variable.

WHAT IS PLACKETT-BURMAN DESIGN?

The Plackett-Burman design is a quantitative experimental design. It is often used in the initial stages of experiments that must analyze a large number of factors. The Plackett-Burman design is used when only the active factors, also called main effects, are of interest to researchers. Interactions between factors are not considered to be significant in this design. The Plackett-Burman design is used to screen out factors that do not significantly affect the outcome. As such, it can reduce the amount of data needed to be collected in future experiments.

The Plackett-Burman design is often compared to the full factorial design. A full factorial design analyzes all possible combinations of factors in an experiment. Using this design, a researcher performs an experimental run for every combination of levels. Using a full factorial design, the number of runs for a two-level experiment with k factors would be $2k$. For example, a two-level experiment with four factors (k = 4) would require sixteen experimental runs:

$$2^4 = 2 \times 2 \times 2 \times 2 = 16$$

Conducting a full factorial design with a large number of factors can quickly become expensive, as the number of runs required grows exponentially with each factor. When deciding to use a full factorial design, it is essential that researchers be aware of the costs that can be incurred. Still, the full factorial design has its benefits. For instance, it is thorough. There is little uncertainty when a researcher tries all combinations of factors.

Factors

	1	2	3	4	5	6	7
run 1	x	x	x	o	x	o	o
run 2	o	x	x	x	o	x	o
run 3	o	o	x	x	x	o	x
run 4	x	o	o	x	x	x	o
run 5	o	x	o	o	x	x	x
run 6	x	o	x	o	o	x	x
run 7	x	x	o	x	o	o	x
run 8	o	o	o	o	o	o	o

Tests

The Plackett-Burman design tests various combinations of factors and performs an experimental run for the levels, testing for the main effects. Like the full factorial design, the Plackett-Burman design analyzes large numbers of factors. It has a resolution of 3, meaning that main effects are not confounded by other main effects, but they are confounded by two-way interactions. A lower resolution indicates a design that is more susceptible to confounding. In a design with a resolution of 4, the main effects are not confounded by two-way interactions, but they are confounded by three-way interactions (interactions involving three variables), and two-way interactions may be confounded by other two-way interactions. By comparison, full factorial designs are sometimes said to have a resolution of "infinity" because they have no confounding. Resolution 3 designs are better suited for screening than for a full research study.

Benefits of the Plackett-Burman Design

The Plackett-Burman design is an excellent method for identifying the most important factors when conducting an experiment. It is especially useful in the early stages of an experiment, or as an exploratory study by a researcher who plans to conduct a more thorough experiment later.

In a Plackett-Burman design, the total number of experimental runs must be a multiple of four. It must also be greater than the number of factors. For example, a researcher may be investigating a variety of factors that can affect the quality of cheese. There are many such factors, including mixing temperature, humidity level, fat content, water content, mixing speed, surface area, and storage temperature. Each of these seven factors could have two levels, high or low. Using a full factorial design, the researcher would need to conduct 128 runs (27). Using a Plackett-Burman design, however, the researcher could conduct just 8 runs (7 + 1 = 8).

The Plackett-Burman design aids researchers in evaluating a large number of factors. It is used to screen out those factors that are not important to the outcome being studied. Thus, it reduces the costs and time that a researcher would spend on analyzing factors that are not relevant to the dependent variable.

Because the Plackett-Burman design acts as a screening process, there is no need to conduct a large number of runs in order to identify the main effects. Once the main effects have been identified, further experiments with higher resolutions can be conducted to determine the influence of interactions between factors and further refine the outcome.

Why Use the Plackett-Burman Design?

The Plackett-Burman design allows researchers to identify the active factors in a study with a relatively small number of runs. It is a highly economical design. Plackett-Burman design is often used in manufacturing, agriculture, industrial engineering, and other fields in which a large number of factors must be analyzed and studies need to be cost efficient.

The Plackett-Burman design is often used in multifactor studies due to its simplicity, time savings, and relatively low cost. However, researchers must keep in mind that because the Plackett-Burman design uses a small number of experiments, the main effects are susceptible to be confounded by two-factor interactions. For this reason, it is best used as a preliminary screening to identify factors of interest for further study.

Pamelyn Witteman, PhD

FURTHER READING

Box, George E. P., et al. *Statistics for Experimenters: Design, Innovation, and Discovery*. 2nd ed., Wiley-Interscience, 2005.

Creswell, John W. *Research Design: Qualitative, Quantitative, and Mixed Methods Approaches*. 4th ed., Sage Publications, 2014.

Field, Andy. *Discovering Statistics Using SPSS*. 3rd ed., Sage Publications, 2009.

Orcher, Lawrence T. *Conducting Research: Social and Behavioral Science Methods*. 2nd ed., Taylor & Francis, 2014.

Patten, Mildred L., and Michelle Newhart. *Understanding Research Methods: An Overview of the Essentials*. 10th ed., Routledge, 2017.

Swanson, Richard A., and Elwood F. Holton III, editors. *Research in Organizations: Foundations and Methods of Inquiry*. Berrett-Koehler Publishers, 2005.

PLACKETT-BURMAN DESIGN SAMPLE PROBLEM

A researcher needs to examine which factors led to a motor vehicle accident. The researcher determines that the accident could have been influenced by eleven different factors: weather, road damage, driver distraction, driver skill, vehicle speed, vehicle condition, driver age, vehicle performance, vehicle instruments, vehicle width, and/or driver health. How many experimental runs would the researcher need to do to determine the main effects of the factors using a Plackett-Burman design?

Answer:

The number of runs for a Plackett-Burman design must be greater than the number of factors. The number of runs must also be a multiple of four. Therefore, the computation begins as follows:

$$11 + 1 = 12$$

The number 12 is divisible by 4, so the researcher need only conduct twelve runs. If (factors + 1) is not divisible by 4—for example, if there were ten factors, giving an answer of 11—then the next higher number that is divisible by 4 would be the number of required runs. In this case, the answer would also be 12.

Positivist Methods

FIELDS OF STUDY

Research Theory; Research Design

ABSTRACT

Positivist methods of research are those that focus on quantitative data and the scientific method. Positivist researchers rely on careful observation and measurement, diligent data recording, and statistical or mathematical analysis. As a philosophy of research, positivism is most often referenced as the opposite of interpretivism in the social sciences. Both approaches are important to different kinds of research.

PRINCIPAL TERMS

- **interpretivism:** an approach to social science research in which the researcher interprets gathered data, using humanistic and qualitative methods.
- **objectivity:** the quality of being true according to external criteria, rather than according to individual perspective.
- **qualitative data:** information that describes but does not numerically measure the attributes of a phenomenon.
- **quantitative data:** information that expresses attributes of a phenomenon in terms of numerical measurements, such as amount or quantity.

- **replication:** the repetition of an experiment or study with the aim of achieving the same results in order to confirm findings or reduce error.
- **subjectivity:** the quality of being true only according to the perspective of an individual subject, rather than according to any external criteria.

Positivism versus Interpretivism

Positivist methods make up one of the two main branches of social research design. They are grounded in a theoretical paradigm that dates back to the ancient Egyptians but emerged as a significant element in Western philosophy and scientific thought during the mid-nineteenth century. In positivism, nature is a vast system of complex forces governed by equally complex laws. "Positive" here is used in the sense not of affirmative, but of definite or real. Scientists are not philosophers conjecturing about the nature of the universe or interpreting it for some meaning. Rather, scientists are observers of the natural world, meticulously recording empirical data, tracking its implications, and searching within it for patterns. This alone can give a reasonable sense of how natural phenomena work, if not why they work that way.

At the core of positivist methods, then, is the perception that a society is like any other natural phenomena. It abides by laws, and events happen for specific causes that have little to do with the individual people within it. Like natural scientists, positivist social scientists gather data, analyze it statistically or mathematically, and use results to cautiously predict future developments. Thus, positivist methods rely on quantitative data to provide work with appropriate dimension. Major goals of positivist research are objectivity and replication. Ideally, every other researcher investigating the same general body of subjects, in the same manner, will come to the same conclusions. In this way, data confirms that social phenomena are measurable and definable.

Positivism is best understood in contrast to interpretivism, its opposite in research theory. Interpretivism rejects the objective analysis of empirical data as a valid paradigm for social research. A more humanistic approach, it attempts to account for subjectivity in social data. Interpretivists view people, and therefore societies, as shaped by deeply individual perspectives that cannot be always be measured scientifically. Indeed, they argue that even the measurements used by scientists are socially constructed and therefore not absolutely objective. Interpretivism relies more on qualitative data, such as that from interviews, case studies, letters, and journals. Researchers themselves become involved with the data, bringing in a human factor of interpretation. Conclusions are intended to be illuminating rather than replicable.

Benefits and Burdens

As with any research, positivist methods aim to uncover the truth and gain factual information specific to the chosen variables in a study. Positivist research has the advantages of following the scientific method. Arguments are evidence-based and deductive. Studies can be replicated and conclusions tested. Such research is very useful for identifying broad patterns and trends.

However, positivist methods have their downsides. They fail to account for anything that cannot be precisely measured. Combined with the lack of attention to individuals, this can prevent researchers from truly understanding the full complexity of social phenomena. Overconfidence in quantitative data can cause researchers to miss nuance or deeper meaning behind the results in an experiment, or even introduce bias. In many forms of social research, some level of interpretivism is necessary.

Quantitative Social Research

Assume there is an interest in the relationship between millennials (those born between the early 1980s and 2000) and the institution of marriage. Marriage is a complex issue, and millennials are a highly diverse group, so there are many potential areas of research. A positivist, however, would only be interested in that which is measurable. For example, they might want to know if millennials are getting married at the same rate as previous generations.

Quantitative data would be gathered using a structured, scientific approach. The researcher would set a large sampling frame, defining a pool of subjects (all millennials in the United States, for example) to investigate. The emphasis would be on broad coverage. The preferred instruments of the positivist approach in social research are often large-scale surveys or questionnaires or official data such as census statistics. Such broad data tends to be representative and reliable and therefore more suited to positivist methods such as statistical analysis.

In this case, data might be collected from state marriage records and compared across age ranges. The study might reveal that millennials in the United States are holding off on marriage by an average of twenty months longer than the generation before. This result could then be used in further positivist research, such as comparison to similar studies in other countries. Assuming the collected data were accurate and the analysis was done correctly, results are clear, measurable facts. The question of whether millennials are holding off on marriage has been answered.

However, no further conclusions can be made with positivist methods. Understanding or analyzing why millennials are holding off on marriage is beyond the scope of the study. Interpretivism would likely be needed to dig deeper into the observed phenomenon and provide critical insights into wide patterns.

Understanding Positivism

Remember that the nature of positivist methods is to form a theory. A theory must be formed using hypotheses with measurable variables. The positivist approach provides quantitative data in a study or experiment in order to examine the specific relationship between chosen variables. This objective viewpoint is not only invaluable to much social research, but also serves as a major part of the foundation of science itself. Therefore, understanding positivism is essential for every researcher. However, researchers must also understand the limits of positivism and the role of subjectivity in scientific study, as represented by interpretivism or qualitative methods.

Joseph Dewey

Further Reading

Aliyu, A., et al. "Positivist and Non-Positivist Paradigm in Social Science Research: Conflicting Paradigms or Perfect Partners?" *Journal of Management and Sustainability*, vol. 4, no. 3, 2014, pp. 79–95, doi:10.5539/jms.v4n3p79. Accessed 11 Apr. 2017.

Comte, Auguste, and J. H. Bridges. *A General View of Positivism*. Routledge, 2015.

Heshusius, Louis, and Keith Ballard, editors. *From Positivism to Interpretivism and Beyond: Tales of Transformation in Education and Social Research*. Teachers College P, 1996.

Riley, Dylan. "The Paradox of Positivism." *Social Science History*, vol. 31, no. 1, 2007, doi:10.1215/01455532-2006-017. Accessed 11 Apr. 2017.

Tonon, Graciela, editor. *Qualitative Studies in Quality of Life: Methodology and Practice*. Springer International, 2015.

Trochim, William M. K. "Positivism & Post-Positivism." *Research Methods Knowledge Base*, 2006, www.socialresearchmethods.net/kb/positvsm.php. Accessed 11 Apr. 2017.

Posttest-Only Design

FIELDS OF STUDY

Experimental Design; Research Design

ABSTRACT

Posttest-only design is the simplest form of research design. At its most basic, a single experimental group receives an intervention, treatment, or test, and a posttest evaluation. The two-group form adds a control group that does not receive the intervention. Posttest-only designs are useful in cases where pretest-posttest design is not possible or practical.

PRINCIPAL TERMS

- **external validity:** the extent to which the results of a study can be generalized to apply to other populations and settings.
- **internal validity:** the extent to which the results of a study can confidently be attributed to the quality being tested, as determined by how well the study was designed to minimize systematic errors.
- **pretest:** a test performed to collect baseline information before application of a treatment or intervention in an experiment.
- **sample size:** the number of subjects included in a survey or experiment.

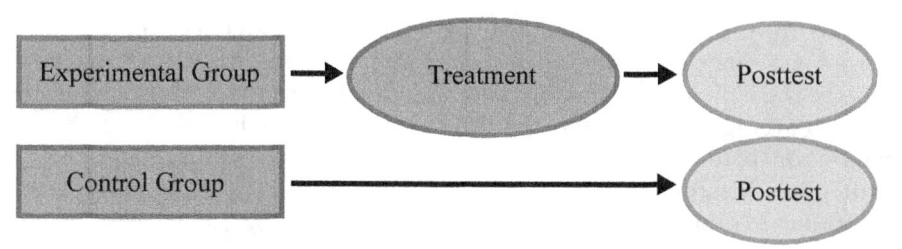

Using Posttests

Posttest-only design is an approach to research that differs somewhat from the more typical pretest-posttest design. In many experiments, researchers administer a pretest to the group being studied. This is then followed by some type of treatment or intervention and then a posttest that is identical to the pretest. This approach is designed to determine whether the intervention caused any change in the group being tested. As its name suggests, a posttest-only design does away with the pretest. It involves the researchers performing the intervention and then administering a test to measure its effects.

Posttest-only research can be conducted on just one group. However, often both an experimental group and a control group are used to improve the experiment's scientific rigor. These groups have equal numbers of members, or sample size, and subjects are randomly assigned to one of the two groups. The experimental group receives the treatment while the control group does not. Afterward, both groups are given a posttest. Any differences between the groups' performance on the posttest are analyzed to see if they are likely to have been caused by the intervention. Statistical analysis is commonly used to determine if a difference is likely to have been caused merely by chance or if it is probable that the difference can be attributed to the treatment. Often the performance of the two groups is represented as a bell curve on a two-dimensional graph. The distance between the two curves shows the variation in performance.

Effects on Validity

Research designs are evaluated on many criteria, two of which are internal validity and external validity. Internal validity is the reliability of the experiment itself, or the degree to which the quality or treatment being tested is likely to be responsible for the change that is observed. External validity is concerned with how reliably the results of the experiment can be generalized to the population from which the sample has been drawn. For example, if a trial shows that taking a new medication caused subjects' cholesterol to drop by 30 percent, it is important to find out if this supports the conclusion that the same results could be expected for others taking the drug. Ideally, a study will have high internal and external validity, meaning that it is accurately measuring the trait of interest and that its results can be generalized to the population at large.

In some cases, the use of a posttest-only design can have an influence on the validity of the study. Pretest-posttest designs are generally more rigorous because of their higher amount of control. However, there are subjects, particularly in the behavioral sciences, with which it is not possible or inadvisable to administer a pretest. This may be because the test involves unpleasant or unsafe experiences, or because the pretest would alter the test subjects too much compared to subjects in a real-world setting. In the latter case, the experimental results would be less generalizable. Such a situation would be an ideal one in which to use a posttest-only design and improve validity. Posttest-only designs can also reduce threats to validity by using a large sample size.

Limited Utility

Posttest-only designs are considered one of the weaker research approaches, so they tend to be used only in situations where an alternate design is not practical. Many examples come from medical literature, in cases where pretests are impossible or unethical. Take the case of a study that compares the recovery rates of smokers and nonsmokers who have contracted pneumonia. Obviously, it would not be appropriate to conduct a pretest by selecting groups of smoking and non-smoking participants, measuring the health of their immune systems, and then causing them to contract pneumonia in order to study their recuperation. This would unethically expose the participants to a high level of potential harm.

Instead, it is more appropriate to employ a posttest-only approach. Participants are selected based on their medical condition (having pneumonia) and non-randomly assigned to smoker and nonsmoker groups. Researchers can then review these patients' records to see how long each person's recovery took. Results might finally be applied to the overall population, making predictions about the effects of smoking on health, the immune system, and resiliency in general.

This approach has a number of potential weaknesses, however, which must be factored in before any conclusions are drawn. The lack of a pretest means that the researchers do not really know whether the study's participants began from comparable situations. Some may have had compromised immune systems at the outset, for example. Essentially, the simple structure of the posttest-only design comes at the price of many other factors that researchers must consider.

An Important Tool

Despite their drawbacks, posttest-only designs are important for researchers to be aware of for several reasons. Their simplicity makes them easy for those new to research design to understand and learn. Their shortcomings help to explain why more complex methodologies have been developed. Finally, and importantly, there are some situations in which posttest-only design may be the only option available. In those scenarios, there may be vital information that the posttest-only design makes accessible. At the same time, researchers must be cautious and proceed methodically to account for the weaknesses of any research design before placing too much faith in the findings. In that sense, posttest-only research is often most useful for its ability to point toward phenomena in need of further study.

Scott Zimmer, JD

Further Reading

Christensen, Larry B., et al. *Research Methods, Design, and Analysis.* 12th ed., Pearson, 2015.

DePoy, Elizabeth, and Laura N. Gitlin. *Introduction to Research: Understanding and Applying Multiple Strategies.* 5th ed., Elsevier, 2016.

Edmonds, W. Alex, and Thomas D. Kennedy. *An Applied Reference Guide to Research Designs: Quantitative, Qualitative, and Mixed Methods.* Sage Publications, 2013.

"Evaluation Designs: Posttest Only." *CYFAR*, U of Minnesota, cyfar.org/evaluation-designs-posttest-only. Accessed 11 Apr. 2017.

Salkind, Neil J., editor. *Encyclopedia of Research Design.* Sage Publications, 2010.

Trochim, William M. K. "Posttest-Only Analysis." *Research Methods Knowledge Base*, 2006, www.socialresearchmethods.net/kb/statsimp.php. Accessed 11 Apr. 2017.

Pretest-Posttest Experimental Research Design

FIELDS OF STUDY

Research Design; Experimental Design; Hypothesis Testing

ABSTRACT

The pretest-posttest experimental design is a simple way of testing the efficacy of a treatment or intervention. It is one of the most common forms of scientific research. Participants in the experiment are divided into two groups. In the pretest, both groups are tested to measure the characteristic of interest. One group then receives a treatment, and then both groups are retested in the posttest. If a difference is observed, it may be due to the treatment.

PRINCIPAL TERMS

- **attribute:** in experimental research, a characteristic or quality possessed by a subject in an experiment and represented by a variable.
- **control group:** the subjects in an experiment who do not receive any intervention or who receive a placebo or sham intervention.
- **dependent variable:** in experimental research, a measurable occurrence, behavior, or element with a value that varies according to the value of another variable.
- **experimental group:** the subjects in an experimental research study who receive the intervention or treatment being tested.
- **independent variable:** in experimental research, a measurable occurrence, behavior, or element that is manipulated in an effort to influence changes to other elements.

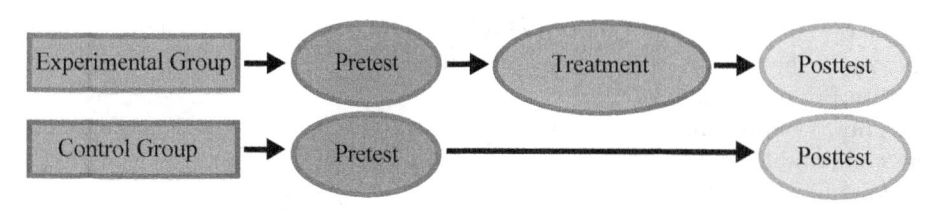

Improving Research Design

Pretest-posttest experimental design is a widely used method of scientific research. At its core, such research attempts to evaluate the effect of an intervention on an attribute of interest. This attribute, however it is measured, is treated as the dependent variable in an equation in which the independent variable is the intervention or treatment being studied. A basic pretest-posttest design uses only one study group. The group is given a pretest, the intervention is applied, and then the group is given a posttest to measure any difference. While this simple form, like even simpler posttest-only designs, has its uses, it is not considered robust and its results are limited.

To be considered a true experimental design, pretest-posttest research must add a control group. The control group does not receive the intervention; its purpose is to demonstrate what would happen if no treatment were given. The experimental group receives the treatment, and both groups are given a posttest. Ideally, subjects are recruited and randomly assigned to either the experimental group or the control group in order to reduce bias.

The experimental group's posttest results can then be compared not only to its pretest results, but also to the control group's posttest and pretest results. Statistical analysis is often used to determine whether results are significant. This better informs the hypothesis that the attribute under study is dependent upon the intervention. The two-group control group pretest-posttest design is the classic controlled experiment.

Advantages and Disadvantages

The two-group control group pretest-posttest experimental design is a benchmark in scientific research. It is relatively simple and easy to set up. Yet it is much more robust than a single-group pretest-posttest design or a posttest-only design. The use of a control group provides greater internal validity (the degree to which the results can be confidently attributed to the intervention). Confounding variables that might skew results are minimized, especially if randomization of the subjects is possible.

Pretest-posttest designs do have drawbacks. Most importantly, the gains in internal validity come at the expense of some degree of external validity. External validity refers to the generalizability of study results, or how the results can be applied to a general population. Applying a pretest to subjects inherently makes them different from the general population, which would not normally get a pretest, thereby potentially reducing external validity. Another major problem, especially in sociology, education, and medicine, is ethical considerations. It may be unethical, impractical, or impossible to fully isolate or randomize subjects, especially if they are people. This can introduce confounding variables or selection bias. In some cases, only a single-group study or a posttest-only design may be possible.

Pretest-Posttest Research in Education

The two-group control group pretest-posttest design is often used in educational research into strategies to improve student performance. Take a scenario in which teachers are seeking a way to help students improve their grades in algebra. One suggestion is to offer daily tutoring after school, but proof is needed that this will have a positive effect. To study the effect of tutoring, teachers would randomly assign students into experimental and control groups. To create a baseline, all of the students would be given an initial assessment to determine their level of proficiency. This is the pretest.

Next, the experimental group would receive daily tutoring to help them with algebra. The control group would not. This would continue for a predetermined time, after which both groups would again be given a test to measure their knowledge of algebra. The average score of each group on this posttest would be compared with their average score on the pretest, as well as with the scores of the other group. If the experimental group's posttest scores show no improvement (or worsen), this suggests tutoring is ineffective. If the experimental group's posttest scores are higher than their pretest scores and the posttest scores of the control group, the results may suggest tutoring is effective. This might be convincing enough evidence to hire tutors or at least to study the tutoring program further.

The researchers would need to be careful to make sure that any changes in scores were due to the tutoring and not to other factors. Statistical analysis of the results can help account for the effects of chance. Factors such as students dropping out of the study or unusually high or low scores can be controlled for. However, others may be difficult or impossible to control. External validity may be compromised, as the pretest itself may have an impact on students' posttest scores or even their motivation. Furthermore, there may be no practical way to prevent students in the experimental group from mingling with students from the control group, potentially skewing results. Using groups from different schools could fix this problem but removes randomization and thus risks selection bias.

USEFUL BUT LIMITED

The pretest-posttest design is one of the most common forms of scientific research. From the education to medicine, pretest-posttest research underlies many fields that have direct influence on daily life. It remains an invaluable tool for countless researchers. Yet though the basic structure of this research design is well-known and thoroughly tested, its limitations mean it must be used with caution. If those limitations and potential errors are understood, they can be accounted for properly. Other more complex and costly designs, such as the Solomon four-group method, can be substituted as appropriate. Having a full understanding of the methods used helps researchers to avoid flawed results and conclusions and members of the public to appreciate the value and limitations of research findings.

Scott Zimmer, JD

FURTHER READING

Christensen, Larry B., Burke Johnson, and Lisa Anne Turner. *Research Methods, Design, and Analysis.* 12th ed., Pearson, 2015.

DePoy, Elizabeth, and Laura N. Gitlin. *Introduction to Research: Understanding and Applying Multiple Strategies.* 5th ed., Elsevier, 2016.

Edmonds, W. Alex, and Thomas D. Kennedy. *An Applied Reference Guide to Research Designs: Quantitative, Qualitative, and Mixed Methods.* Sage Publications, 2013.

"Evaluation Designs: Pretest/Posttest." *CYFAR*, U of Minnesota, cyfar.org/common-evaluation-designs-continued-0. Accessed 12 Apr. 2017.

Knapp, Thomas R. "Why Is the One-Group Pretest-Posttest Design Still Used?" *Clinical Nursing Research*, vol. 25, no. 5, 2016, pp. 467–472, doi:10.1177/1054773816666280. Accessed 12 Apr. 2017.

Salkind, Neil J., editor *Encyclopedia of Research Design.* Sage Publications, 2010.

PROBABILISTIC SAMPLING

FIELDS OF STUDY

Sampling Techniques; Sampling Design

ABSTRACT

Probabilistic sampling is a class of methods by which researchers select individuals or groups of individuals from a larger population in order to study them. These methods are called "probabilistic" because they are designed to ensure that every member of the population has a chance, or probability, of being included in the sample to be studied. Probabilistic sampling methods include simple random sampling, stratified random sampling, systematic random sampling, and cluster sampling.

PRINCIPAL TERMS

- **cluster sampling:** a probabilistic sampling method in which the population being studied is divided into subsets called "clusters" based on a shared characteristic unrelated to the research question, such as geographic location, and then multiple clusters are selected at random for comprehensive examination.
- **random sampling:** one of several probabilistic sampling methods in which individuals within

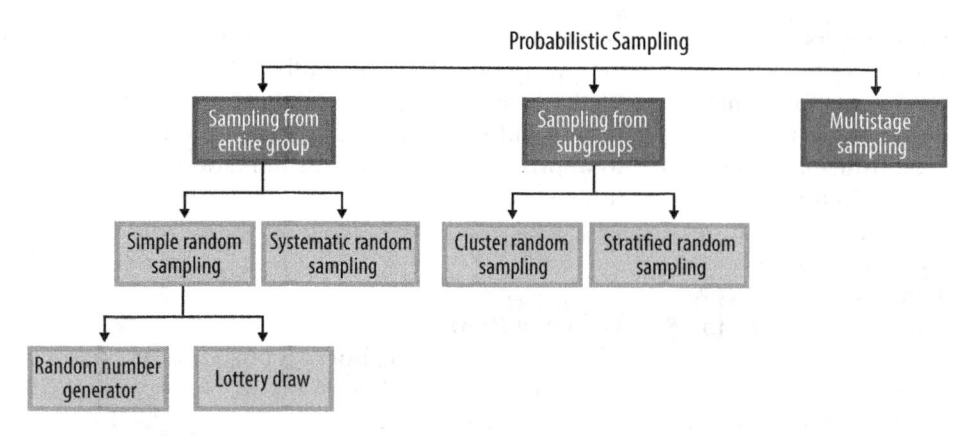

the population being studied are selected for examination at random, so that each member of the population has a chance (often an equal chance) of being included.

STUDYING GROUPS WITH PROBABILISTIC SAMPLING
When researchers study large groups, or populations, they typically do not study every member of the population. In most cases, doing so would be too time consuming and expensive. Instead, researchers use a process called sampling in order to select a representative portion of the population to study. They then use the data they gather from this sample to estimate characteristics of the larger population.

The sampling process begins by identifying the population to be examined. Next, a sampling frame must be selected. A sampling frame is a source that identifies all members of the population to be studied. For example, take a polling firm trying to find out which candidate had the greatest support among eligible voters in a national election. The population being studied would be all eligible voters. The sampling frame might be the voting registration rolls from each voting district.

Once the researchers have chosen the sampling frame, they then decide which sampling method to use. Sampling methods can be divided into two types, probabilistic and non-probabilistic. A probabilistic sampling method is one that uses some form of random selection. In probabilistic sampling, every member of the population has a calculable nonzero probability of being included in the sample. Non-probabilistic methods do not use random selection.

Probabilistic sampling methods include simple random sampling, stratified random sampling, systematic random sampling, and cluster sampling. In simple random sampling, each member of the population has an equal chance of being included in the sample. For example, each member could be assigned a unique number, and then a random number generator could be used to select which members to include in the sample.

Stratified random sampling divides the population into representative subgroups, or strata, based on mutually exclusive criteria, such as age, education level, or town of residence. No individual can be assigned to more than one stratum. Individuals are then chosen at random from each stratum, in numbers proportionate to the makeup of the population as a whole. For example, if the members of one stratum make up 30 percent of the overall population, and the researchers are choosing a total of one hundred individuals to study, then thirty people will be chosen from that stratum.

In systematic random sampling, all members of the population are first sorted into an ordered list. Then members are selected from the list at regular intervals. For example, the researchers might order the names of all population members alphabetically, then choose every tenth name on the list. The size of the interval is determined based on the sample size desired.

Cluster sampling first divides the population into subgroups called clusters, usually based on geographic location. A random sample of these clusters is then taken, and every member of each cluster in the sample is studied.

With all statistical sampling methods, the possibility remains that the sample will not accurately represent the entire population. This is known as sampling error. For example, take a population with an equal number of men and women. It is possible that a randomly selected sample might contain only men or only women, thus misrepresenting the population. In general, the larger the sample size, the lower the sampling error and the greater the likelihood that the sample will accurately represent the entire population. Appropriate sample sizes can be determined statistically using formulas, tables, and charts.

Comparing Sampling Methodologies

Simple random sampling is straightforward and relatively easy to implement. However, it may not be the best choice when the population contains smaller subgroups that need to be represented for a complete understanding of the research question. In such cases, stratified random sampling would be a better choice. This method can be used to ensure that the subgroups are properly represented. Systematic random sampling can be useful if a population is particularly large and other random sampling methods would be too laborious or time consuming. Cluster sampling is typically used when a population is widely dispersed geographically and members of the population cannot be reached remotely. In such cases, cluster sampling reduces the travel necessary to reach all members of the sample.

Note that sampling error can be introduced at any point in the sampling process. For example, researchers might fail to identify the correct population or choose the wrong sampling frame. Alternatively, a number of the chosen sample might decline to participate in the study when contacted.

In some situations, it may be preferable to use non-probabilistic sampling methods. For example, if a researcher wants to determine all possible answers to a question, a non-probabilistic method such as heterogeneity sampling might be more appropriate.

Probabilistic Sampling in Action

A researcher for a large international corporation is trying to determine the level of customer satisfaction among customers living in the European Union. This is a large population that includes millions of members, so contacting each customer is not practical. Therefore, selecting a sample from the larger population is necessary. Customers selected will be contacted via telephone or e-mail, so researchers will not need to visit each customer in person. Because of this, cluster sampling will not be used, as it would provide no apparent benefit. Systematic random sampling is not necessary, because the researchers do not anticipate any difficulty locating and contacting randomly chosen subjects.

Simple random sampling could be used. However, there are a relatively small number of customers in some EU nations, so simple random sampling may fail to represent those customers effectively. Therefore, stratified random sampling might be a better choice. It ensures that each member nation will be represented in a manner proportional to its population.

The Power of Probabilistic Sampling

Probabilistic sampling allows researchers to effectively study large groups that would be difficult to study accurately in any other way. It can be used to answer important research questions in fields as varied as advertising, political polling, medical research, engineering, and the physical and social sciences. Improving computer and communication technologies allow ever larger amounts of data to be collected, stored, and retrieved. Probabilistic sampling allows researchers to make effective use of such data to solve problems, test theories, and better understand the subjects they choose to study.

Maura Valentino, MSLIS

Further Reading

Blair, Edward, and Johnny Blair. *Applied Survey Sampling.* Sage Publications, 2015.

Fuller, Wayne A. *Sampling Statistics.* John Wiley & Sons, 2009.

Grosof, Miriam Schapiro, and Hyman Sardy. *A Research Primer for the Social and Behavioral Sciences.* Academic Press, 1985.

Kish, Leslie. *Survey Sampling.* 1965. John Wiley & Sons, 1995.

Thompson, Steven K. *Sampling.* 3rd ed., John Wiley & Sons, 2012.

Weisberg, Herbert F., et al. *An Introduction to Survey Research, Polling, and Data Analysis.* 3rd ed., Sage Publications, 1996.

Proofs

FIELDS OF STUDY

Research Theory

ABSTRACT

A proof is a series of structured arguments that uses formal logic to establish the validity of a statement, typically a mathematical statement. Proofs use deductive reasoning to establish validity beyond doubt. Proofs proceed step by step, with each step providing either a statement of an accepted axiom or theorem or an argument that is logically derived from the step preceding it.

PRINCIPAL TERMS

- **deductive reasoning:** a type of reasoning in which general rules are applied to a specific situation in order to draw a narrower conclusion; if the premises on which the conclusion is based are true, the conclusion is also deemed to be true.
- **inductive reasoning:** a type of reasoning in which knowledge of specific situations is used to draw a broader and more general conclusion; if the premises on which the conclusion is based are true, the conclusion is deemed to be highly probable but not certain.
- **mathematical proof:** a logical and valid argument that demonstrates the truth of a mathematical statement.
- **rule of inference:** an established argument structure that can be used to justify a step in a formal proof.

Mathematical Proofs

A proof is a specially formed argument or series of arguments that provides logically sound evidence for a statement. Proofs use deductive reasoning to proceed through a series of statements, each of which is either an established axiom or a logical derivation from the preceding statement. Proofs can be constructed using any formal system of abstract thought. However, the most common types of proofs are mathematical proofs.

The conclusion, or final statement, of a proof is called a theorem. An axiom, or postulate, is one of a number of specific predefined assertions that are held to be true. For example, all right angles are equal to one another, and a straight line can be drawn between any two points. Much of what is self-evident to the layperson about mathematics is not logically provable and must instead be made axiomatic. A set of axioms used to logically derive theorems is called an axiomatic system. According to Kurt Gödel's incompleteness theorems, no axiomatic system containing basic arithmetic is capable of proving all the true statements about the natural numbers within that system.

The reasoning employed in mathematical proofs—and, by extension, the certainty inherent in their conclusions—differs from that used in scientific research. Proofs are deductive, save for rare exceptions that establish degree of certainty rather than certainty beyond doubt. In contrast, scientific research often makes use of inductive reasoning. (Mathematical induction should not be mistaken for inductive reasoning, as it is, confusingly, a form of deductive reasoning used in formal proofs.) In many cases, a scientist will form a hypothesis by extrapolating from prior observations to a more general proposed truth. Then an experiment is conducted in an attempt to disprove that hypothesis. If extensive, repeated testing fails to disprove a hypothesis, it may be accepted as a scientific theory. However, unlike a mathematical statement, a hypothesis can never be proved true with complete certainty; it can only repeatedly fail to be proved false.

The Challenge of Proofs

There are multiple possible ways of proving any given (valid) statement. However, all proofs must proceed from the premises (statements that can be assumed to be true, such as axioms or accepted theorems) to the conclusion using rules of inference. Additionally, most mathematicians have a sense of aesthetics that finds certain styles of proof more elegant than others. A shorter proof is not always better than a longer proof, depending on the steps involved.

Constructing a proof can be difficult. It is easy to make a simple mistake, such as skipping a necessary logical step or including an operation that results in division by zero. There are countless axioms and proven theorems that may be drawn on. This means much knowledge is required to construct meaningful proofs and do so well, much less elegantly. Large proofs pose their own difficulties. While software can be used to assist in the verification of large proofs (in some cases consisting of hundreds of thousands of statements), this verification is post facto and still requires the mathematician to chart a vast conceptual course.

The Elegance of Proofs

Proofs are one of the most frequently cited examples of mathematical beauty. Paul Erdős, a prolific and influential mathematician of the twentieth century, imagined God keeping the most elegant proofs in a book (although Erdős was actually agnostic). Though both mathematical beauty in general and the elegance of proofs in particular are subjective concepts, as with more conventional forms of art, there are certain masterpieces about which a consensus has formed.

Elegant proofs usually have certain elements in common. A proof that consists only of axioms qualifies for one kind of elegance. Also elegant is a proof that can surprise the reader in some way. In this sense, appreciation of mathematical beauty is like any other kind of artistic appreciation, informed by the viewer's familiarity with the art form. Greater understanding typically imparts heightened appreciation. While anyone can appreciate the beauty of a painting, art connoisseurs who are well versed in color theory and recognize the artist's influences are able to appreciate aspects of that painting that will escape the casual viewer.

For example, Euler's identity, named for Swiss mathematician Leonhard Euler, is widely considered an exemplar of mathematical beauty. The equation can be appreciated by people with only a high school mastery of mathematics. Its elegance derives in part from the fact that it links so many disparate and important mathematical concepts in such a short statement:

$$e^{i\pi} + 1 = 0$$

However, just as a painting can be appreciated on more than one level, learned mathematicians consider the true elegance of Euler's identity to lie in its proof. Polls of working mathematicians have named it the most beautiful proof in mathematics. A neurological study of mathematicians' experience of mathematical beauty found that they responded to the proof in the same way that non-mathematicians respond to other forms of art, such as music and poetry.

The Necessity of Proofs

Both proofs themselves and the standards applied to them are important to modern mathematics and science. Proofs are important to mathematics for the same reason that scientific theories are important to the sciences: the more things that are proved, the greater the body of knowledge becomes. Understanding those proofs, by extension, means having a better understanding of the theorems they prove, and of the workings of mathematics in general. Furthermore, although the testing of scientific hypotheses and the construction of proofs follow different reasoning processes, both make similar demands in terms of rigor, attention to detail, and internal consistency. Finally, a solid understanding of mathematics is important for research work in most sciences. Historically, the professionalization of the sciences has meant the quantification of scientific work, including the adoption of quantitative research in the social sciences.

Bill Kte'pi, MA

Further Reading

Agler, David W. *Symbolic Logic: Syntax, Semantics, and Proof.* Rowman & Littlefield Publishers, 2013.

Andrews, Peter B. *An Introduction to Mathematical Logic and Type Theory: To Truth through Proof.* 2nd ed., Kluwer Academic Publishers, 2002.

Barker-Plummer, Dave, et al. *Language, Proof, and Logic.* 2nd ed., CSLI Publications, 2011.

Chartrand, Gary, et al. *Mathematical Proofs: A Transition to Advanced Mathematics.* 3rd ed., Pearson, 2013.

Lakatos, Imre. *Proofs and Refutations: The Logic of Mathematical Discovery.* 1976. Edited by John Worrall and Elie Zahar, Cambridge UP, 2015.

Polya, George. *How to Solve It: A New Aspect of Mathematical Method.* 1945. Foreword by John H. Conway, expanded Princeton Science Library ed., Princeton UP, 2014.

Shoenfield, Joseph R. *Mathematical Logic.* 1967. A K Peters, 2001.

Wolf, Robert S. *Proof, Logic, and Conjecture: The Mathematician's Toolbox.* W. H. Freeman, 1998.

Prospective Cohort Design

FIELDS OF STUDY

Research Design; Sampling Design

ABSTRACT

Prospective cohort design is an approach to research that tracks a group of subjects over time, usually for years or even decades. The purpose of this tracking is to determine which of the participants develop, or fail to develop, the outcomes that are the subject of the research. Ideally, this will allow inferences to be made about what factors lead to the development of these outcomes.

PRINCIPAL TERMS

- **cohort:** a group of purposefully chosen research subjects who share one or more defining characteristics.
- **population:** the entire pool of individuals or other subjects that a researcher wishes to obtain information about.
- **posttest-only research:** a research design in which no pretest is administered prior to the treatment.

STUDIES OVER TIME

The goal of prospective cohort research is to observe the eventual outcomes of defined characteristics or suspected risk factors that members of a cohort, or sample group, already possess or are exposed to at the outset of the research. These characteristics or risk factors may be innate, such as a genetic predisposition to high cholesterol, or they may be acquired, such as exposure to a potentially carcinogenic chemical. This sample group is often compared to a control group of individuals, who are selected for observation because they have a similar background to members of the cohort but do not have or have not been exposed to the specific trait or risk factor being studied.

A distinguishing feature of prospective cohort studies is that the study begins after the characteristic or risk factor has been identified or the exposure has occurred, but before the subjects have developed the condition or outcome being studied. The goal is to monitor any outcomes that may develop in future. The researcher collects the same, specific set of data about each member of the cohort. This information serves as a baseline for future comparisons.

Prospective cohort studies are also called longitudinal studies because they track participants over an extended period of time. The participants are chosen intentionally, rather than through random sampling, because they share characteristics or past exposures of concern. These studies are particularly useful for research that seeks to better understand how a condition develops. They can help determine whether there are any qualities that could be useful in predicting who will develop the condition and who will not. The ultimate purpose of this type of research is often to identify specific populations that may be at higher risk for developing certain conditions.

PROSPECTIVE, NOT RETROSPECTIVE

Cohort designs differ from other research approaches in that they do not involve a treatment or intervention being applied to a group of subjects. Because of this, ethical concerns are less of a factor.

There are two main types of cohort designs, prospective and retrospective. A prospective cohort study begins before the development of the outcome of interest. It is a type of pretest-posttest research because the researcher is able to collect baseline information from the subjects before any changes are observed. One advantage of this method is that the researcher can control what initial data is collected, and therefore can more confidently establish a link between a characteristic or exposure and an outcome. In contrast, retrospective cohort studies begin after cohort members have developed the outcome of concern, and historical research is required to determine what shared exposures or risk factors could have caused it. Retrospective studies are a type of posttest-only research, because the researcher is not able to gather baseline information from the subjects prior to the outcome.

Prospective cohort design gives researchers more control over the participants involved and the methods used for data collection. They can also study more than one outcome at a time. However, because there is no randomization involved in sample selection, there is a greater risk of sampling bias affecting

Independent variable | **Dependent variable**

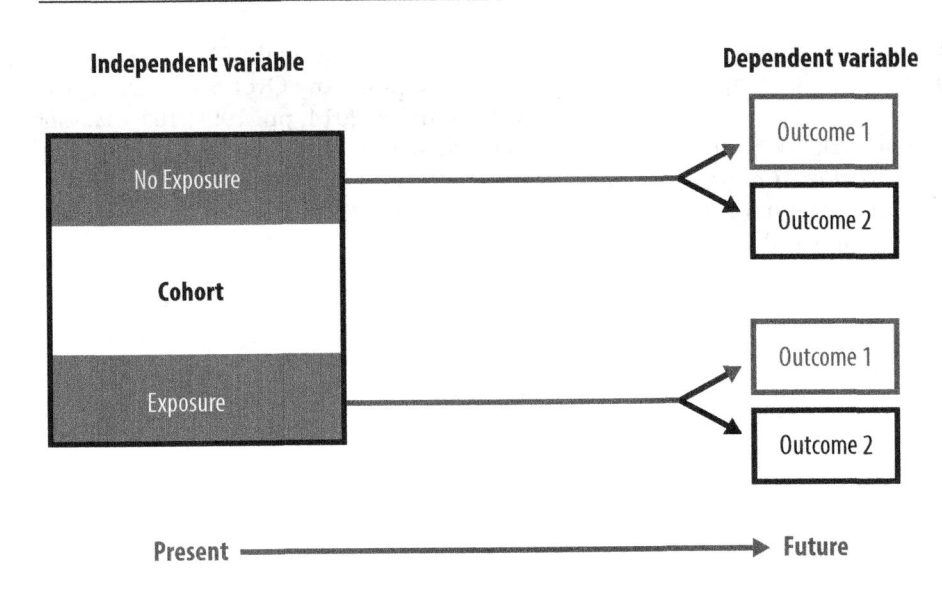

Cohort: No Exposure → Outcome 1, Outcome 2
Cohort: Exposure → Outcome 1, Outcome 2

Present → Future

the results. Another disadvantage of prospective cohort research is that the study must continue for a long enough period to allow some subjects to develop the condition. Such studies can last for years or even decades, and can therefore prove costly. There are also inevitably participants who drop out, die, move, and so forth, all of which threaten the validity of the study.

Use in Epidemiology

A familiar form of prospective cohort research is the kind that produces population-level predictions of medical conditions such as heart disease. It has become common knowledge that low levels of exercise and a high-fat diet lead to higher levels of cholesterol in the blood. This increases the risks of heart attacks, strokes, and high blood pressure.

However, there once was a time when none of this had been established. Doctors and scientists had theories about the possible connections that might exist between diet, exercise, and heart disease. In order to confirm them, they had to conduct prospective research. They recruited cohorts of participants, gathered from them basic information about their health and lifestyle, and then tracked their health over many years. Eventually, these studies allowed researchers to look for correlations between lifestyle factors and heart disease. Similarly, in the mid-nineteenth century, cases of lung cancer began steadily increasing at the same time that the number of Americans smoking had increased. Some scientists used prospective cohort studies to argue that there was a link between the two. E. Cuyler Hammond's 1952–55 and 1959–72 prospective studies, which followed cohorts of men and women and studied their smoking habits, provided sufficient evidence to identify smoking as a cause of lung cancer.

Potential of Prospective Cohort Studies

An understanding of prospective cohort research is important. A great deal of what is known about public health has been learned using this method. When trying to figure out a potential cause for a condition, the researcher can have more confidence in their results if they can observe the subjects from exposure to development. If carried out successfully, prospective cohort studies can reasonably determine incidence rates and potential risk factors for specific conditions or diseases. Even a single prospective study could save millions of lives by providing information that could lead to behavioral changes among large numbers of people.

Scott Zimmer, JD

Further Reading

Breeze, Elizabeth, et al. "Harnessing the Power of Cohort Studies for Dementia Research." *Journal of Public Mental Health*, vol. 14, no. 1, 2015, pp. 8–17.

Dos Santos Silva, Isabel. *Cancer Epidemiology: Principles and Methods*. International Agency for Research on Cancer, 1999.

Jarde, Alexander, et al. "Q-Coh: A Tool to Screen the Methodological Quality of Cohort Studies in Systematic Reviews and Meta-Analysis." *International Journal of Clinical and Health Psychology*, vol. 13, no. 2, 2013, pp. 138–46. *Academic Search Complete*, search.ebscohost.com/login.aspx?direct=true&db=a9h&AN=87706691&site=ehost-live. Accessed 14 Apr. 2017.

Lenney, Warren, et al. "The Problems and Limitations of Cohort Studies." *Breathe*, vol. 10, no. 4, 2014, pp. 307–11.

Mendes, Elizabeth. "The Study That Helped Spur the US Stop-Smoking Movement." *American Cancer Society*, 9 Jan. 2014, www.cancer.org/latest-news/the-study-that-helped-spur-the-us-stop-smoking-movement.html. Accessed 14 Apr. 2017.

Thiese, Matthew S. "Observational and Interventional Study Design Types: An Overview." *Biochemia Medica*, vol. 24, no. 2, 2014, pp. 199–210. *Academic Search Complete*, search.ebscohost.com/login.aspx?direct=true&db=a9h&AN=96680415&site=ehost-live. Accessed 14 Apr. 2017.

Q

Quadrat Sampling

FIELDS OF STUDY

Sampling Techniques; Sampling Design

ABSTRACT

Quadrat sampling is a method in which subjects in a specified area or volume are measured in order to estimate distribution over a larger area. It is commonly used by environmental researchers for inventorying the plant and/or animal life of an ecosystem or habitat. Using quadrat sampling, researchers section off an ecosystem into carefully measured units (most often in the shape of a squares) called quadrats and gather precise data on the biological life within each quadrat. Researchers can then make estimates of the ecological profile of the larger area based on those data.

PRINCIPAL TERMS

- **sampling area:** in quadrat sampling, a predetermined location that is selected within the larger environment under study and then sectioned off into smaller units that are randomly selected for examination.
- **sampling point:** the individual observed units, or subjects, (most often plant, insect, or animal life) within a sampling area.

The Need for Quadrat Sampling

Quadrat sampling is a cost-efficient method for studying the flora or fauna of an ecosystem or the geography of a large area. Once the ecosystem or location under study has been identified, researchers can then use random sampling to select a sampling area. Researchers then divide the sampling area into quadrats. Quadrats are premeasured units of the same size and shape. These units have traditionally been squares. However, quadrats can be any shape as long as they are uniform. The area or volume of the quadrat must be known.

The researchers closely examine each quadrat and record their observations. From the data collected, conclusions can be drawn about the larger environment. For example, researchers using quadrat sampling may study the distribution, variety, and abundance of certain plants, animals, soil types, rock formations, or minerals in a certain area.

Increasing awareness of the impact of development on ecosystems has promoted the need for accurate metrics of the environment. Field ecologists use quadrat sampling to measure an area's biological diversity. Rain forests, open fields, wetlands, seashores, off-shore underwater ecosystems, and agricultural tracts of land can be analyzed using quadrat sampling. Any environmental system, from the polar ice caps to a suburban backyard, can be subject to careful analysis to determine its biodiversity.

Counting every sampling point within an ecosystem is simply not practical or even realistic. For instance, counting all the snails in a state-protected bog or systematically inventorying all the clover stems in an industrial dairy pasture would be nearly impossible to complete with a reasonable degree of accuracy. Quadrat sampling represents a workable, standardized system for determining the makeup of an ecosystem.

Sectioning Off an Ecosystem Using Quadrats

Field ecologists developed a system for measuring the ecological profile of large ecosystems using a simple quadrat. Traditionally, a quadrat is a square frame. Quadrats are often made of plastic or wood. However, quadrats can also be created using wires or strings. A quadrat can measure virtually any size depending on the research project. However, the size of the quadrat can affect sampling precision. Thus, quadrat size is often determined by the subject under study. For small study subjects such as insects or lichen, a quadrat that measures 10 square centimeters may be selected. If studying shrubs, the researcher may divide the sampling area into quadrats that measure several square meters.

The researcher begins by determining the specific dimensions of the area under study. Then, the researcher sections off that area into sampling areas of equal size. The number of sampling areas and their dimensions depend on the size of the total area under study, the number of researchers available, and the time constraints of the project. The researcher can then number each sampling area. A random number table or a random number generator can be used to select which sampling areas will be examined. Random sampling reduces bias and improves the accuracy of the data compared to haphazard sampling. The researcher then places the quadrat frame down in the selected areas to begin the sampling process.

Nonambulatory Organisms

In the case of nonambulatory organisms, such as plants or trees, the researcher will count the number of specific organisms under study that are present in each selected sampling area. Some sampling projects, however, want a more general census of a certain plant within the quadrats. In this case, the researcher may simply record quadrats in which the organism of interest covers 50 percent or more of the quadrat's area. Whatever the goal, the process is not complicated. Researchers simply get down on their hands and knees and actually count the number of sampling points within each quadrat. The work is time-consuming and physically demanding. Observations must be careful and the record-keeping meticulous.

In addition to recording the number of sampling points, researchers may also note specific patterns of growth and distribution. For example, the researchers may record whether the flora is regularly spaced, clumped together, or randomly dispersed throughout the quadrat. Using this method, the researcher can provide a fairly accurate environmental profile of the entire system. The researcher can multiply the number of sampling points within each sampling area by the total number of sampling areas.

Ambulatory Organisms

Quadrat sampling is best suited to nonambulatory organisms. However, it is occasionally used to study slow-moving organisms, such as snails. When quadrat sampling involves faster-moving ambulatory species, the work is essentially the same. However, instead of counting organisms themselves, the researcher may look for evidence of animal activity. They may record footprints, claw or teeth marks, eggs or nests, patches of fur, dental remains, or entrails of prey. From that evidence, the researchers can determine the presence or absence of a species in the area under study.

Joseph Dewey

Further Reading

Brawer, James, Jerrold Zar, and Carl N. von Ende. *Field and Laboratory Methods for General Ecology.* 4th ed., McGraw-Hill, 1997.

Fox, Gordon A., et al., editors. *Ecological Statistics: Contemporary Theory and Application.* Oxford UP, 2015.

Gotelli, Nicholas J., and Aaron M. Ellison. *A Primer of Ecological Statistics.* 2nd ed., Sinauer, 2012.

Karban, Richard, et al. *How to Do Ecology.* 2nd ed., Princeton UP, 2014.

Smith, Thomas M., and Robert Leo Smith. *Elements of Ecology.* 9th ed., Pearson, 2014.

Weathers, Katherine, et al., editors. *Fundamentals of Ecosystem Science.* Harcourt/Academic Press, 2012.

Qualitative Research

Fields of Study

Research Theory; Research Design

Abstract

Qualitative research focuses on collecting observations, narratives, interviews, and other nonnumerical data, usually without the framework of hypothesis testing. The overall goal is to add to the body of knowledge about the area of inquiry rather than to gather evidence in support of explaining an observed phenomenon. It is particularly associated with disciplines that study human behavior.

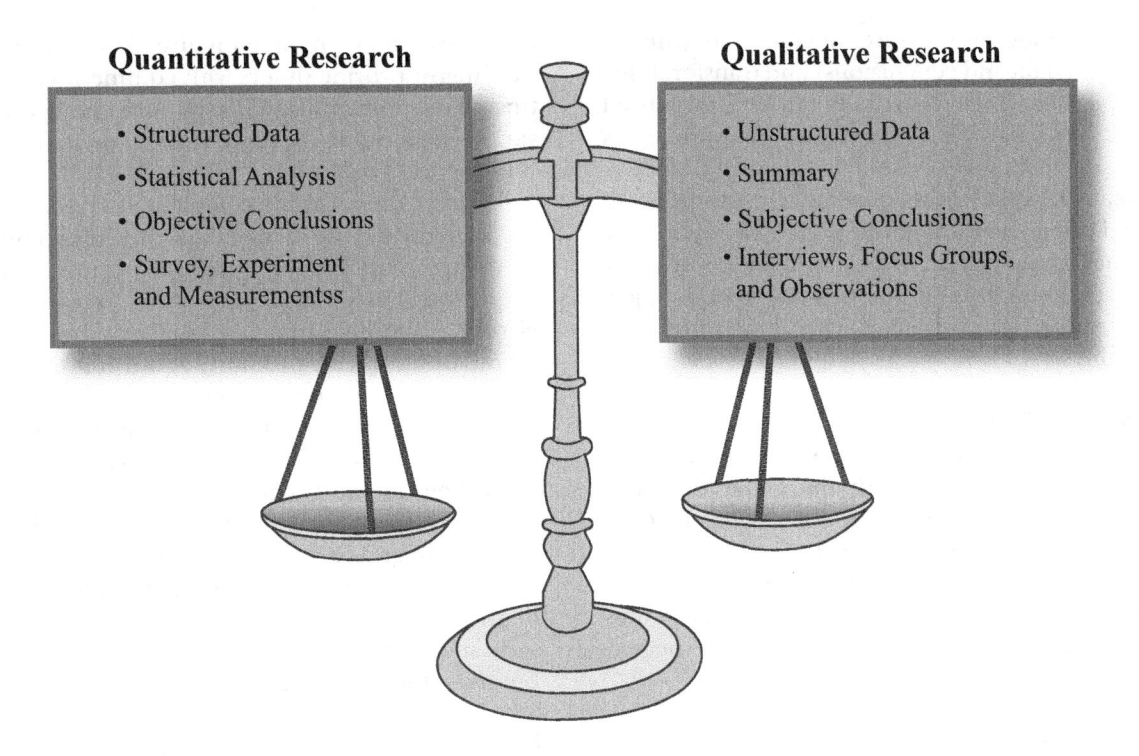

Quantitative Research
- Structured Data
- Statistical Analysis
- Objective Conclusions
- Survey, Experiment and Measurementss

Qualitative Research
- Unstructured Data
- Summary
- Subjective Conclusions
- Interviews, Focus Groups, and Observations

PRINCIPAL TERMS

- **bias:** an inclination toward a particular perspective or preconceived notion; also, in statistics, refers to a method of collecting or evaluating data that inherently and consistently produces results that inaccurately reflect the characteristics of the population or phenomenon being studied.
- **case study:** a research method in which selected individuals or small groups are observed and examined thoroughly.
- **credibility:** in qualitative research, the quality of trustworthiness or ability to inspire belief; analogous to internal validity in quantitative research.
- **quantitative data:** information that expresses attributes of a phenomenon in terms of numerical measurements, such as amount or quantity.
- **subjectivity:** the quality of being true only according to the perspective of an individual subject, rather than according to any external criteria.
- **transferability:** the degree to which the outcome of a research project can be generalized to contexts outside the project.

QUALITATIVE RESEARCH METHODS

Qualitative research focuses on subjective and often unstructured data, using exploratory methods to increase understanding of a subject. By contrast, quantitative data is most directly connected to the scientific method and the testing of hypotheses that have been formulated to explain observed phenomena. The two approaches naturally differ in their methods. Quantitative research is more likely to involve experimentation and statistical analysis. Qualitative research tends to rely on observation and data collection and interpretation. Where quantitative data is numerical or otherwise mathematically measurable, qualitative data emphasizes interviews, stories, and other subjective information. A common form of qualitative data is the case study, which collects relevant information on a particular individual or group. Case studies are used in fields as varied as ethnography, legal education, social work, business, and medicine.

Due to the subjectivity of its data, qualitative research is judged by different criteria. While quantitative research aims for objective measurement, qualitative research seeks trustworthiness. Instead of internal and external validity, qualitative research is judged by analogous measures of credibility (the believability of data and conclusions) and transferability (the degree to which the results can be generalized to other contexts). Many qualitative methodologies also seek to make subjective data more quantitative. For instance, coding sorts qualitative data into standardized categories, which can then be compared or statistically analyzed. An example is when a teacher reads unique student essays and, through both objective and subjective critique, assigns them numerical grade values.

Benefits and Burdens

One major drawback of qualitative research is that the subjective nature of both the generation and interpretation of data leaves it susceptible to bias. Though this is widely acknowledged, it can remain difficult to control for. Types of bias, and the severity of its effects, vary by area of inquiry, but meta-studies often reveal systemic research bias in unexpected areas.

Other disadvantages of qualitative research also exist. Many studies rely on unique information and therefore can never be reproduced exactly. Observations can be difficult to present in a clearly transferable form, even with coding, limiting their use as results.

Qualitative research nevertheless provides access to information that quantitative research cannot. No amount or quality of numerical data can reproduce the information in a well-conducted interview, narrative history, or in-depth observation. Qualitative research can provide much greater detail and nuance than quantitative research, making it invaluable when an inquiry does not have a yes or no answer. It tells a story, even if no hypothesis is proven. Furthermore, qualitative research is generally simpler to design, can use a smaller sample size, and requires less time and cost than quantitative research. For most researchers, qualitative and quantitative methods are used side by side.

Qualitative Research in the Real World

While qualitative research is used often in many fields, a simple example comes from customer service. Take a restaurant that seeks insight into how satisfied customers are. Some quantitative data may be available, such as the number of customers and the frequency of different orders over a given period, but these give an incomplete picture. As a first step, the staff might provide diners with comment cards to be turned in anonymously. These comments, expressing various subjective opinions, form qualitative data that can then be analyzed in various ways.

Perhaps the most basic analysis would be to identify general trends in the data. If eighty comment cards are received and sixty are positive, the staff could conclude that customers are happy overall. However, the data may not be that simple. A comment could remark positively about one menu item and negatively about another. Deeper analysis might reveal that everyone who orders steak loves it, but most people who order pasta are disappointed.

Coding might be used to better break down complex data. This could be done by specifying response categories on the comment cards or by the staff sorting responses into set categories. Comments might be coded as satisfied, unsatisfied, or neutral, with subcategories for food, service, and facilities. Entering the information into a spreadsheet would allow for quantitative analysis of the qualitative data. Meanwhile, notable comments or issues might be explored with further qualitative research by holding a focus group or interviews with specific customers.

The Need for Qualitative Research

Qualitative research may appear less conclusive than quantitative research, in which experiments either prove or disprove hypotheses. Yet neither form of research is better or worse. They are different and often highly complementary. Qualitative research is exploratory, usually seeking less to prove specific theories than to deepen knowledge and better inform researchers. Not everything can be explained by numbers and statistics, least of all human behavior. For this reason, qualitative research stands as a pillar of scientific inquiry, well suited to questions about motivations, opinions, decision-making processes, and the like.

Bill Kte'pi, MA

Further Reading

Corbin, Juliet, and Anselm L. Strauss. *Basics of Qualitative Research: Techniques and Procedures for Developing Grounded Theory.* 4th ed., Sage Publications, 2015.

Marshall, Catherine, and Gretchen B. Rossman. *Designing Qualitative Research*. 6th ed., Sage Publications, 2016.

Merriam, Sharan B., and Elizabeth J. Tisdell. *Qualitative Research: A Guide to Design and Implementation*. 4th ed., Jossey-Bass, 2016.

Miles, Matthew B., A. Michael Huberman, and Johnny Saldaña. *Qualitative Data Analysis: A Methods Sourcebook*. 3rd ed., Sage Publications, 2014.

Patton, Michael Quinn. *Qualitative Research and Evaluation Methods: Integrating Theory and Practice*. 4th ed., Sage Publications, 2015.

Seidman, Irving. *Interviewing as Qualitative Research: A Guide for Researchers in Education and the Social Sciences*. 4th ed., Teachers College P, 2013.

Quantitative Research

FIELDS OF STUDY

Research Theory; Research Design

ABSTRACT

Quantitative research collects and analyzes measurable data through specialized techniques. It may involve experimental research that seeks evidence to support a hypothesis about a phenomenon, or descriptive research that collects more information about the subject of study.

PRINCIPAL TERMS

- **bias:** an inclination toward a particular perspective or preconceived notion; also, in statistics, refers to a method of collecting or evaluating data that inherently and consistently produces results that inaccurately reflect the characteristics of the population or phenomenon being studied.
- **descriptive research:** research intended to produce an accurate description of the current state of its subject, whether a phenomenon or a population, without addressing when, why, or how the current state came to be.
- **experimental research:** research conducted by manipulating one variable and observing any resulting changes in one or more presumed dependent variables in order to identify any cause-and-effect relationships among them.
- **inference:** a step in reasoning that develops a conclusion from observations, whether by deduction, induction, or abduction.
- **qualitative data:** information that describes but does not numerically measure the attributes of a phenomenon.

EMPIRICAL INVESTIGATION

Quantitative research focuses on empirical investigation of an area of inquiry, through the collection, measurement, and analysis of quantifiable data. Specialized techniques are often used to manipulate that data in order to reveal specific types of information. This research may be either experimental or descriptive. Experimental research uses inference to try to establish a causal link between two or more variables in support of a hypothesis. In contrast, descriptive research simply collects more accurate information about the subject being studied. Because quantitative research involves less contact with subjects than does qualitative research, there is generally less chance of bias, though that risk is present in all research.

Experimental quantitative research follows the scientific method. Based on prior observation of the phenomenon being studied, a hypothesis is proposed to explain that phenomenon, usually in some variation of "if p, then q." An experiment is designed to test the hypothesis. At least one variable is manipulated, with the expectation that it will effect a change in another variable. Empirical data are collected during the experiment and analyzed to determine if they bear out the claims of the hypothesis. Experimental research can be classified according to where it takes place: in a controlled lab setting or in the real world (a field experiment).

Descriptive research can similarly be classified into subtypes, such as correlational and causal-comparative. Correlational research aims to establish a relationship between at least two quantifiable variables and analyze that relationship to make predictions. Descriptive research that seeks to establish a causal link between quantifiable variables through observation and analysis is causal-comparative. While

Quantitative Research

- Structured Data
- Statistical Analysis
- Objective Conclusions
- Survey, Experiment and Measurementss

Qualitative Research

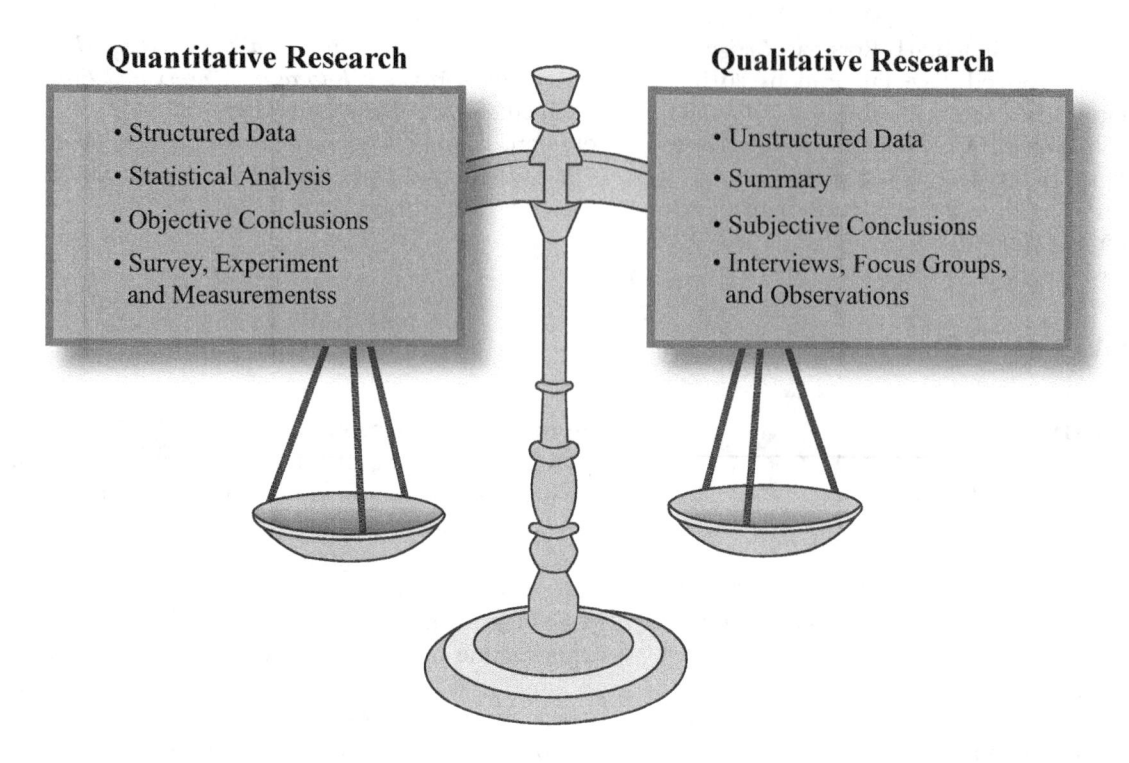

- Unstructured Data
- Summary
- Subjective Conclusions
- Interviews, Focus Groups, and Observations

it is harder to establish causation without experimentation, experiments are not always possible, particularly in the social sciences.

STRENGTHS AND DRAWBACKS

The main strength of quantitative research is its rigor. Quantitative research produces large amounts of data that tends to be more objective than qualitative data. There are many quantitative methodologies that are clear and reliable, and produce results that are likely to be generalizable. Experimental research lends itself best to quantitative research, making it standard in many fields. Research funding also tends to be most available to research that can clearly explain its results in quantifiable ways.

There are results that quantitative research simply cannot produce that qualitative research can, however. Not all areas of inquiry can be approached in terms of numbers and statistics. Qualitative research, which deals with narratives and other more subjective data, complements quantitative research and adds the details that raw numbers cannot. Another drawback of quantitative research is that the apparent objectivity of numbers can lead researchers to assume results are bias-free. Yet quantitative methods are susceptible to sampling bias and other systematic errors that are often difficult to find and correct. Finally, quantitative research is time consuming, statistically intensive, and costly.

THE US CENSUS

While quantitative research is most often associated with experimental research, the decennial US census is an example of how important descriptive quantitative research can be. A population census, it is intended to establish a count as accurate as possible of people living in residences in the United States. This includes citizens, noncitizen legal residents and long-term visitors, and unauthorized immigrants, as well as those without conventional housing. Most US citizens living outside the country are left out of the count, unless they are federal employees or their dependents.

Over the census's history, the level of detail it collects about the US population has varied. The Constitution requires at least a geographically accurate count by state. This is used to apportion representation in Congress and electoral votes. Historically, the census has also gathered demographic data, including individual traits (such as age, gender, and race) and household traits (such as how many adults and children live in each household). This information informs state and federal programs that take demographics into account, from education to social assistance. For privacy reasons, individual data are not publicly available for seventy-two years, but aggregate statistical data are released once tabulated. These provide social researchers and demographers with invaluable information about the US populace as a whole.

Raw census data are considered relatively accurate and vital to further research, both quantitative and qualitative. However, elements of the data are not fully objective. Most notably, census definitions of race and ethnicity have changed over the years, demonstrating their subjective nature. Partly due to controversies over such categories on past questionnaires, the planning stages of the 2020 census included the largest research project ever undertaken on the ways Americans self-identify by race and ethnicity. Such efforts seek to allow quantitative research on otherwise qualitative information.

The Foundation of Science

Quantitative research is the foundation of modern science. Even before a rigorous scientific method with hypothesis testing was devised, the collection of empirical data was synonymous with scientific activity in many cultures. Quantitative methods remain particularly associated with the physical sciences. Yet beginnings of quantitative research in the social sciences marked their professionalization, and quantitative methodologies continue to be vital to nearly every realm of science. Along with its qualitative counterpart, quantitative research is fundamental to the pursuit of knowledge.

Bill Kte'pi, MA

Further Reading

Creswell, John W. *Research Design: Qualitative, Quantitative, and Mixed Methods Approaches.* 4th ed., Sage Publications, 2014.

Hoy, Wayne K., and Curt M. Adams. *Quantitative Research in Education: A Primer.* 2nd ed., Sage Publications, 2016.

Martin, William E., and Krista D. Bridgmon. *Quantitative and Statistical Research Methods: From Hypothesis to Results.* Jossey-Bass, 2012.

O'Dwyer, Laura M., and James A. Bernauer. *Quantitative Research for the Qualitative Researcher.* Sage Publications, 2014.

Vogt, W. Paul. *Quantitative Research Methods for Professionals in Education and Other Fields.* Pearson, 2006.

Wrench, Jason S., et al. *Quantitative Research Methods for Communication: A Hands-On Approach.* 3rd ed., Oxford UP, 2016.

Random Sampling Error

FIELDS OF STUDY
Sampling Design; Sampling Techniques; Experimental Design

ABSTRACT
Random sampling errors are the result of differences between random samples and the populations from which they are taken. When these errors are statistically significant, they skew the results of research. Various methods exist to reduce their effect, mainly through improved sampling techniques. Undetected sampling errors can mislead researchers as to the accuracy of the study's results.

PRINCIPAL TERMS

- **type I error:** a mistaken conclusion that an experiment has provided support for the experimental hypothesis (i.e., that a relationship exists between the variables under study), leading to the incorrect rejection of the null hypothesis.
- **type II error:** a mistaken conclusion that an experiment has provided support for the null hypothesis (i.e., that no relationship exists between the variables under study), leading to the incorrect rejection of the experimental hypothesis.

Sampling Errors in Experimentation
Random sampling errors occur when there is a difference between the measured traits of a sample and the actual incidence of those traits in the total population from which the sample was drawn. When only a single simple random sample is taken, random sampling error is all but inevitable, unless the population is homogenous with respect to the trait in question. There is no reason to think that a random selection of subjects from the population will possess the same distribution of characteristics as the population as a whole.

There are various strategies for reducing sampling error. The true danger of sampling error occurs when it is undetected. An undetected sampling error will mislead researchers about the results of their research.

Sampling errors are often discussed in terms of their relationship to the null hypothesis. In experimental research, the null hypothesis is the prediction that there is no relationship between the independent and dependent variables under study. It encompasses all other possible outcomes not stated in the research hypothesis. In other words, the null hypothesis holds that the experiment's hypothesis is wrong. A **type I error** is the mistaken belief that the experimental hypothesis is correct. Type I error describes the incorrect rejection of the null hypothesis. A **type II error** is the mistaken belief that the null hypothesis is correct. Type II error describes the incorrect acceptance of the null hypothesis.

Type I and type II errors are relevant across all fields of research. Less universal are type III errors. A type III error is the correct rejection of the null hypothesis for the wrong reason. Type III errors occur when researchers develop accurate conclusions based on faulty evidence, due for instance to a random sampling error. Alternately, some researchers have used type III error to refer to a conceptual error unrelated to the data or its interpretation. These researchers consider type III errors to describe having solved the wrong problem or having found the right answer based on the wrong questions. Although the results of a type III error are correct, due to an error of vision or conception, the results will not be useful for whatever larger goal the researchers had intended.

Reducing Sampling Error
The simplest way of reducing sampling error is to increase the size of the sample. The larger the sample, the greater the probability that the sample will have parameters in common with the larger population, and the closer it will come to accurately representing

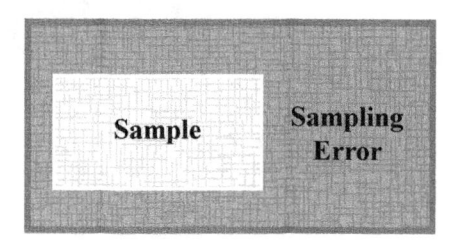

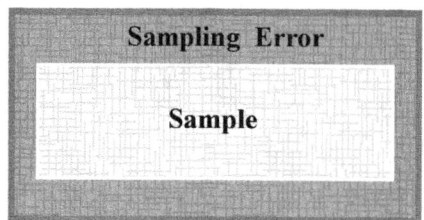

that population. A sample that is equal to the size of the entire population would contain no errors. A sample consisting of just one person would have the maximum possible margin of error. In experimental design, individual members of a sample are called replicates, and increasing the sample size to avoid sampling error is called replication. The repetition of the experiment on larger groups of people increases the validity of the results.

Another common way to reduce random sampling error is to use more sophisticated random sampling techniques. Instead of simple random samples, in which every member of a population has an equal probability of being selected for study, more sophisticated samples can be constructed in stages using one or more methods. Stratified random sampling divides the population into groups according to specific traits and then samples from those groups in proportion to their representation in the population. Cluster sampling divides a population into clusters based on shared characteristics, most often geographical area, and randomly selects some of the clusters, which are either surveyed in their entirety or randomly sampled a second time. Systematic random sampling lists all members of the population in order and then selects individuals for the sample at constant, predetermined intervals.

The Danger of Sampling Error

Research depends first and foremost on the accuracy of its data. In many cases, obtaining that data requires randomly sampling from a larger population in order to create a data set of manageable size. These random samples introduce the possibility of errors, which must be managed and corrected in order to ensure the integrity of the research and the validity of the results.

Bill Kte'pi, MA

Further Reading

Banerjee, Amitav, et al. "Hypothesis Testing, Type I and Type II Errors." *Industrial Psychiatry Journal*, vol. 18, no. 2, 2009, pp. 127–31.

Biemer, Paul P., et al., editors. *Total Survey Error in Practice.* John Wiley & Sons, 2017.

Cochran, William G. *Sampling Techniques.* 3rd ed., John Wiley & Sons, 1977.

Shoukri, Mohamed M. *Measures of Interobserver Agreement and Reliability.* 2nd ed., CRC Press, 2010.

Smith, Gary. *Essential Statistics, Regression, and Econometrics.* 2nd ed., Academic Press, 2015.

Strang, Kenneth D., editor. *The Palgrave Handbook of Research Design in Business and Management.* Palgrave Macmillan, 2015.

Thompson, Steven K. *Sampling.* 3rd ed., John Wiley & Sons, 2012.

RANDOM SAMPLING ERROR SAMPLE PROBLEM

Consider a hypothetical research scenario. Researchers are testing the hypothesis that people who eat a high-protein breakfast in the morning will experience less weight gain over time than people who eat a sweet breakfast. The study monitors the eating habits of survey respondents over a period of time. The researchers then randomly select subjects from each group to be weighed. Due to an insufficiently large sample, the sampled subjects from the sweet group show net weight loss. However, overall the sweet group gained more weight on average than the high-protein group. What type of error is this?

Answer:

This is a type II error. The results from the sampled subjects contradict the experimental hypothesis (that the sweet group will gain more weight), which would cause the researchers to incorrectly reject their experimental hypothesis and accept the null hypothesis.

Randomization

FIELDS OF STUDY

Sampling Design; Experimental Design

ABSTRACT

Randomization refers to any of several chance-based methods that are used in selecting and assigning subjects for experiments. The use of randomization helps limit selection bias and sampling errors. When used correctly, randomization produces research results that are more likely to be generalizable.

PRINCIPAL TERMS

- **random assignment:** the process of dividing the subjects in an experiment into the experimental or control groups using methods that are based on chance.
- **random sampling:** one of several probabilistic sampling methods in which individuals within the population being studied are selected for examination at random, so that each member of the population has a chance (often an equal chance) of being included.

Random Chance in Experiments

Randomization is the process of generating a result by chance rather than by choice. When using random methods, all possible choices have an equal probability of being selected. Various methods are used for randomization. Mechanical methods such as shuffling cards, flipping a coin, or picking selections out of a hat are relatively reliable, especially for randomizing a complete sequence. Computer software is also easy to come by for more sophisticated randomization. Randomization is one of the most important methods in sampling and experimental design. In many research studies, certain aspects must be randomized for the results to be considered valid. Other elements of research designs can be made stronger through randomization. Randomization can protect against specific kinds of error and the effects of bias. Two key forms of randomization in experiment design are random sampling and random assignment. Both are chance-based methods that are used to determine the involvement of participants in an experiment. Random sampling is used to select participants for the experiment from a larger population, and random assignment is used to assign those participants to either the control group or the experimental group.

Random sampling should not be confused with haphazard sampling, which is a type of nonprobability sampling. Haphazard or convenience sampling selects participants based on availability, such as by taking the first available volunteers. With this method, there is no reason to think that the sample is actually representative of the larger population. Furthermore, some studies need to either eliminate or include extraneous variables. This means that although participants should be randomly selected, the experimenters may also require that they represent a demographically diverse group, or a demographically similar one.

Randomization Strategies

How randomization is used in a study is determined by the study's purpose. When deciding on sampling and assignment methods, the key consideration is how the experiment results should be generalized. In many cases, medical research needs to be generalized as broadly as possible. In these cases, the best randomization strategy is a random sampling method that will ensure that the research participants are demographically diverse. A demographically diverse sample represents a variety of ages, ethnicities, income levels, health conditions, and other potentially confounding factors. A good method for ensuring diversity is stratified random sampling, also called proportional random sampling or quota random sampling. In stratified random sampling, the population of potential participants is first divided into mutually exclusive groups according to the characteristics being studied. Simple random samples are then taken from each subgroup in proportion to their representation in the larger population. Simple random samples by themselves are more vulnerable to sampling error and are labor intensive when performed on large populations. Stratified random samples help protect against this type of error.

Simple Randomization	Units are randomly assigned to treatment groups and have equal likelihood of assignment.
Unequal Allocation Randomization	Units are randomly assigned to treatment groups; groups may not be equal in size, but must fall within an acceptable ratio.
Stratified Randomization	Units are pregrouped into strata based on some intrinsic characteristic; groups are then randomly assigned to treatments.
Blocked Randomization	Units are randomly assigned to small and equal groups; groups are randomly asigned to treatments.

Researchers may also want to sample from demographically similar groups. For instance, a social science experiment may examine what news articles are shared by retired baby boomers social media. This would require an entirely different strategy of random sampling. The results of this study need to be generalized to a population of Americans born between 1946 and 1964. Thus, the participant group clearly needs to be limited to participants who share that description. Stratified random sampling could be used to proportionally represent gender, race, and education level within that age range. Because both social-media usage and news access could be affected by geographic location, the researchers would draw from a geographically diverse group of participants rather than participants who are local to the experimenters.

Cluster sampling is a randomization technique that can be useful for simplifying the process of random sampling over large geographic areas. In cluster sampling, the relevant population is first divided into clusters based on mutually exclusive criteria, such as geographic location. All members of a population who fit the criteria make up a single cluster. Simple random sampling is then used to select several clusters at random, and every individual in each cluster is surveyed, measured, or counted. The main drawback to this technique is the time and labor involved. There is also a possibility of sampling error, which is greater if the number of clusters sampled is low or if the impact of geography is pronounced. However, because coordinating participants over a large geographic area can be expensive and time consuming, cluster sampling can be beneficial in many situations.

THE NEED FOR RANDOMIZATION

In the sample problem, randomization was used in order to make the results of the research generalizable beyond the group of students in the study. If stratified random sampling is used, randomly selected participants in separate groups could still be assured to be similar enough in respects outside the area of inquiry to eliminate the effect of extraneous variables. This is the key benefit of random sampling over haphazard sampling, which can be subject to bias and has a higher risk of sampling errors.

Bill Kte'pi, MA

FURTHER READING

Alferes, Valentim R. *Methods of Randomization in Experimental Design*. Sage Publications, 2012.

Chow, Shein-Chung, and Jen-Pei Liu. *Design and Analysis of Clinical Trials: Concepts and Methodologies*. 3rd ed., John Wiley & Sons, 2014.

Donner, Allan, and Neil Klar. *Design and Analysis of Cluster Randomization Trials in Health Research*. 2000. John Wiley & Sons, 2010.

Dugard, Pat, et al. *Single-Case and Small-n Experimental Designs: A Practical Guide to Randomization Tests*. 2nd ed., Routledge, 2012.

Kirk, Roger E. *Experimental Design: Procedures for the Behavioral Sciences*. 4th ed., Sage Publications, 2013.

> **RANDOMIZATION SAMPLE PROBLEM**
>
> Imagine a hypothetical research scenario. Researchers want to study the effects of eating breakfast on the performance of first-semester students in a foreign-language class. Participants are drawn from hundreds of such classes across multiple colleges and universities. Participants have volunteered for the study and answered a questionnaire about their breakfast habits and course enrollment. Over the course of a semester, the performance of students who do not eat breakfast will be compared with the performance of students who do eat breakfast. The researchers hypothesized that students who eat breakfast will have a better average performance.
>
> In this research scenario, where can (or must) the researchers use randomization?
>
> **Answer:**
> Because students have already indicated their breakfast habits, random assignment to the two groups (breakfast or no breakfast) is not possible. However, random sampling of participants from the volunteer pool is possible. Stratified random sampling can be used to reduce the impact of extraneous variables. For instance, the experimental groups and control groups can be constructed through stratified random sampling to ensure that members of both groups have the same grade-point averages. Such sampling can also ensure that the two groups have the same mix of majors, classes, socioeconomic status, and proficiency in other foreign languages.

Montgomery, Douglas C. *Design and Analysis of Experiments*. 8th ed., John Wiley & Sons, 2013.

Rosenberger, William F., and John M. Lachin. *Randomization in Clinical Trials: Theory and Practice*. 2nd ed., John Wiley & Sons, 2016.

Randomized Complete Block Designs

FIELDS OF STUDY

Research Design; Sampling Design

ABSTRACT

The randomized complete block (RCB) design is used to test multiple new treatments in a small area or among a small population. Researchers begin by dividing the space or population into smaller, essentially uniform parts. They then randomly test each treatment on each part to accurately compare the effects.

PRINCIPAL TERMS

- **replicate:** an experimental run that tests the same set of treatment levels as another run within the same experiment, or an experimental subject that receives the same set of treatment levels as another subject in the experiment.
- **spatial effects:** environmental elements, such as sunlight, that affect the area under study and could therefore confound experimental results.
- **subject:** the variable that is to be studied to determine the effect of a treatment; could be a person, an animal, or an object, depending on the experiment.
- **treatment:** the process or intervention applied or administered to members of an experimental group.

Accounting for Variations

The randomized complete block (RCB) design is an experimental design used when the researcher knows or suspects that there will be variations among the experimental units. Before randomly assigning treatments to the experimental units, the researcher will need to isolate and account for these differences as much as possible. This helps ensure that any observed effects can more accurately be linked to the treatment. To do this, researchers break up the experimental units into more homogenous

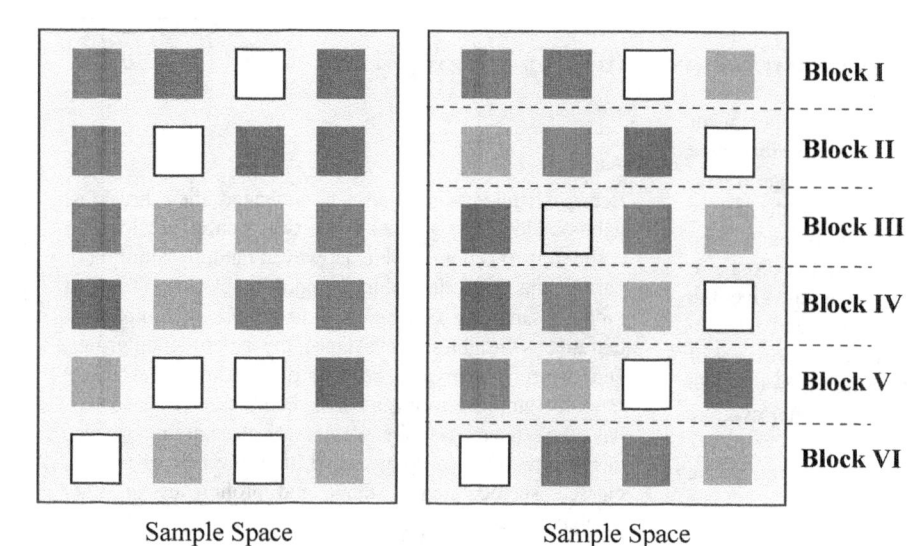

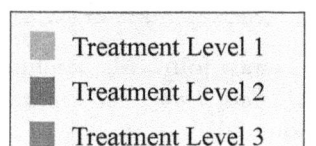

Treatment Level 1
Treatment Level 2
Treatment Level 3
No Treatment Level

subgroups, or blocks, according to the different characteristics or environmental factors. The different treatments are then randomly applied equally to subjects in each block. Each treatment is used once in each block.

Agricultural tests typically use the RCB design to determine how best to grow new crops. Farmland is often a varied space. Some parts of a field have more sunlight, wind, or wildlife than other parts. These differences are called spatial effects. If they are not accounted for in a research study, they can lead to invalid results. To account for spatial effects, researchers divide a field into smaller units of crops known as blocks. Then they apply each treatment, such as a type of fertilizer, in each block. The blocks serve as replicates, in that they all receive the same set of treatment levels. To be sure that the results were not accidental or caused by spatial effects, researchers will replicate each treatment under study in several locations.

When to Use RCB Design

RCB designs are more precise than completely randomized designs, particularly for field experiments. Greater precision leads to results with greater external validity. In a completely randomized design, the researcher assumes that the experimental units being studied are essentially homogenous. Therefore, a treatment can be randomly applied to any unit of subjects in the entire sample. However, in an RCB design, treatments are applied to specific blocks of subjects. RCB design is preferred by researchers who are concerned that variations among the units, commonly spatial effects, will confound their study. This design is not limited to agricultural studies, however. It is also used in fields such as biology and human psychology where animal and human populations are studied.

Theoretically, there is no limit to the number of replicates that can be used or the number of different treatments that can be applied. However, real-world limitations, such as the size of a field or the budget of a research team, often restrict the number of replicates used in a single study. Additionally, researchers must be careful not to apply too many treatments to one replicate, as this will increase the number of things that can go wrong during the experiment. Testing a large number of treatment would require larger blocks, which would decrease homogeneity. In cases where larger blocks are needed, an incomplete block design may be more suitable.

Medical Studies and RCB Designs

Researchers often rely on RCB designs when testing new medications to ensure that certain variations are considered when assessing the effectiveness of the treatment. Common variations that researchers want to control in these studies are age and gender. People of different ages and genders react to medications in different ways. Researchers need to ensure that men and women of different ages are well represented in both the experimental and control groups. An RCB design can be used to divide the subjects into blocks according to these characteristics. This way, the experiment would control for age and gender, allowing the researcher to make a more accurate analysis of the effect of the medication that is also more externally valid.

USE IN MANY FIELDS

Because of its flexibility, greater precision, and minimal bias, the RCB design has been used in many different types of experiments. It is especially helpful for establishing control and increasing power in field experiments. Therefore, it is regularly used by farmers, ranchers, and other agriculturalists to increase the yield of their crops and livestock. Medical researchers have also used RCB designs to study human and animal diseases, such as testing out a new antibiotic. Additionally, food scientists have used this method to test the best way to make and preserve foods.

Allison Hahn, PhD

FURTHER READING

Alferes, Valentim R. *Methods of Randomization in Experimental Design.* Sage Publications, 2012.

Christensen, Larry B., et al. *Research Methods, Design, and Analysis.* 12th ed., Pearson, 2014.

Clewer, Alan G., and David H. Scarisbrick. *Practical Statistics and Experimental Design for Plant and Crop Science.* John Wiley & Sons, 2001.

Creswell, John W. *Research Design: Qualitative, Quantitative, and Mixed Methods Approaches.* 4th ed., Sage Publications, 2014.

Festing, Michael F. W. "Randomized Block Experimental Designs Can Increase the Power and Reproducibility of Laboratory Animal Experiments." *ILAR Journal,* vol. 55, no. 3, 2014, pp. 472–76. *MEDLINE Complete,* search.ebscohost.com/login.aspx?direct=true&db=mdc&AN=25541548&site=ehost-live. Accessed 31 May 2017.

Levy, Paul S., and Stanley Lemeshow. *Sampling of Populations: Methods and Applications.* 4th ed., John Wiley & Sons, 2008.

RANDOMIZED COMPLETE BLOCK DESIGNS SAMPLE PROBLEM

Researchers want to test the effects of four new fertilizers on daisy flowers. They have forty pots of flowers to test, and they will conduct the test in a large room that has ten tables. The researchers are concerned that some parts of the room are warmer than others, and some tables are closer to the windows than other tables. How should they organize the study? Specifically, how should they divide the space, and how many blocks should they have? Also, how should they assign the treatments to experimental units within each block?

Answer:

First, determine the number of blocks to be used in the study. A natural division is to make each table into one block, so there will be ten different blocks in this study. The forty flowerpots can then be sorted onto those tables, putting four flowerpots (experimental units) on each table. This will allow the researcher to test each fertilizer treatment on one flowerpot in each block. The treatment must be randomized, so the researchers may use a computer program, or pull numbers out of a bag, to assign the numbers 1, 2, 3, and 4 to the flowerpots in each block. Each number represents one of the four different fertilizer treatments. This way, no two flowerpots on the same table will receive the same treatment.

Regression Discontinuity Design

FIELDS OF STUDY

Experimental Design; Research Design; Statistical Analysis

ABSTRACT

The regression discontinuity (RD) design is a quasi-experimental, pretest-posttest comparison design. It is distinguished by its use of cutoff values, rather than random assignment, to assign participants to groups. One advantage of this feature is that the participants most in need of the program or treatment being tested will have access to it, and will not be randomly assigned to a control group.

PRINCIPAL TERMS

- **before-and-after graph:** a graph that depicts the relationship between two variables, typically either independent and dependent variables or pretest and posttest scores, before and after a treatment for both the experimental group and the control group.
- **cutoff point:** a designated value used to divide a range of numbers into two relevant categories; in a regression discontinuity design, the numbers are the participants' pretest scores, and the categories are the experimental group and the control group.
- **sample size:** the number of subjects included in a survey or experiment.
- **treatment:** the process or intervention applied or administered to members of an experimental group.

BASICS OF REGRESSION DISCONTINUITY

The regression discontinuity (RD) design is a quasi-experimental pretest-posttest design. It differs from randomized and other quasi-experimental designs in how subjects are assigned. Assignment is based on a predetermined cutoff point or threshold. This aids in estimating the local average effect of the treatment in situations where randomization is not feasible.

There are several variations on RD design. The basic design is a two-group pretest-posttest design, in which participants are assigned to groups based on their pretest scores. The scores are evaluated against the cutoff point, which is established prior to the pretest. It is usually based on either the resources available or the degree of need indicated by each score. The participants whose scores indicate they are most in need of treatment are assigned to the experimental group. Those whose scores indicate they are least in need of treatment are assigned to the control group. The treatment is then given to the experimental group, after which both groups are given a posttest.

After the treatment, RD analysis is conducted. This is done by plotting the pretest and posttest scores on a before-and-after graph, with pretest scores on the x axis and posttest scores on the y axis. A regression line is calculated for each group. If the treatment had a significant effect, there will be a "jump," or discontinuity, between the regression lines at the cutoff point on the x axis.

AN UNCOMMON DESIGN

RD design was first implemented in the mid-1970s to help educators evaluate compensatory education programs funded by the US Elementary and Secondary Education Act of 1965. It is not a commonly used design, in part because researchers may consider its criteria to be overly restrictive. It is also often confusing to inexperienced researchers. Most research designs prefer the average characteristics of the control groups and the experimental or treatment groups to be as similar as possible. However, the cutoff criterion of RD design means that the differences between control and experimental groups are maximized.

When properly executed, the RD design can result in an unbiased estimate of the treatment effect. In this regard, it is almost equivalent to a randomized design measuring a treatment effect. It also helps to eliminate ethical issues that can arise as a result of randomization. The use of the cutoff point in group assignment means that the participants most in need of the treatment or program being tested are not at risk of being randomly assigned to the control group. However, because RD design is not randomized, it is more prone to error than a true experimental design, and therefore less efficient. An RD design must

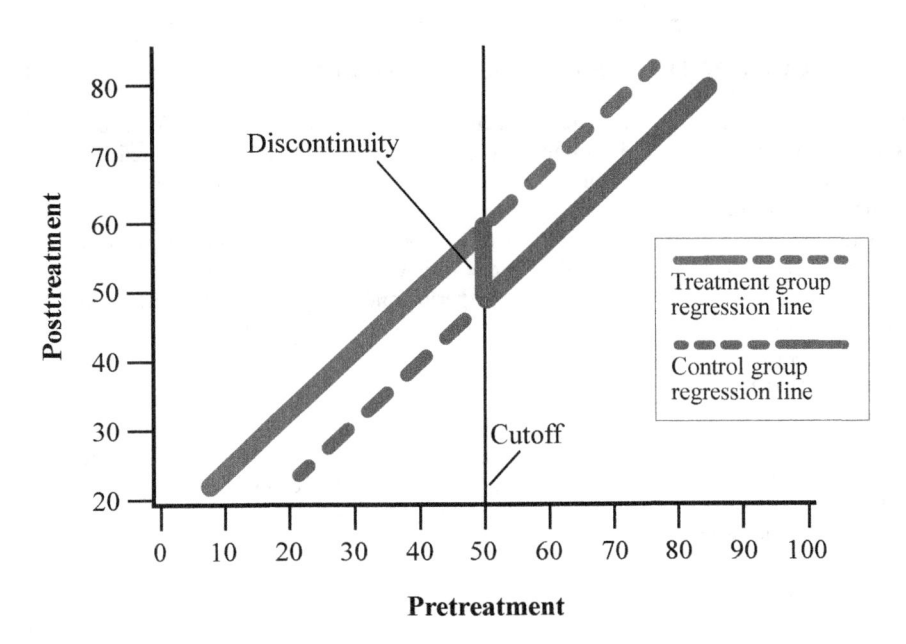

[Figure: Graph showing Posttreatment (y-axis, 20–80) vs Pretreatment (x-axis, 0–100), with a Treatment group regression line and Control group regression line meeting at the Cutoff (x=50) with a Discontinuity at that point.]

have a larger sample size than a randomized design in order to have the same statistical power.

REGRESSION DISCONTINUITY ANALYSIS

Once the pretest and posttest scores are obtained for all participants, RD analysis is conducted to estimate the effects of the program or treatment being studied. This requires terms to indicate the pretest score (x), the posttest score (y), and a variable to signify whether or not the treatment was given (z, which would take a value of either 0 or 1). Before the scores are plotted on the before-and-after graph, the pretest scores are transformed by subtracting the cutoff point from each one. This sets the cutoff point to $x = 0$ on the before-and-after graph. The experimental group's scores are on one side of the y axis and the control group's scores are on the other side.

Note that which group is on which side of the y axis depends on what the pretest measures. Whether a score above or below the cutoff point indicates that a participant is most in need of intervention. If the participants who score below the cutoff point are determined to be most in need of intervention, and those who score above the cutoff point are least in need, then the experimental group will be to the left of the y axis and the control group will be to the right. If the opposite is true—that is, if higher scores indicate greater need—then the control group will be on the left, and the experimental group will be on the right.

Once all scores are plotted, regression lines are calculated for both groups, each one cutting off at the y axis. Any treatment effect will be indicated by a discontinuity at this point. The two lines may have the same slopes, indicating a direct, additive effect from the treatment. Different slopes indicate an interaction effect. If both slopes are the same, the equation for both regression lines will have the basic format

$$y = \beta_0 + \beta_1 x + \beta_2 z$$

where y is posttest score, β_0 is the y intercept (where the line crosses the y axis; also the cutoff point), β_1 is the coefficient of the transformed pretest score x, and β_2 is the mean difference caused by the treatment z. (For the control group, z is equal to 0, and thus that element is not present in the equation.) Other elements may be added to the equation, but this is its simplest form.

REGRESSION DISCONTINUITY DESIGN IN PRACTICE

Although RD design is not widely used, it has important implications for use in both education and the clinical sciences. It is especially useful in quality control programs. Care must be taken to correctly model the relationship between treatment and outcome effects or a biased result can occur. Contamination of the treatment effect can occur if other, non-measured treatments occur simultaneously in the sample group. However, if such care is taken, and if sample sizes are sufficiently large to produce statistically significant results, RD design can be a powerful tool for study in cases where random assignment is not feasible.

Mandy McBroom-Zahn

FURTHER READING

Cappelleri, Joseph C., and William M. K. Trochim. "An Illustrative Statistical Analysis of Cutoff-Based

> **REGRESSION DISCONTINUITY DESIGN SAMPLE PROBLEM**
>
> A hospital administrator wishes to implement a quality-improvement program to reduce cases of hospital-acquired anemia. The program involves training staff members to use pediatric phlebotomy tubes rather than full-size tubes to draw blood. However, the program is too expensive to implement in all departments. Instead, the hospital will implement the program in a subgroup of departments, which will serve as the experimental group, and all other departments will be the control group. Each department is equivalent to a single "participant." The pretest that will be used to assign departments to either the experimental group or the control group will be the number of cases of hospital-acquired anemia per department staff member in each department over the past year. A cutoff point is determined, and all departments whose number of cases per department employee is above that point will receive the program. After the program is completed, a posttest will be conducted to see if it had any effect. Then each department's pretest and posttest scores are plotted on a before-and-after graph, with the cutoff point transformed to be equal to $x = 0$ and the experimental and control groups on either side of the y axis, and regression lines are drawn for both groups. Which side of the y axis (i.e., the cutoff point) will the experimental group's regression line be on?
>
> **Answer:**
>
> The experimental group's regression line will be on the right side of the y axis, because the experimental group consisted of departments whose number of cases per employee was *higher* than the cutoff point.

Randomized Clinical Trials." *Journal of Clinical Epidemiology*, vol. 47, no. 3, 1994, pp. 261–70.

Hedges, Larry V., and Jennifer Hanis-Martin. "Can Non-randomized Studies Provide Evidence of Causal Effects? A Case Study Using the Regression Discontinuity Design." *Education Research on Trial: Policy Reform and the Call for Scientific Rigor*, edited by Pamela Barnhouse Walters et al., Routledge, 2009, pp. 105–24.

Hocking, Ronald R. *Methods and Applications of Linear Models: Regression and the Analysis of Variance.* 3rd ed., John Wiley & Sons, 2013.

Jacob, Robin Tepper, et al. *A Practical Guide to Regression Discontinuity.* MDRC, Aug. 2012. *MDRC,* www.mdrc.org/publication/practical-guide-regression-discontinuity. Accessed 8 May 2017.

Trochim, William M. K., and Joseph C. Cappelleri. "Cutoff Assignment Strategies for Enhancing Randomized Clinical Trials." *Controlled Clinical Trials*, vol. 13, no. 3, 1992, pp. 190–212.

Van Leeuwen, Nikki, et al. "Regression Discontinuity Design: Simulation and Application in Two Cardiovascular Trials with Continuous Outcomes." *Epidemiology*, vol. 27, no. 4, 2016, pp. 503–11.

Regression Modeling and Analysis

FIELDS OF STUDY

Statistical Analysis

ABSTRACT

Regression analysis is a statistical method for evaluating a relationship between a dependent variable and one or more independent variables. Regression analysis enables researchers to devise a model to account for the data observed and then determine how well the model fits that data.

PRINCIPAL TERMS

- **least squares:** a method for determining the best fit for a regression line that minimizes the sum of the squares of the distances between the line and each data point.
- **relationship:** also called "correlation"; a statistical method used to measure and describe the strength and direction of the relationship between an independent and dependent variable.
- **R-squared:** the amount of variability in a data set that is accounted for by the regression equation.

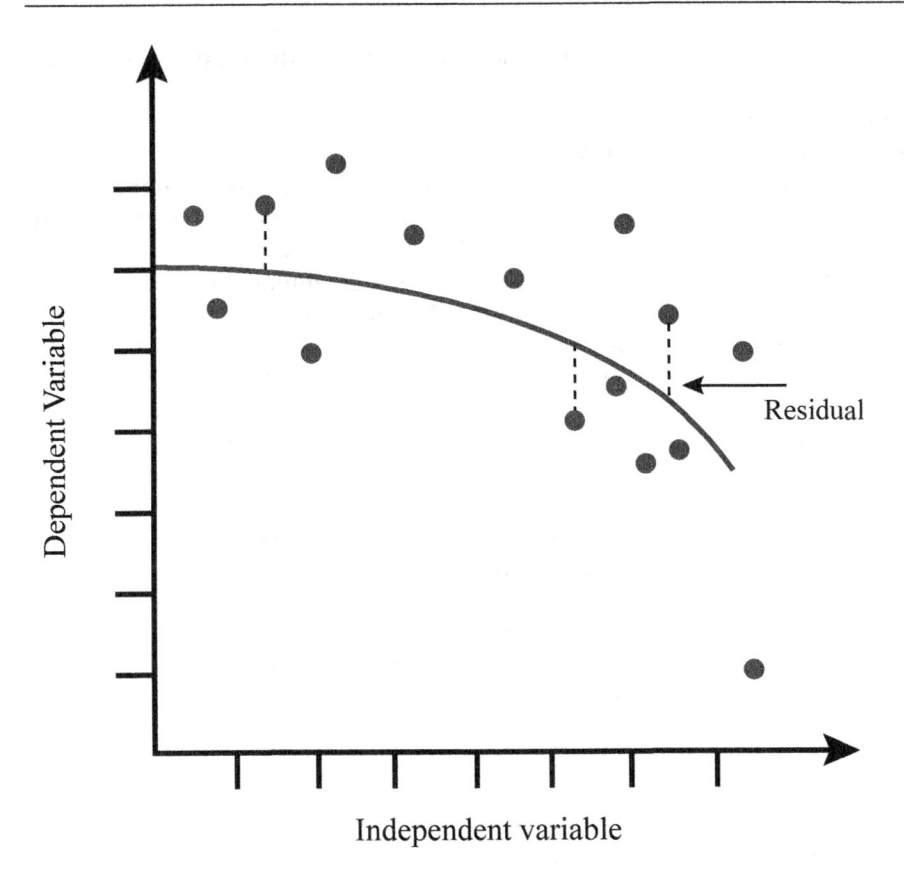

STATISTICAL MODEL

There is a series of steps to be taken to identify the correct regression method to be used. First, the data points for the response variable should be plotted on a scatterplot and the basic shape evaluated. Basic questions that can be asked of a regression model are whether there is a relationship between a response variable, y, and a predictor variable, x, and, if so, what the strength and direction of that relationship is. In each model, the least squares method is used to obtain the estimates of the y-intercept and the slope.

The simple linear regression formula is

$$y = \beta_0 + \beta_1 x + \varepsilon$$

where β_0 is the y-intercept, β_1 is the slope, and ε is the estimated value for random prediction errors, or "residuals." The slope is usually the parameter of interest. (Note that in regression, the terms "linear" and "nonlinear" refer to the form of the function of the parameters, not the line that the function produces. Thus, a linear regression function may produce a curve.)

Another measurement that is used in regression modeling is the coefficient of determination, r^2. This statistic allows one to examine the closeness of fit of the sample regression equation to the values of the response variable. R-squared is calculated as

$$r^2 = SSR / SSTO \text{ or } r^2 = 1 - (SSE / SSTO)$$

where SSR is the regression sum of squares (explained variation), $SSTO$ is the total sum of squares (total variation), and SSE is the error sum of squares. The largest value that r^2 can be is 1, which means that all the variation in the response variable is explained by the regression line. Conversely, if the value is 0,

- **variable:** in research or mathematics, an element that may take on different values under different conditions.

OVERVIEW

Regression analysis is a type of statistical analysis that can help determine the presence of a relationship between a dependent variable and one or more independent variables. Dependent variables are also called "response variables" or "outcome variables"; independent variables are also called "predictor variables" or "explanatory variables."

The type and complexity of the regression model chosen depends on several properties of the data. These include whether the data are continuous or discrete, if multiple predictor variables are to be measured against a response variable, and the distribution of data for the relationship between the two variables. Regression models rely on the least squares method for obtaining the best-fitting line and whether all the data in the response variable can be explained by the line using the coefficient of determination, r^2.

then none of the variation in y is explained by the regression line.

The multiple linear regression model is written such that a linear relationship is presumed to exist between some response variable (y) and k predictor variables, $x1, x2, \ldots xk$, as follows:

$$y = \beta_0 + \beta_1 x_1 + \beta_2 x_2 + \ldots + \beta_k x_k + \varepsilon$$

When a scatterplot indicates that the relationship between a predicator and the residuals might be non-linear, polynomial regression is used. The polynomial regression equation takes the form below:

$$y = \beta_0 + \beta_1 x + \beta_2 x^2 + \ldots + \beta_k x^k + \varepsilon$$

The highest power in the polynomial, k, determines the degree of the model. For example,

$$y = \beta_0 + \beta_1 x + \beta_2 x^2 + \varepsilon$$

would be a second-degree model because the highest power is 2.

STEPWISE REGRESSION FOR BEST-FITTING MODEL
When there are many predictors to evaluate, a procedure known as stepwise regression is used to arrive at a single model for the data. First, all predictor variables that could be responsible for the response are listed. Next, two significance levels, α_1 and α_2, are specified for entering a predictor into the stepwise model and removing it, respectively. Then, predictors are entered one by one, in ascending order by P value for the t-test, provided that P value is under α_2. Once more than one predictor has been entered, the significance level for the preceding predictor(s) is rechecked. If the new t-test P value at a slope of 0 exceeds α_2, the predictors with those higher P values are removed. If the new P value does not exceed α_2, all models that include the entered predictors are tested for fit (regressed). This procedure continues until no predictor can enter without exceeding α_2. At that point, the final model has been found.

Stepwise regression is limited, however, by several factors. Among them are the variables chosen, the potential for errors when testing predictors, and the fact that models other than the final one might be equally useful.

LIMITATIONS
Regression is a powerful tool for examining the relationship between variables. When there are multiple predictor variables, however, problems can arise. These include multicollinearity (several predictor variables are linear functions of each other), confounding (impure effects), and interaction between uncorrelated predictors. Each has specific diagnostic methods to find it and treat it to improve the statistical model. Also, successive error values can be correlated with one another (autocorrelation). False positives or negatives as well as omitting variables can also cause a researcher to select the wrong regression model.

APPLICATIONS
Regression is used in many different applications across many fields. It is commonly used to test the hypothesis that there is a relationship between a dependent variable and one or more predictor variables. For example, in medicine, it might be used to test a predicted relationship between blood pressure, weight, and stress.

Mandy McBroom-Zahn

FURTHER READING
Daniel, Wayne W., and Chad Lee Cross. *Biostatistics: A Foundation for Analysis in the Health Sciences.* John Wiley & Sons, 2014.
Darlington, Richard B., and Andrew F. Hayes. *Regression Analysis and Linear Models: Concepts, Applications, and Implementation.* Guilford Press, 2017.
Haslam, S. Alexander, and Craig McGarty. *Research Methods and Statistics in Psychology.* 2nd ed., SAGE Publications, 2014.
Lomax, Richard G., and Debbie L. Hahs-Vaughn. *An Introduction to Statistical Concepts.* 3rd ed., Routledge, 2012.
Muller, Keith E. *Applied Regression Analysis and Other Multivariable Methods.* Brooks, 2013.
Rothman, Kenneth J., et al. *Modern Epidemiology.* Wolters Kluwer Health, 2015.

> **SAMPLE PROBLEM**
>
> In the table below are the weights and heights of a random sample of eleven nutritionally deficient children.
>
> | Child ID | Weight | Child ID |
> |---|---|---|
> | 1 | 64 | 57 |
> | 2 | 71 | 59 |
> | 3 | 53 | 49 |
> | 4 | 67 | 62 |
> | 5 | 55 | 51 |
> | 6 | 58 | 50 |
> | 7 | 57 | 48 |
> | 8 | 56 | 42 |
> | 9 | 51 | 42 |
> | 10 | 76 | 61 |
> | 11 | 68 | 57 |
>
> Given that weight is the response variable and height is the predictor variable, which of the following models is the best-fitting model?
>
> a. $y = \beta_0 + \beta_1 x + \varepsilon$; $r^2 = 0.432$
> b. $y = \beta_0 + \beta_1 x + \beta_2 x^2 + \varepsilon$; $r^2 = 0.830$
> c. $y = \beta_0 + \beta_1 x_1 + \ldots + \beta_k x_k + \varepsilon$; $r^2 = 0.78$
>
> **Answer:**
>
> The strong r^2 value for answer b., a second-order polynomial equation, suggests that this model best fits the data.

REGRESSION POINT DISPLACEMENT

FIELDS OF STUDY

Statistical Analysis; Quasi-experiment; Research Design

ABSTRACT

Regression point displacement (RPD) is a type of statistical analysis that is used to analyze the difference in outcomes between experimental and control groups. RPD analysis is usually used in regression point displacement designs (RPPDs), which are most often used in community research. In an RPDD, the entire sample serves as the experimental group, and the results from the sample are compared to a large number of other, similar populations, rather than to just one control group.

PRINCIPAL TERMS

- **statistical analysis:** the collection, examination, description, manipulation, and interpretation of quantitative data.
- **treatment:** the process or intervention applied or administered to members of an experimental group.
- **variable:** in research or mathematics, an element that may take on different values under different conditions.

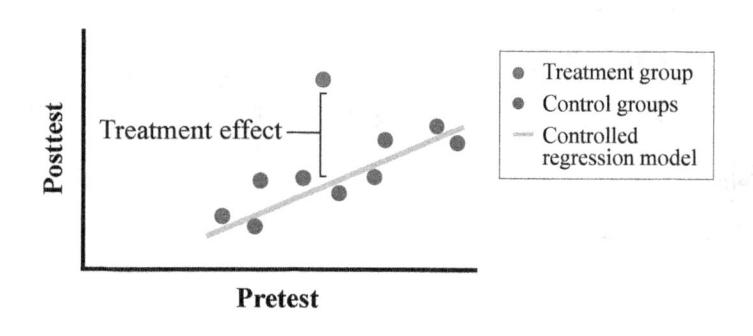

Community-Based Research

Regression point displacement (RPD) is a type of statistical analysis that is applied to the results of a research study involving experimental and control groups. It is used to analyze the difference in outcomes between groups after the experimental group has received a treatment and the control group has not. If the difference is significant, then the null hypothesis is rejected, and the treatment is deemed to have had an effect.

RPD analysis is usually used in a regression point displacement design (RPPD). RPPD is a quasi-experimental design that is most often used in community research. It may be used if a researcher wants to know the effect of a certain program on a sample population. In RPDD, the entire sample serves as the experimental group, and the results from the sample are compared to a large number of other, similar populations, not just one control group. This is considered more robust than a comparison against one control sample.

To conduct RPD analysis, all members of the sample must first be given a pretest. A pretest is a measurement of some variable of interest to the study, usually the dependent variable, that is taken before the study begins. After the study is complete, a second measurement, or posttest, is taken. All study participants must have pretest and posttest scores, as they will be plotted against each other on an *x-y* graph, with pretest scores on the *x*-axis and posttest scores on the *y*-axis. The same information is plotted for the control groups. A regression line is fitted to the control data. Then, the distance, or displacement, of the experimental data from the regression line is evaluated for significance.

Benefits and Drawbacks

RPDD is often used when there are only enough resources to give a treatment to a limited population. To complete RPD analysis for this design, values of the dependent variable must be available both for the group being studied and for the multiple control groups to which the experimental group will be compared. For example, a university studying the outcomes of a program would need data from a sample of other universities without a similar program for comparison. It is more robust to use RPD analysis than to use the mean of the other populations as a single comparison point.

One possible drawback of RPD analysis is that the comparison data may be too scattered to allow for an accurate regression line to be calculated. For example, a curved line may fit the data better than a straight line. In such cases, the direction of the discrepancy must be considered. If the less accurate straight regression line reduces the apparent effect of the treatment but still shows a significant difference, this can be used to show a conservative estimate of the treatment effect. However, if the straight regression line exaggerates the apparent effect, using it to estimate the treatment effect would be misleading.

Pretest and posttest scores are required from all populations used in the analysis for it to be accurate. If the researcher cannot obtain pretreatment data from both the treated population and the comparison populations, RPD analysis cannot be used. Similarly, if posttreatment data cannot be obtained from all populations, it cannot be used. For example, if a researcher wants to test the effect of a county-based program to educate teenagers about drunk driving, they might compare the rate of accidents involving intoxicated teenagers in the county with the program to the rates in surrounding counties. To do so, the researcher would need to obtain rates from all of the counties of interest (the experimental one as well as the control ones) for both before and after the program takes place. If those figures are not available, RPD analysis cannot be used.

RPD Analysis in Action

To expand on the example above, imagine that a state wants to reduce the number of driving accidents involving alcohol. An educational program for young drivers is developed, but there are too few resources to implement the program statewide. Instead, a pilot program is implemented in all high schools in one county. The rate of accidents involving teenagers and alcohol per year is determined for each county in the state, including the one that will receive the education. After the program is implemented, similar data is gathered for the following year.

These pre- and post-program accident rates are the pretest and posttest scores. The scores of all of the counties are plotted on an *x-y* graph, with the pretest scores on the *x*-axis and the posttest scores on the *y*-axis. The scores of the county that ran the program are also plotted, but this plot point is noted differently so that it can be distinguished from the others. A regression line is computed and drawn through the plot points of the control counties. Then RPD analysis is conducted on this data to determine if the vertical displacement of the notated point, representing the county that received the program, from the regression line is statistically significant. If it is, then the null hypothesis is rejected, and the program can be said to have had a significant effect.

The Importance of RPD

It is important in any experimental study to understand which research design is best suited for which fields of study and what types of statistical analysis are best suited for said design. RPDD is most commonly used in education, health care, and military research, due to the low costs of testing. This design is particularly well suited for use in determining causation in large populations.

Pamelyn Witteman, PhD

Further Reading

Creswell, John W. *Research Design: Qualitative, Quantitative, and Mixed Methods Approaches.* 4th ed., SAGE Publications, 2014.

Linden, Ariel, et al. "Evaluating Program Effectiveness Using the Regression Point Displacement Design." *Evaluation & the Health Professions*, vol. 29, no. 4, 2006, pp. 407–23, doi:10.1177/0163278706293402. Accessed 19 May 2017.

Orcher, Lawrence T. *Conducting Research: Social and Behavioral Science Methods.* 2nd ed., Taylor & Francis, 2014.

Patten, Mildred L., and Michelle Newhart. *Understanding Research Methods: An Overview of the Essentials.* 10th ed., Routledge, 2017.

Swanson, Richard A., and Elwood F. Holton III, editors. *Research in Organizations: Foundations and Methods of Inquiry.* Berrett-Koehler Publishers, 2005.

Trochim, William M. K. "Other Quasi-Experimental Designs." *Research Methods Knowledge Base*, www.socialresearchmethods.net/kb/quasioth.htm. Accessed 19 May 2017.

Trochim, William M. K. "Regression Point Displacement Analysis." *Research Methods Knowledge Base*, www.socialresearchmethods.net/kb/statrpd.php. Accessed 19 May 2017.

REGRESSION POINT DISPLACEMENT SAMPLE PROBLEM

A researcher is studying the size of fish in a lake before and after an industrial accident that released toxic chemicals into the lake. Data on the average size of the fish in that lake and several other regional lakes is available from before the accident and from two years after the accident. How would the researcher determine if the industrial accident influenced the size of the fish in the lake after two years?

Answer:

The average size of the fish in each lake would be graphed, with the average size before the industrial accident plotted on the *x*-axis and the size two years after the accident plotted on the *y*-axis. Next, a regression line would be calculated and drawn for all points on the graph except for the one representing the lake in which the accident occurred (the experimental point). As the regression line is created from the points representing the lakes that did not have the accident, their R^2 (pronounced "R squared") values, or the squares of their correlation coefficients, must be high—equal to or close to 1—with respect to the regression line. Then, RPD analysis is used to determine if the vertical displacement of the experimental point from the regression line is significant and, thus, if the null hypothesis can be rejected. For the null hypothesis to be rejected, the experimental point must have a lower R^2 value than the other points. This analysis can be done manually, or it can be easily performed using statistics software.

Trochim, William M. K., and Donald T. Campbell. *The Regression Point Displacement Design for Evaluating Community-Based Pilot Programs and Demonstration Projects.* Unpublished manuscript, 1996. *Research Methods Knowledge Base,* www.socialresearchmethods.net/research/RPD/RPD.pdf. Accessed 19 May 2017.

Vogt, W. Paul. *Quantitative Research Methods for Professionals.* Pearson, 2007.

Wyman, Peter A., et al. "Designs for Testing Group-Based Interventions with Limited Numbers of Social Units: The Dynamic Wait-Listed and Regression Point Displacement Designs." *Prevention Science,* vol. 16. no. 7, 2015, pp. 956–66, doi:10.1007/s11121-014-0535-6. Accessed 19 May 2017.

Replication, Manipulation, and Randomization

FIELDS OF STUDY

Research Theory; Experimental Design; Sampling Design

ABSTRACT

Replication, manipulation, and randomization are three factors that are important considerations in experimental design and sampling design. Though not mandatory in every research project, when applicable and available, each offers advantages that strengthen the reliability of the results, especially in quantitative experimental research.

PRINCIPAL TERMS

- **control group:** the subjects in an experiment who do not receive any intervention or who receive a placebo or sham intervention.
- **experimental research:** research conducted by manipulating one variable and observing any resulting changes in one or more presumed dependent variables in order to identify any cause-and-effect relationships among them.
- **generalization:** the act of applying a characteristic of a sample group to a larger population, or the result of such an action.
- **quantitative data:** information that expresses attributes of a phenomenon in terms of numerical measurements, such as amount or quantity.

Sampling and Experimental Design

Replication, manipulation, and randomization are key interrelated concepts in experimental research, particularly in experimental and sampling design.

Replication is the repetition of the experiment's treatment combinations. It is not the repeated measurement of the same sample (which is also an important aspect of experimental design), but rather the administration of the same treatment combination to more than one subject. For example, in a clinical trial of a proposed new medication, the medication would be administered to an experimental group consisting of many people, rather than just one patient.

Manipulation refers to the change introduced by experimenters. An independent variable is manipulated with the expectation that it will effect change in one or more dependent variables. This may be done in a laboratory experiment or a field experiment, and usually involves quantitative data. Such experiments divide their participants into groups. Experimental groups receive the treatment in which the variable is manipulated. The control group is not manipulated and serves as a benchmark of comparison.

Randomization is the process of making a sequence or selection unpredictable, statistically unbiased, and without intentional pattern or deliberate choice. In experimental design and sampling, randomization can be important in many areas. These include selecting participants for the experiment, assigning them to control or experimental groups, and taking samples from data sets. Randomization is one of the tools that can assist in generalization of an experiment's results.

Strengths and Drawbacks

Replication is an important step in demonstrating the generalizability of research. If there are major differences in response to the treatment between individual subjects, the same must be expected when the treatment is conducted outside the research. But if the variation among subjects is statistically insignificant, this strengthens the case for generalization. Furthermore, replication helps eliminate the effects of bias, statistical anomalies, and observational errors. The drawback is that it increases the scale and complication of the research.

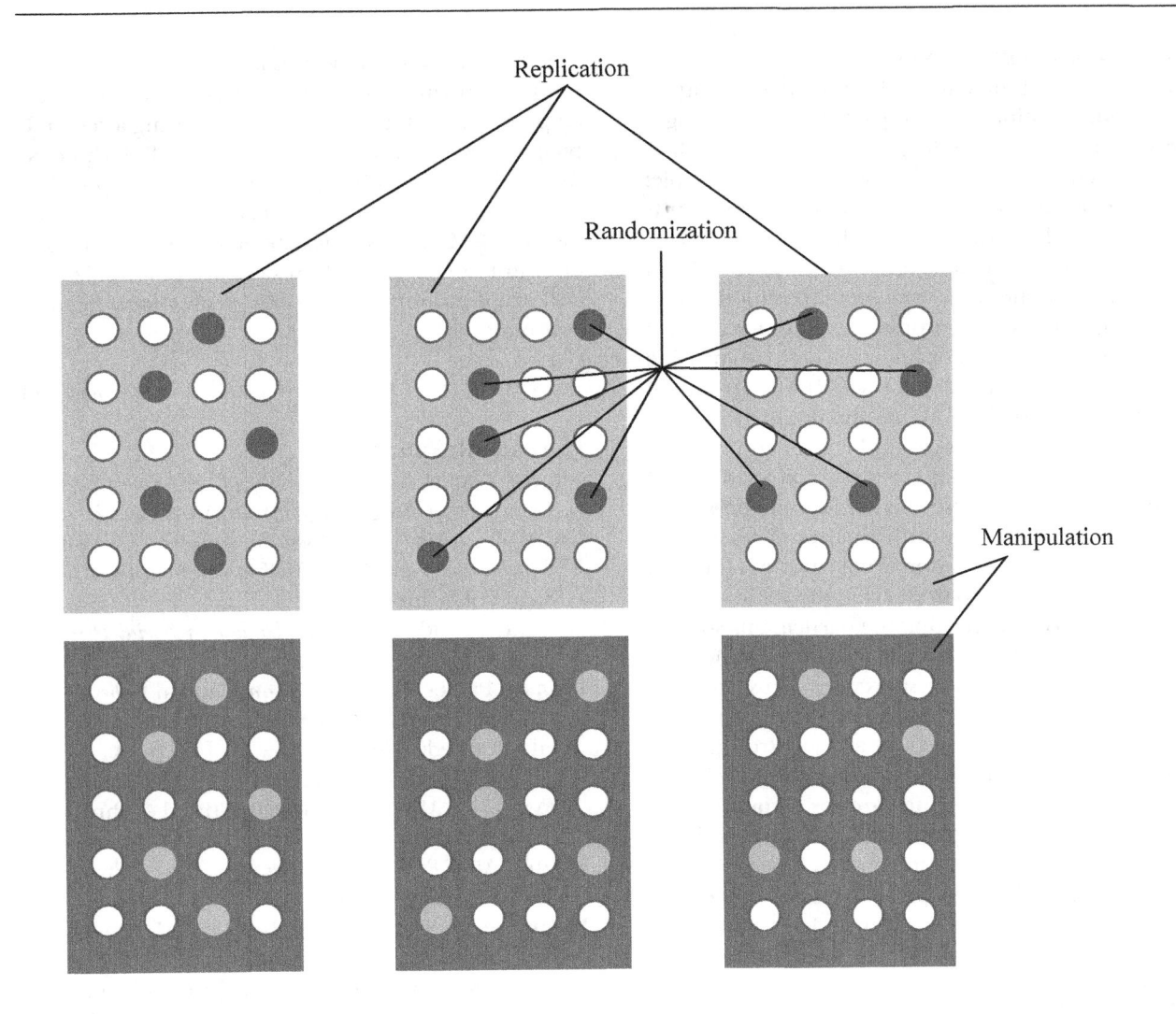

By some definitions, not all experimental research requires manipulation of a variable. Natural experiments are experimental research insofar as they seek to test a hypothesis, and thus are not purely descriptive. However, rather than introduce a change, researchers in these experiments observe naturally occurring phenomena that appear to demonstrate causal relationships. Typically, this is because the phenomena either cannot be induced by manipulation or cannot be recreated in the laboratory except in computer models, which are themselves based in part on such field observations. Natural experiments have the advantage of offering research possibilities, often with both replication and randomization, that would not otherwise be available. The drawback is the challenge to generalization. When researchers are not in control of manipulation and cannot directly observe hypothesized effects, the causal and predictive claims of their research will be subject to scrutiny.

Randomization is one of the strongest tools for reducing selection bias, where applicable, and for reducing confounding effects. Confounding is the unwanted effect of unaccounted-for variables on the causal relationship under study. This can cast doubt on the results of the study. When no causal relationship is being studied, randomization is less important. For instance, in purely descriptive research, randomization is of little use.

A Survey Sample Example

As an example of the ways replication, manipulation, and randomization can impact research design, consider a survey sample project about breakfast-cereal-purchasing habits. The survey is very simple: Do the participants surveyed purchase Bran Puffs or Honey Flakes? Conducting this survey without replication would mean gathering responses from only one household or one neighborhood. With replication, the survey would be conducted on a large number of participants, enough to provide a statistically significant data set. If the group of participants is sufficiently large and demographically diverse, the researchers may be able to uncover trends of preferences according to data collected about the participants. Because this version of the survey is just an opinion survey, with no hypothesis and no control group, manipulation is not a factor. However, randomization is key to preserving the validity of the results, primarily in selecting the participants so as to avoid selection bias on the part of the researchers.

Another version of the survey could use replication, manipulation, and randomization to divide the participants into experimental and control groups. In this version, participants are randomly selected and randomly assigned to groups. The control group receives the simple survey already described, but the experimental group receives a push poll instead. Although the push poll includes the question on the simple survey—"Do you prefer to purchase Bran Puffs or Honey Flakes?"—it asks another question first: "Are you aware of recent news stories alleging unhealthy side effects of Bran Puffs cereal?" This push poll question is the manipulated variable, testing the hypothesis that the experimental group will show a less favorable opinion of Bran Puffs.

Strategies of Research Design

Replication, manipulation, and randomization are all powerful tools to consider in designing a research project. A project need not make use of all three, as they will not always be appropriate or even possible, but they do need to be considered before being dismissed. Manipulation is the defining characteristic of most experimental research. Randomization and replication are strong protections against confounding.

Bill Kte'pi, MA

Further Reading

Alferes, Valentim R. *Methods of Randomization in Experimental Design*. Sage Publications, 2012.

Creswell, John W. *Research Design: Qualitative, Quantitative, and Mixed Methods Approaches*. 4th ed., Sage Publications, 2014.

De Vaus, David A. *Research Design in Social Research*. Sage Publications, 2001.

Dugard, Pat, et al. *Single-Case and Small-n Experimental Designs: A Practical Guide to Randomization Tests*. 2nd ed., Routledge, 2012.

Goodwin, C. James, and Kerri A. Goodwin. *Research in Psychology: Methods and Design*. 8th ed., John Wiley & Sons, 2017.

Kazdin, Alan E. *Research Design in Clinical Psychology*. 5th ed., Pearson, 2017.

Montgomery, Douglas C. *Design and Analysis of Experiments*. 8th ed., John Wiley & Sons, 2013.

Rosenberger, William F., and John M. Lachin. *Randomization in Clinical Trials: Theory and Practice*. 2nd ed., John Wiley & Sons, 2016.

Retrospective Cohort Design

Fields of Study

Experimental Design; Research Design

Abstract

A retrospective cohort design is a type of observational study that measures historical variables in a specific group of individuals that have the same exposure to determine risk of developing a certain disease or other outcome of interest. This study type can help investigators ask new questions and determine the sample size for a larger prospective study.

Principal Terms

- **cohort:** a group of purposefully chosen research subjects who share one or more defining characteristics.

Independent variable — **Dependent variable**

Exposure ← Outcome 1 / Outcome 2

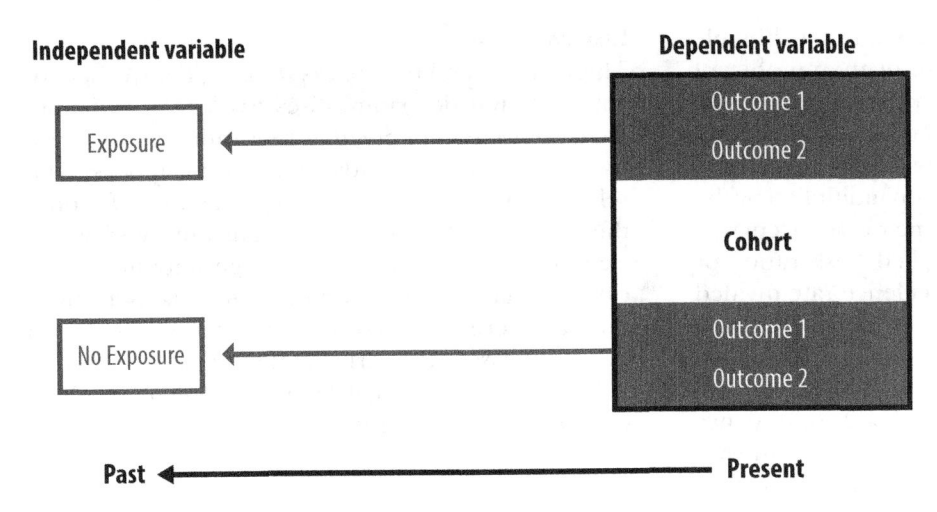

Cohort

No Exposure ← Outcome 1 / Outcome 2

Past ← Present

- **population:** the entire pool of individuals or other subjects that a researcher wishes to obtain information about.
- **posttest-only research:** a research design in which no pretest is administered prior to the treatment.

Cohorts and Cohort Studies

A cohort is defined as members of a population who share a common trait. That trait is most often the experience of the same event during a specific period of time. For example, a cohort may consist of people who were born in the same year or who graduated in the same year. In medical or public health research, a cohort may consist of people who undergo the same procedure or are exposed to the same potentially disease-causing agent. The sampling strategy for a cohort study may be based on population or on an exposure of interest.

Cohort studies uncover information that might otherwise go unnoticed. For instance, both cohort and case-control studies are common research designs in epidemiology. A case-control study starts with a group of individuals who have a disease and selects a group of individuals from the same population who do not have the disease to serve as a control group. It investigates both to determine what past exposures are most closely associated with their current health status. Such a study can be used to determine the odds ratio (the probability of an event occurring divided by the probability of it not occurring) of an individual developing the disease based on certain exposures. It can only examine one possible outcome of those exposures, however, and it cannot estimate the incidence of the disease in the larger population. In contrast, a cohort study based on exposure begins before any disease has developed, so it has the potential to observe multiple outcomes. It can also be used to estimate the incidence rate among the larger exposed and unexposed populations.

A cohort research design is defined temporally (based on point in time measured). Specifically, it is either prospective or retrospective. A prospective cohort study is a type of pretest-posttest research that begins before the outcome of interest has taken place. Again using the exposure example, a prospective cohort study would identify a cohort based on exposure status, gather baseline information before disease has developed, and then follow the subjects over a period of time to find out if they develop a disease in the future. The study would also track any changes in exposure status over time. A retrospective cohort study is a type of posttest-only research that begins after some of the subjects have developed the outcome of interest. In a retrospective cohort study, a researcher uses historical data to identify other members of the cohort from a time before the disease developed, determine their exposure status then, and assesses the ultimate outcome. Rather than following subjects over time, this type of cohort study looks solely at events that have occurred in the past.

At first glance, the retrospective cohort design might seem inferior to the prospective cohort design. Unwanted extraneous factors (confounders) cannot be controlled in a retrospective study, and the researcher must rely entirely on historical records or individuals' imperfect memories. However, the retrospective cohort design offers several advantages over the prospective design. It is more cost-effective, much less time-consuming, and better for studying rare exposures or examining multiple outcomes at the individual level.

Statistics: Measures of Association

Measurements of interest in a retrospective cohort design include the incidence rate of disease (or other outcome), incidence density ratio, and attributable risk. Incidence rate is a measure of the number of new cases developed over a certain time interval. The numerator is the number of new cases in a given cohort or population. The denominator is the total number in the cohort or population multiplied by the number of years over which the new cases occurred. Incidence density ratio, also called "risk ratio" or "relative risk," is the exposed incidence rate divided by the unexposed incidence rate. Attributable risk is the unexposed incidence rate subtracted from the exposed incidence rate.

To help calculate these items, a 2 × 2 contingency table can be constructed and the cells filled in with appropriate subject numbers. The 2 × 2 contingency table should appear as follows:

Limitations

There are several drawbacks to using a retrospective cohort study to determine disease risk in a population. First, they are not ideal for examining very rare diseases, although odds ratios can approximate relative risk in some very rare diseases. Second, the quality of data available might not be ideal for examining a disease outcome. Large amounts of data can be obtained from national data sets, but unless the data were collected in a way that allows one to directly measure the variable of interest, they may be of limited value. Third, it can be difficult to match an exposed group to a control group.

Exposed	Disease	No Disease	Total
Yes	a	b	a + b
No	c	d	c + d
Total	a + c	b + d	a + b + c + d

Applications

Determining the risk of developing diseases based on exposure is a useful and desirable undertaking for medical investigators. There are many different applications for the retrospective cohort design, from calculating risk of chronic diseases to determining the source of a foodborne illness or a communicable disease outbreak. Calculating incidence is often done initially to determine the disease burden or other outcome so that planning can be performed to prevent or stop a disease process from occurring. For example, calculating the incidence rate of lung disease among those exposed to asbestos and comparing it to the incidence rate among the unexposed population via relative risk or attributable risk would provide compelling evidence that asbestos should not be used in new construction.

Mandy McBroom-Zahn

Further Reading

Agresti, Alan. *Categorical Data Analysis*. 3rd ed., John Wiley & Sons, 2013.

Aschengrau, Ann, and George R. Seage III. *Epidemiology in Public Health*. 3rd ed., Jones & Bartlett Learning, 2014.

Friis, Robert H., and Thomas A. Sellers. *Epidemiology for Public Health Practice*. 5th ed., Jones & Bartlett Learning, 2014.

Kelsey, Jennifer L., et al. *Methods in Observational Epidemiology*. 2nd ed., Oxford UP, 1996.

Mantel, Nathan, and William Haenszel. "Statistical Aspects of the Analysis of Data from Retrospective Studies of Disease." *Journal of the National Cancer Institute*, vol. 22, no. 4, 1959, pp. 719–48.

Riegelman, Richard K. *Studying a Study & Testing a Test: Reading Evidence-Based Health Research*. 6th ed., Lippincott Williams & Wilkins, 2013.

Rothman, Kenneth J., et al. *Modern Epidemiology*. 3rd ed., Lippincott Williams & Wilkins, 2008.

Sedgwick, Philip. "Retrospective Cohort Studies: Advantages and Disadvantages." *The BMJ*, vol. 348, no. 7943, 2014, doi:10.1136/bmj.g1072. Accessed 11 Apr. 2017.

> **SAMPLE PROBLEM**
>
> At a reception, the caterers offered two choices for dinner, steak and fish. However, they did not properly store the fish. Out of seventy guests who ordered the fish for dinner, twenty people developed signs of food poisoning. Out of sixty guests who ordered the steak, five developed signs of food poisoning. Assuming that having eaten the fish constitutes exposure, construct a 2×2 contingency table and calculate the incidence of food poisoning among the exposed population and the unexposed population, the incidence density ratio, and the attributable risk.
>
> **Answer:**
>
> | Exposed? | Symptoms | No Symptoms | Total |
> |---|---|---|---|
> | Yes | 20 | 50 | 70 |
> | No | 5 | 55 | 60 |
> | Total | 25 | 105 | 130 |
>
> To calculate the incidence rate among the exposed population, divide the number of people who were exposed and developed symptoms by the total number of people exposed:
>
> $$\text{incidence rate}_{exposed} = 20 / 70 = 0.286$$
>
> Repeat this calculation for the unexposed population:
>
> $$\text{incidence rate}_{unexposed} = 5 / 60 = 0.0833$$
>
> The incidence ratio was 28.6 percent among the exposed population and 8.33 percent among the unexposed population.
> To calculate the incidence density ratio, divide the exposed incidence rate by the unexposed incidence rate:
>
> $$\text{incidence density ratio} = 28.6 / 8.33 = 3.433$$
>
> To calculate the attributable risk, subtract the unexposed incidence rate from the exposed incidence rate:
>
> $$\text{attributable risk} = 28.6 - 8.33 = 20.17$$
>
> The attributable risk was 20.17 percent.

Sample Frame Error

FIELDS OF STUDY

Sampling Design; Sampling Techniques; Research Theory

ABSTRACT

Sample frame errors are caused by poor sampling framework. This error can occur in both qualitative and quantitative studies that attempt to study a small portion of a population and then infer that these study findings apply to a larger population. When this smaller group is selected from a source that incompletely or inaccurately represents the larger population, usually in a way that biases the study, there has been a sample frame error.

PRINCIPAL TERMS

- **inference:** a step in reasoning that develops a conclusion from observations, whether by deduction, induction, or abduction.
- **population:** the entire pool of individuals or other subjects that a researcher wishes to obtain information about.
- **sample group:** the portion of a larger population that is selected to participate in an experiment.
- **sampling framework:** the method or group of methods and techniques that determine which participants will be chosen for study.

SAMPLING FRAMEWORKS

Sample frame errors occur when an improper sample frame is selected for a research study. A sample frame is the source from which the participants or subjects of a study (the sample group) are chosen. It is meant to include all members of the population of interest, so that the sample group will accurately reflect the larger population. Poor sample frame selection occurs when a researcher is developing the study's sampling framework, which outlines not only the source of the sample group but also the method by which it will be selected.

The goal of studying a sample rather than the entire population is to save time and money. However, the study must be carefully designed to produce results that describe the entire population. Randomization of sample selection, or random sampling, is considered one of the best ways to reduce bias. It is intended to ensure that there is as little difference as possible between the measured traits of the sample and the actual occurrence of those traits in the total population. However, the benefits of random sampling will be voided if the wrong sample frame is used in the study. Sample frame errors negatively affect a study by preventing researchers from delivering findings, or inferences, from the study results that are actually reflective of the population of interest. This type of error often occurs when the researcher does not fully understand the target population.

PROBLEMS WITH SAMPLE FRAMES

While an ideal sample frame contains all elements of the population, such a frame may not always be available. In these cases, researchers will choose a frame that closely approximates the overall population but may exclude some elements. For example, if the population of interest is all people living in a certain city, a researcher may select their sample group from a telephone directory for that city. The vast majority of residents will be listed in the directory, but inevitably some residents will not be.

Using a narrowed frame to identify a sample can cause problems for researchers. A list might include the same individual twice, or it might have been created in a biased way. Additionally, it might contain old information, or it might not be inclusive enough. A common problem in online research is coverage error. This is a specific kind of sample frame error that arises from the fact that certain demographic groups, such as the elderly and people with lower incomes, have lower rates of Internet access compared to other groups. These groups are then underrepresented in the sample frame because they cannot participate in online research. Coverage error also occurs when individuals have the ability to respond to online surveys but still prefer to respond in person or on paper.

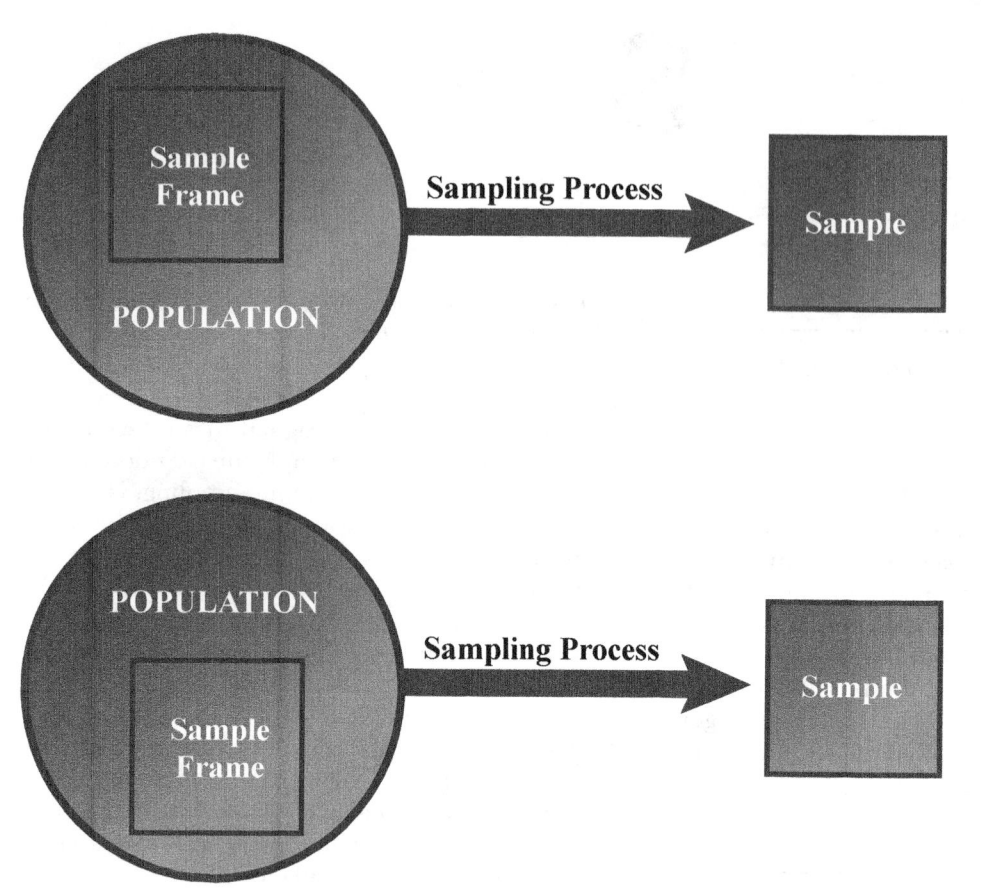

Dangers of Sample Frame Error

A historic sample frame error occurred during the American presidential election of 1936. Researchers wanted to predict who would win the election. It would have been impossible to interview every single voter, so they decided to study a specific subset of the target population. The researchers used telephone directories and car registries as their sample frame. They took names from telephone directories and car registries throughout the country, surveyed a random sample that they drew from those names, and based their findings on the results. However, because cars and telephones were still considered luxuries at the time and were mostly owned by the wealthy, the sample was heavily biased toward rich people, who were more likely to vote Republican. This produced a sample frame error. A survey of the rich could not accurately describe how poorer Americans would vote. As a result, the researchers predicted that Republican candidate Alf Landon would win the election, when in fact Democratic candidate Franklin D. Roosevelt won with the largest electoral margin (523–8) in US history.

Occurrence in Many Fields

Sample frame errors can occur in many fields. For example, sociologists are careful to consider all of the ways that a sample frame error could occur when designing their studies. This includes accounting for demographic information such as gender, race, ethnicity, and class. Such considerations ensure that the population that the study seeks to address is the population that is surveyed. Medical researchers are also careful to ensure that their studies address groups that are commonly missing from surveys, such as members of the military, college students living in dormitories, and prisoners. The locations of each of these populations and the limitations put on their access to information require that researchers make an extra effort to ensure that they are included in the study. If they are not, the study report must clearly indicate that the findings do not speak to the experiences of such groups.

Allison Hahn, PhD

> **SAMPLE PROBLEM**
>
> Several researchers are studying how work stress affects family dynamics. They want to find out if parents with high-stress jobs spend less time with their children than do parents with low-stress jobs. The researchers have divided the parents' careers into two sets, high-stress career types (such as EMTs, stockbrokers, and lawyers) and low-stress career types (such as massage therapists, bakers, and hair stylists). They plan to interview a sample from each group and measure the amount of time each study participant spends with his or her children in a week. What is one way that a sample frame error might occur in this study? How might this sample frame error be prevented?
>
> **Answer:**
>
> Societal gender norms are one reason why this study could experience a sample frame error. The positions identified as high stress are more commonly held by men, and the low-stress careers are more commonly held by women. Additionally, in many communities, women are expected to spend more time with their children, regardless of their career choice or stress level. As such, the study will most likely find that women spend more time with children than men. The study, by attempting to compare all individuals from high-stress jobs to all individuals from low-stress jobs, could falsely evaluate the impact of stress load on the time spent with children due to a gender bias built into the sample frame. This bias could be prevented if the study restricted itself to only studying women working in each career, or if the high-stress and low-stress jobs were redefined to avoid gender bias.

FURTHER READING

Biemer, Paul P., et al., editors. *Total Survey Error in Practice.* John Wiley & Sons, 2017.

Blair, Edward, and Johnny Blair. *Applied Survey Sampling.* Sage Publications, 2015.

Braithwaite, Dejana, et al. "Using the Internet to Conduct Surveys of Health Professionals: A Valid Alternative?" *Family Practice*, vol. 20, no. 5, 2003, pp. 545–51. MEDLINE Complete, search.ebscohost.com/login.aspx?direct=true&db=mdc&AN=14507796&site=ehost-live. Accessed 30 Apr. 2017.

Christensen, Larry B., et al. *Research Methods, Design, and Analysis.* 12th ed., Pearson, 2014.

Crespi, Irving. *Pre-election Polling: Sources of Accuracy and Error.* Russell Sage Foundation, 1988.

DiGaetano, Ralph. "Sample Frame and Related Sample Design Issues for Surveys of Physicians and Physician Practices." *Evaluation & the Health Professions*, vol. 36, no. 3, 2013, pp. 296–329.

Levy, Paul S., and Stanley Lemeshow. *Sampling of Populations: Methods and Applications.* 4th ed., John Wiley & Sons, 2009.

Litwin, Mark S. *How to Measure Survey Reliability and Validity.* Sage Publications, 1995.

Sampling Design vs. Experimental Design

FIELDS OF STUDY

Research Theory; Sampling Design; Experimental Design

ABSTRACT

Sampling design and experimental design together comprise the planning or conceptual stage of research work, and attention to them rewards researchers with more reliable results in the execution stage. From an experimenter's perspective, sampling design is an aspect of experimental design; however, it is applicable to nonexperimental research projects as well.

PRINCIPAL TERMS

- **cluster sampling:** a probabilistic sampling method in which the population being studied is divided into

	More samples	**Fewer samples**
More treatments and controls	Multiple replications within a control group and multiple treatment levels **Highest statistical power**	Few replications within a control group and multiple treatment levels **Medium statistical power**
Fewer treatments and controls	Multiple replications within a single treatment and a single control group **Medium statistical power**	Case study: single sample, one or no treatments, one or no controls **Lowest statistical power**

Experimental design is the design of laboratory or field experiments in which evidence is collected in support of a hypothesis. In the case of true experiments, this requires the manipulation of a variable in order to observe the subsequent effects on other variables. Experiments are usually divided into control groups and experimental groups for the purposes of comparison. Random sampling is used to determine which participants are assigned to which group. Sampling design is the researcher's method for studying a population. For instance, in cluster sampling, the population under study is divided into groups based on common characteristics from which samples are randomly selected.

Sampling can be used in numerous types of research, such as quantitative, qualitative, descriptive, and experimental. Sampling design considers whether the results of the research need to be generalized. Research that seeks only to add to the store of information about the exact things that it will study—rather than a representative sample of a larger population—is not concerned with this generalization. Otherwise, the sampling design is informed by the differences between the population the results need to be generalized to and the population the researchers have access to. The type of sampling method and sample size are considered to reduce sampling error. Experimental design needs to keep in mind the concerns of sampling design. Statistical control should be used to minimize bias and the effects of confounding and extraneous variables on the data.

subsets called "clusters" based on a shared characteristic unrelated to the research question, such as geographic location, and then multiple clusters are selected at random for comprehensive examination.
- **control group:** the subjects in an experiment who do not receive any intervention or who receive a placebo or sham intervention.
- **experimental group:** the subjects in an experimental research study who receive the intervention or treatment being tested.
- **random sampling:** one of several probabilistic sampling methods in which individuals within the population being studied are selected for examination at random, so that each member of the population has a chance (often an equal chance) of being included.
- **variable:** in research or mathematics, an element that may take on different values under different conditions.

Sampling Design and Experimental Design

Sampling design and experimental design are both important stages of research. Having a clear experimental design helps a researcher determine the appropriate sampling method to use to get the intended outcome. There are many methods of sampling, not all of which are appropriate to every kind of experiment. Furthermore, not every type of research that includes sampling is experimental in nature.

Design Considerations

Unlike the execution of a research project, the design does not consume research materials, lab time, or travel expenses. It does not even need research participants. It takes only the time and mental energies of the researchers. Time and care put into choosing the right sampling methods, and the right experimental design if applicable, pay off in fewer wasted resources when the research project begins. This also allows for greater chances of successful research results.

The greatest burdens of design are that it can be time-consuming and must deal with practicality. The ideal sampling method is not always the available one. For example, a graduate student planning a research project may have access only to participants they can reach locally or online. They may also have a short window of time in which to conduct research. Even a professional researcher has a budget to consider. This budget impacts the scale of sampling and may affect experimental design choices such as whether to pursue a lab experiment or a field experiment.

Gender Bias in Work-Life Balance

A University of Plymouth study sought to explore the different ways fathers and mothers are treated when they seek part-time work. Previous studies had shown that fathers were more likely to be hired for full-time work than mothers. The new study sought to find if this were also true for part-time work. This hypothesis was formed in response to social trends showing that more mothers were seeking full-time employment while more fathers were reducing their work time in order to spend more time at home. The study included an online survey sent to about one hundred managers who were asked to rate hypothetical job applicants, a focus group of managers and working parents, and interviews conducted with managers and parents.

The results found that the hypothetical applicants who were mothers were more likely to be rated highly than the fathers, despite having the same qualifications, when both were seeking part-time work. Focus groups and interviews confirmed that managers were comfortable with mothers who did not want full-time work. However, they viewed men seeking part-time work with suspicion.

The survey, focus group, and interviews were part of the experimental design in this study. The experimental component was the manipulated variable of the gender of the hypothetical applicant. The sampling design was the researchers' methods of choosing which managers to survey. This involves whether they were randomly distributed across region and industry, for instance. It also involves whether male and female managers were equally represented, in order to account for the possibility of manager gender having a confounding effect.

The Importance of Design

Any successful research has two stages: conception and execution. While problems can occur in execution, they can be guarded against, and their effects mitigated, in the conception stage. Problems in conception, however, cannot be similarly fixed in execution. More often, small errors compound themselves into larger ones. For these reasons, sampling design and experimental design are the most important parts of a research project to ensure accurate, representative results.

Bill Kte'pi, MA

Further Reading

Christensen, Larry B., et al. *Research Methods, Design, and Analysis.* 12th ed., Pearson, 2014.

Cochran, William G. *Sampling Techniques.* 3rd ed., Wiley, 1977.

Goodwin, C. James, and Kerri A. Goodwin. *Research in Psychology: Methods and Design.* 8th ed., Wiley, 2017.

Kotu, Vijay, and Bala Deshpande. *Predictive Analytics and Data Mining.* Morgan Kaufmann, 2014.

Lohr, Sharon L. *Sampling: Design and Analysis.* 2nd ed., Brooks/Cole, 2010.

Strang, Kenneth D., editor. *The Palgrave Handbook of Research Design in Business and Management.* Palgrave Macmillan, 2015.

Thompson, Steven K. *Sampling.* 3rd ed., Wiley, 2012.

Williams, Alan. "Fathers Face Negative Bias over Quest for Work-Life Balance, Study Suggests." *Phys.org*, Omicron Technology, 22 Feb. 2017, phys.org/news/2017-02-fathers-negative-bias-quest-work-life.html. Accessed 10 Apr. 2017.

Sampling Design: Randomness and Interspersion

FIELDS OF STUDY

Sampling Design; Research Theory

ABSTRACT

Randomness and interspersion are related concepts in research design and sampling. Most sampling methods used in research are random in order to reduce bias. Interspersion is often considered in combination with randomization to avoid confounding and preserve the appropriate distribution of samples or experiment treatments as needed for the nature of the research being conducted.

PRINCIPAL TERMS

- **cluster sampling:** a probabilistic sampling method in which the population being studied is divided into subsets called "clusters" based on a shared characteristic unrelated to the research question, such as geographic location, and then multiple clusters are selected at random for comprehensive examination.
- **distribution:** a mathematical function that defines all possible values of a variable according to either the probability of their occurring or the frequency with which they occur.
- **interspersion:** the distribution of samples throughout an area (target population).
- **random sampling:** one of several probabilistic sampling methods in which individuals within the population being studied are selected for examination at random, so that each member of the population has a chance (often an equal chance) of being included.
- **stratified random sampling:** a probabilistic sampling method in which the population to be studied is divided into mutually exclusive subgroups, or strata, and then random samples are taken from each stratum in numbers proportional to that stratum's share of the larger population.

Randomness and Interspersion

In sampling design, the greatest challenge is negotiating between the demands of selecting a design that is manageable for the research project and avoiding the effects of sampling error. The goal of sampling is to study a sample of a population that is sufficiently representative of that population in respects relevant to the research being conducted. This way, the results of the research conducted on the sample will be more generalizable to the original, total population.

Sampling error is the difference in the sampling distribution of a characteristic between the sample and the population from which it was taken. For instance, any given college campus is unrepresentative of the human population as a whole with respect to age. This has been raised as a source of sampling error in a number of social science and psychological experiments conducted using college students as the sole participants. While sufficiently large samples protect against error, they are impossible to obtain in most cases. They may be out of reach due to the amount of time, money, or other resources such sampling would demand. Sampling methods and statistical controls have been developed to guard against error instead.

Random sampling is a powerful tool for guarding against selection bias and other errors. In simple random sampling, every member of the population has an equal chance of selection for study. Cluster sampling divides the population into clusters, usually defined geographically, and then randomly selects some of the clusters for inclusion. Stratified random sampling sorts the population into groups according to certain traits and then randomly samples from those groups in proportion to their representation in the population. Sampling methods are regularly combined.

Interspersion is another means by which to avoid sampling error. Researchers usually strive to make sure that the samples studied are well distributed, or interspersed, across the area of the target population. Random samples do not necessarily guarantee good interspersion. Imagine an experiment on a deck garden that is partially in shade. If samples are taken completely at random, with every plant having an equal chance of being sampled, they may come entirely from plants in the shade or entirely from plants in the sun. Cluster or stratified sampling can be used to ensure interspersion in the sample, including

Systematic Sampling

Random Sampling

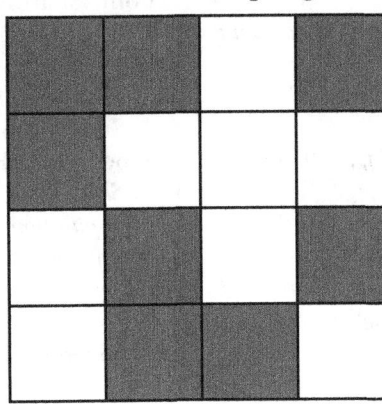

both sun and shade plants. Interspersion can also be achieved through deliberate, nonrandom sampling, if appropriate to the study.

Randomization and interspersion are also used in experimental design to avoid selection bias and other sources of experimenter error. For instance, participants are often randomly assigned to experimental or control groups. In studies where treatments are administered to particular locations rather than to subjects in a laboratory, the interspersion of treatments is an important part of experimental design.

STRENGTHS AND DIFFICULTIES

Different sampling methods lend themselves best to different kinds of research. Interspersion without random sampling is a type of nonprobability sampling called purposive sampling. For example, heterogeneity sampling involves selecting a deliberately diverse group of participants without concern to keeping that diversity in proportion to the population from which it was sampled. This is appropriate only to certain kinds of research and would skew the results of most others. However, it is sometimes used for specific purposes in market research and psychological studies to get a target sample quickly.

The most difficult aspect of sampling or experimental design is knowing which method to use. In many experiments, randomness is best viewed as a means to achieve interspersion. Therefore, sampling methods and methods of administering treatments should be chosen accordingly. In some cases, to get greater interspersion, the researcher may choose to use systematic sampling rather than simple random sampling. This method distributes samples throughout the target area by selecting them first at random and then by regular intervals. Simple random samples by themselves are more prone to random sampling error unless they are sufficiently large.

ANIMAL BEHAVIOR EXPERIMENTS

In a 2003 paper, R. Haven Wiley wrote about randomness and interspersion in animal behavior experiments. He pointed out that when dealing with the natural world, no ideal experimental design exists. Instead, the challenge of design is to manage compromises. The researcher must identify the challenges, acknowledge the problems introduced when adopting an imperfect method, and justify its inclusion by pointing out its strengths relative to alternatives. The importance of interspersion in treatments is underscored by a birdsong playback experiment Wiley recounts. Recordings of birdsong are played in a natural environment in order to observe birds' responses. If these playbacks are not sufficiently interspersed, a subject's response to the playback can influence a neighboring subject's response. This in turn has ramifications on the use of treatment and control groups. In the playback experiment, interspersion is necessary to keep subjects sufficiently segregated. Random assignment is used to assign them to control or treatment groups.

DESIGN STRATEGY

Wiley's description of the birdsong experiment shows the relationship between interspersion and randomization in the research design stage. While both guard against error, they do so in different ways. They can also be at odds with one another. For instance, simple randomization can act against interspersion. As the Wiley example shows, interspersion is more important in some work than in others. In experimental design, it is most important when insufficiently interspersed treatments are at risk of contamination. Interspersion here is critical to the work's external validity. This is the criterion that measures whether the research results can be generalized adequately.

Bill Kte'pi, MA

FURTHER READING

Agresti, Alan. *An Introduction to Categorical Data Analysis.* 2nd ed., Wiley-Interscience, 2007.

Biemer, Paul P., et al., editors. *Total Survey Error in Practice.* Wiley, 2017.

Chaudhuri, Arijit. *Modern Survey Sampling.* CRC Press, 2014.

Cochran, William G. *Sampling Techniques.* 3rd ed., Wiley, 1977.

Daniel, Johnnie. *Sampling Essentials: Practical Guidelines for Making Sampling Choices.* Sage, 2012.

Lohr, Sharon L. *Sampling: Design and Analysis.* 2nd ed., Brooks/Cole, 2010.

Rosenberger, William F., and John M. Lachin. *Randomization in Clinical Trials: Theory and Practice.* 2nd ed., John Wiley & Sons, 2016.

Seber, G. A. F., and Mohammad M. Salehi. *Adaptive Sampling Designs: Inference for Sparse and Clustered Populations.* Springer, 2013.

Thompson, Steven K. *Sampling.* 3rd ed., Wiley, 2012.

Wiley, R. Haven. "Is There an Ideal Behavioural Experiment?" *Animal Behaviour*, vol. 66, 2003, pp. 585–88.

SAMPLING FRAMEWORK

FIELDS OF STUDY

Sampling Design; Sampling Techniques

ABSTRACT

"Sampling framework" refers to the tools used to define the population from which samples are drawn for study and the chosen methods for sample selection. Its purpose is to guide the sampling and data collection processes. It is a key part of the research design, as researchers have a wide array of sampling techniques from which to choose, according to their research question and goals.

PRINCIPAL TERMS

- **attribute:** in experimental research, a characteristic or quality possessed by a subject in an experiment and represented by a variable.
- **parameter:** in statistics, a quantity that defines a certain characteristic of a population, such as mean, median, mode, or standard deviation.
- **sample frame error:** sampling errors that occur when the sample frame does not represent the total population adequately.
- **variable:** in research or mathematics, an element that may take on different values under different conditions.

What Is a Sampling Framework?

It is often difficult or impossible to survey all members of a population for a research study. Sampling is therefore essential to research. Sampling allows researchers to draw conclusions about a large population from a limited number of observations. However, the members of the sample group must be representative of the entire population. Because researchers often lack a definitive list of all members of the population under study, a sampling framework helps to guide the sampling process. A sampling framework outlines the members and the attributes of the population of interest. It also specifies the sampling methods that will best answer the research question. A comprehensive sampling framework helps researchers avoid sample frame errors and other forms of bias.

"Sampling framework" refers to the sampling methods selected for a research project. It encompasses the sample frame, which attempts to list and identify the members of the population of interest as comprehensively and accurately as possible. A sampling framework guides the selection of subjects from the population for further examination. A comprehensive sampling framework ensures the most accurate results. In general, the best sample is selected randomly. Random sampling techniques ensure that each member of the target population has an equal probability of being selected. If a sample group is not truly representative of the population, it is considered biased. Any results derived from a biased study will be unreliable.

When determining the sampling framework for a study, researchers have a wide array of sampling methods from which to choose. They often select a combination of sampling techniques based on the research question. Sampling methods can be random or nonrandom. Random sampling schemes are the best strategy for ensuring a sample is representative. In some cases, a researcher might decide to use a nonrandom sampling method. For instance, for a case study, a researcher may purposefully select individuals who best represent the phenomenon under study.

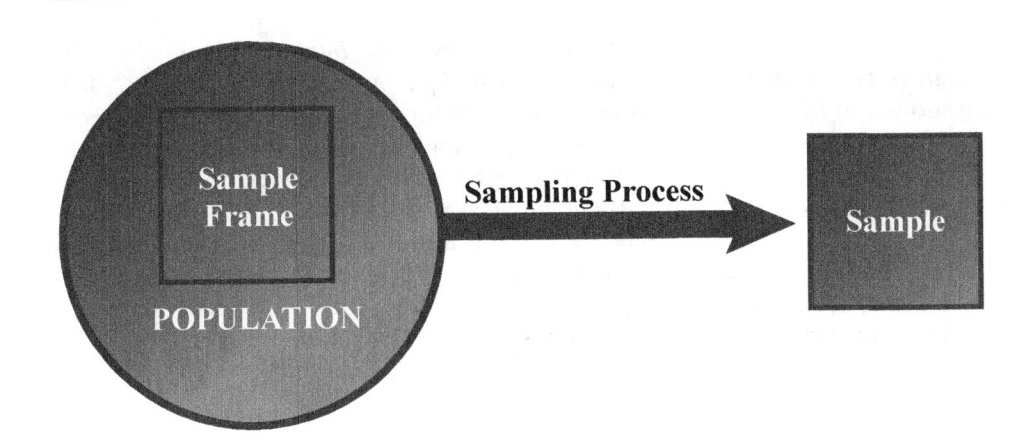

Selecting a Sampling Framework

The design of a study should determine the methods for sampling and data collection at the outset. In research, the main goal is to obtain a sample whose attributes match those of the desired target population as closely as possible. This helps ensure representativeness and relevancy. The relevancy of particular variables change based on the purpose of the study. For instance, a study of health outcomes might take into consideration members' weight and age. In a study of diversity, factors such as race, ethnicity, and gender would be of relevance. For instance, if the population of interest consists of 60 percent women, then the sample should also consist of 60 percent women to be truly representative. Research parameters serve as guidelines for the selection of the sample group. These parameters measure the particular characteristics of a population and help ensure a sample is representative.

The selected sampling framework depends on the attributes and parameters of the target population as well as the researchers' goals and limitations. To obtain accurate information and avoid sample frame errors, a sampling framework must consider the selection of variables for study, sample size, sample frame, sampling process, and sampling validation process. A sample frame must also be well organized and comprehensive. This ensures that the sample frame is accurate, providing identifiers for each individual to avoid duplication and containing accurate contact information.

Ethical considerations are also part of a sampling framework. If the participation of individuals includes identifying features, the sampling framework should list methods to safeguard anonymity and the secure storage of data. It may also be necessary to obtain the informed consent of participating individuals.

Considerations and Applications

Researchers must consider that a sampling framework that was effective for one study or one population may be inadequate for others. For instance, obtaining a representative sample of migrants, who move often, presents unique challenges. Methods such as systematic sampling would be challenging because a comprehensive sample frame of migrants living in an area may not exist. In these cases, researchers may resort to less popular sampling techniques, such as the snowball approach. Snowball sampling relies on respondents to provide names and contact information of others for future study. Snowball sampling is not ideal because it is nonrandom. However, random approaches in this instance could lead to scant inclusion of members of the target population.

In general, the preferred techniques obtain a random sample. However, sometimes participants are selected purposefully. For example, convenience sampling is practical and inexpensive. A research group might want to know whether theatergoers liked a play, so they may stand outside a theater and ask members of the audience for their opinion as they leave. This approach may miss people who are in a hurry or who are shy. However, it is a quick, easy, and cost-effective way for researchers to collect data.

Trudy Mercadal, PhD

Further Reading

Creswell, John W. *A Concise Introduction to Mixed Methods Research*. SAGE Publications, 2014.

Galesic, Mirta, et al. "A Sampling Framework for Uncertainty in Individual Environmental Decisions." *Topics in Cognitive Science*, vol. 8, no. 1, 2016, pp. 242–58.

Kershaw, P., et al. "The Use of Population-Level Data to Advance Interdisciplinary Methodology: A Cell-through-Society Sampling Framework for Child Development Research." *International Journal of Social Research Methodology*, vol. 12, no. 5, 2009, pp. 387–403.

Onwuegbuzie, Anthony J., and Nancy L. Leech. "Sampling Designs in Qualitative Research: Making the Sampling Process More Public." *The Qualitative Report*, vol. 12, no. 2, 2007, pp. 238–54.

Sabo, Roy, and Edward Boone. *Statistical Research Methods: A Guide for Non-Statisticians*. Springer, 2013.

Vigneswaran, Darshan. "Residential Sampling and Johannesburg's Forced Migrants." *Journal of Refugee Studies*, vol. 22, no. 4, 2009, pp. 439–59.

SAMPLING FRAMEWORK SAMPLE PROBLEM

A researcher wants to determine the prevalence of people who wear eyeglasses across genders. The researcher decides to track the presence or absence of eyeglasses on people entering and exiting a department store. The store has two entrances: entrance A and entrance B. However, the research team decides to watch only entrance A due to limited staffing.

Can you see how this research design will result in a sample frame error?

Answer:

Some people will enter the store through entrance A and exit through entrance B, causing them to be counted once. Others will enter through entrance B and exit through entrance A, also being counted once. There will be those who enter and exit through entrance A, resulting in duplication. Those who enter and exit through entrance B will never be counted. This study also failed to provide sampled individuals with unique identifiers. Such identifiers could be used to avoid duplication.

Furthermore, the researchers should have carefully considered whether the patrons of the department store adequately represent the population of interest. If the department store sells high-end goods, then the study might overrepresent wealthy people. If the department store only sells women's clothing, then the study might underrepresent men.

Sampling vs. Census

FIELDS OF STUDY

Sampling Design; Sampling Techniques; Statistical Analysis

ABSTRACT

Censuses and sample surveys are two common techniques used to gather data. Conducting a census requires data to be collected from every member of the population being surveyed. Sample surveying allows data to be collected from a subset of the population.

PRINCIPAL TERMS

- **convenience sampling:** a non-probabilistic sampling method in which members of the sample group are drawn from a population based on their availability and ease of access; also called accidental sampling or haphazard sampling.
- **simple random sampling:** a probabilistic sampling method in which members of the sample group are drawn from the population at random, so that each member of the population has an equal chance of being included.

Target Population → **Census Population** (Sample)

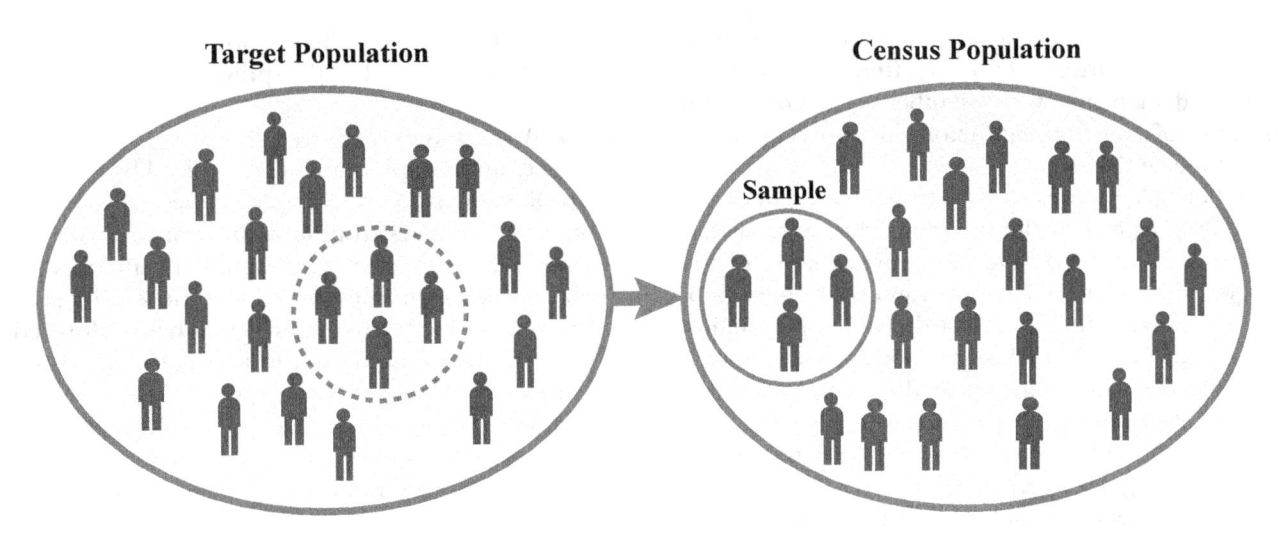

- **voluntary response sampling:** a non-probabilistic sampling method in which members of the population of interest are made aware of a survey or study and can choose whether or not to participate.

Overview of Censuses and Sampling

A population is a set of elements that have at least one common trait and are studied for statistical purposes. A census compiles target statistics from each individual in the population. A sample only collects data from a subset of the population.

Since ancient times, censuses have been used by leaders to understand important statistics about their people. The first census in the United States was conducted in 1790 and resulted in a population count of 3,929,214 people. Since then, the US government has conducted a census every ten years to get an official count of the number of people in the country, as well as descriptive information about them. The government also collects more detailed data on a sample of residents. It began conducting the Current Population Survey in 1940 and the American Community Survey in 2005.

There are two basic ways to collect data about a population. One way is to compile target statistics from every single member of the population. This process is known as a census. A census is defined by its comprehensiveness. The other way is to collect data from a subset of the population, known as a sample. Findings can then be extrapolated to the larger population. A sample survey is more practical than a census, because it does not require data from every member in the population. However, it has a greater margin of error.

There are many ways to select a sample from a population. Sampling methods can be broadly characterized as either probabilistic or non-probabilistic. A probabilistic sampling method is one that involves an element of random chance. Every member of the population has a probability of being selected, and that probability can be calculated for any given individual. A non-probabilistic sampling method is one in which either some members of the population have no probability of being selected, or the probability cannot be calculated for every individual.

One of the easiest sampling methods is convenience sampling. Convenience sampling takes little effort from the sampler, because it involves collecting data from whoever happens to be nearby or otherwise readily available at the time of sampling. It is a non-probabilistic method because it is impossible to calculate the odds of any given member of the population being selected. Another method is simple random sampling. This is the most basic form of probabilistic sampling. In this method, a subset of individuals is chosen at random from the entire population in such a way that every individual has an equal chance of being selected. A simple random sample is much more likely than a convenience sample to be representative of the population as a whole.

Advantages and Disadvantages

Censuses and sample surveys each have advantages and disadvantages. Because a census collects data from every individual in a population, it provides greater accuracy than a sample survey when done correctly. Unfortunately, it is often impossible to collect data from every member of a population because of constraining factors such as population size, time, and cost.

When a census is not possible, researchers use sampling instead. If the sample is chosen in such a way that it is demographically representative of the larger population, researchers can use the data gathered to make estimates about the whole population with a fair degree of accuracy. However, sampling is less effective when studying smaller populations, because larger samples are required to obtain statistically meaningful results. For example, if a researcher were studying a population of one hundred people, they would need a sample size of eighty in order to achieve a standard 95 percent confidence level and 5 percent margin of error.

Another challenge of sampling is avoiding sampling bias. Sampling bias occurs when not all members of the population have an equal chance of being chosen for the sample, and so it is not truly representative. A common type of sampling bias is self-selection bias. This arises when individuals can choose whether or not to be part of the sample, as in voluntary response sampling. Self-selection bias can lead to inaccurate results in any type of study, but it is particularly problematic in opinion polls and surveys, because those who choose to participate tend to be the ones who feel most strongly about a topic, and those with less firmly held viewpoints are underrepresented.

Examples of Census and Sample Surveying

If a basketball coach wanted to know the average shoe size of the team, it would be best to take a census. Typically, a basketball team has around twelve players at any given time. It would take minimal time and effort to collect information about the shoe sizes of all twelve players. A sample would not be the best choice, because in such a small population, the margin of error would be so great that the results would be all but meaningless.

Alternatively, for a scientist researching how a specific bug spray affects mosquitoes, a sample would be more appropriate. The scientist could gather a group of mosquitoes to observe how the spray affects each of them. A census would be impossible, as the scientist would have to observe how all mosquitoes in the world are affected by the spray.

Real-World Application

Censuses and sample surveys are utilized by a variety of professionals. For example, college professors often want feedback from students about a particular course. They may ask a small sample of students for their opinions, or they may have the entire class submit feedback. Once the professor has performed either the sample survey or the census, they can then use the feedback to improve the course.

Knowing when to use a census and when to use a sample survey is important. Both techniques have distinct advantages and disadvantages. Using the appropriate technique enables researchers to make the best possible predictions about the populations they are studying.

Brandon Chupp, Austin Huff, and Daniel Showalter, PhD

Further Reading

Bennett, Jeffrey O., et al. *Statistical Reasoning for Everyday Life.* 4th ed., Pearson Education, 2014.

SAMPLING VS. CENSUS SAMPLE PROBLEM

A statistics teacher wants to compare the average SAT score of the students in their class to the national average SAT score. There are thirty students in the statistics class. Should the teacher conduct a census or a sample survey of the students in order to calculate their average SAT score? Assume that the national average can be looked up online and that all of the students in the class have already taken the SAT.

Answer:

The teacher has direct access to the students in the class, and the number of students is not prohibitively large, so it would not be very difficult to gather information from every student. In addition, because the population is small—only thirty individuals—any sample would have a large margin of error. Thus, a census is the best choice.

Bryan, Thomas. "Basic Sources of Statistics." *The Methods and Materials of Demography*, edited by Jacob S. Siegel and David A. Swanson, 2nd ed., Elsevier Academic Press, 2004, pp. 9–41.

Durrett, Rick. *Elementary Probability for Applications*. Cambridge UP, 2009.

Olofsson, Peter. *Probabilities: The Little Numbers That Rule Our Lives*. 2nd ed., John Wiley & Sons, 2015.

"Statistical Language: Census and Sample." *Australian Bureau of Statistics*, Commonwealth of Australia, 3 July 2013, www.abs.gov.au/websitedbs/a3121120.nsf/home/statistical+language++census+and+sample. Accessed 11 Apr. 2017.

Triola, Mario F. *Elementary Statistics*. 12th ed., Pearson Education, 2014.

Significance Levels

FIELDS OF STUDY

Statistical Analysis; Hypothesis Testing; Variance Analysis

ABSTRACT

The significance level of a study is a statistically determined threshold that indicates the percent chance of the study's null hypothesis being found false when in fact it is true. Significance level is represented by the Greek letter alpha (α). Determining whether a study is statistically significant is a crucial part of statistical hypothesis testing. Significance levels help establish whether results are due to chance. However, statistical significance is not the same as practical significance.

PRINCIPAL TERMS

- **alpha (α):** variable representing the likelihood of a type I error, which occurs when the null hypothesis is rejected despite being true.
- **confidence interval:** a range of values within which the value of an unknown population parameter has a given probability of falling.
- **critical region:** the set of values for which the null hypothesis is rejected.
- **error:** in statistics, a situation in which the null hypothesis is found true despite being false, or found false despite being true.
- ***P* value:** the probability of a test result occurring by chance when assuming that the null hypothesis is true.

OVERVIEW OF SIGNIFICANCE LEVELS

One of the most important steps in statistical hypothesis testing is determining the validity of the null hypothesis. The null hypothesis is the prediction that no significant difference will be found between the control and experimental groups in a study. The null hypothesis can only be rejected if a result is statistically significant. This is established by setting a significance level, represented by the Greek letter alpha (α). The significance level indicates the probability of a type I error, in which the null hypothesis is rejected despite being true. As with other probabilities, the significance level can range from zero to one.

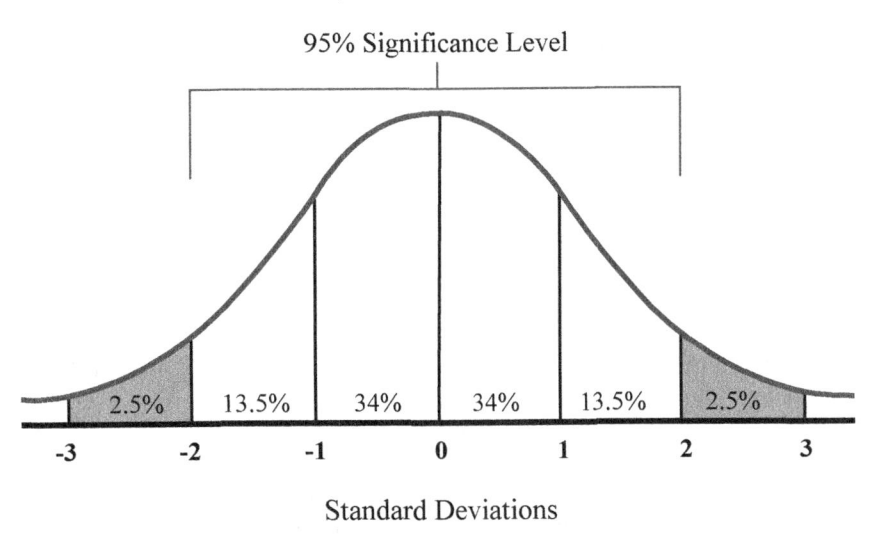

Alpha is ordinarily established before data for a study is collected. This helps researchers avoid intentionally or unintentionally setting alpha to align with sampling results. Once the data is collected, it is displayed on a probability distribution plot, a two-dimensional graph with an *x*-axis and a *y*-axis. Data points tend to be distributed in a bell-shaped curve known as the normal distribution. The critical region is the area on the extreme end of the graph, farthest from the null hypothesis on the x-axis. It can be allocated to just one side of the graph, in what is known as a one-sided test. However, it is most often allocated to both ends of the graph, for a two-sided or two-tailed test. If alpha is set at 0.05, for example, the critical region is the 5 percent of the distribution on the extreme of each tail of the graph (2.5 percent on either side).

Every data set also has a *P* value, which represents the probability of a test result indicating a difference by chance when no difference exists. Like alpha, this value ranges from zero to one. If the *P* value is equal to or less than alpha, the null hypothesis is rejected. If it is greater than alpha, the null hypothesis is not rejected. The rest of the graph includes the area under the bell curve and between the two boundaries of the significance level. This region is known as the confidence interval. It represents the range of probable values for an unknown parameter.

Setting Alpha

A result is said to be statistically significant if its *P* value is less than the significance level established at the outset of the study. Of course, the same data will show different statistical significance when using different alpha values. For example, if alpha is 0.05 and the *P* value is 0.03, the result is statistically significant and the null hypothesis is rejected. However, if alpha is 0.01 with the same *P* value, the result is not statistically significant and the null hypothesis is retained. This is why it is important to set the significance level at the beginning.

Most researchers use a standardized alpha value of 0.05. This is due to the precedent set by pioneer statistician R. A Fisher (1890–1962). It provides a convenient balance between the chances of introducing a type I error and a type II error. A significance level of 0.05 means there is a 5 percent chance that the data will indicate a difference even though no difference is present (a type I error). However, alpha may be raised or lowered depending on the study. The other most common significance level is 0.01. This lowers the chance of a type I error to 1 percent, though it increases the chance of a type II error, in which the null hypothesis is accepted despite being false. This may be preferred for sensitive studies where a high confidence level is needed. The confidence level plus alpha always equals one.

Significance Levels in Action

An experiment researching the effects of math tutoring on student test scores is a useful example for understanding significance levels. Researchers plan to collect scores before and after tutoring and compare them to find the difference. The null hypothesis is that tutoring has no effect on scores. By setting the significance level at 0.05, the researchers accept a 5 percent chance that the data could indicate a difference even if the tutoring has not caused a difference. If the comparison of the test scores shows that, on average, scores after the tutoring were nine points higher than they were before the tutoring, then the next question is to determine if this change is statistically significant.

Making this determination requires calculation of the *P* value of the experimental results. This can be done by graphing the data and the assumed distribution, or by using a formula for the specific type of test. The *P* value is then compared with the significance level. If the *P* value is greater than 0.05, the results are statistically significant. This means the observed change in scores is more likely to have resulted from the tutoring than to have been the product of random chance. If the *P* value is less than or equal to 0.05, the result is not statistically significant and the null hypothesis holds: tutoring likely did not cause the differences in scores.

The Significance of Statistical Significance

Statistical significance is often a confusing subject for many people. It is important to remember that it does not have the same meaning as the word "significance" in normal use; that is, it does not imply importance or meaningfulness. A statistically significant result may or may not be practically significant for various reasons. Setting a significance level is somewhat arbitrary and does not guarantee differentiating between studies with a real effect and those without. Errors are still possible due to the effects of chance. However, significance levels are a useful way of controlling and quantifying error. As a crucial part of statistical hypothesis testing, they are an important tool for researchers of all kinds.

Scott Zimmer, JD

FURTHER READING

De Winter, Joost, and Riender Happee. "Why Selective Publication Of Statistically Significant Results Can Be Effective." *PLOS ONE*, vol. 8, no. 6, 2013, doi:10.1371/journal.pone.0066463. Accessed 10 Apr. 2017.

Gibbs, N. M. "Errors in the Interpretation of 'No Statistically Significant Difference.'" *Anaesthesia and Intensive Care*, vol. 31, no. 2, 2013, pp. 151–3. *MEDLINE Complete*, search.ebscohost.com/login.aspx?direct=true&db=mdc&AN=23577371&site=eds-live. Accessed 10 Apr. 2017.

Sharpe, Donald. "Your Chi-Square Test Is Statistically Significant: Now What?" *Practical Assessment, Research & Evaluation*, vol. 20, no. 8–12, 2015.

Education Source, search.ebscohost.com/login.aspx?direct=true&db=eue&AN=108545923&site=eds-live. Accessed 10 Apr. 2017.

Van Assen, Marcel A. L. M., et al. "Why Publishing Everything Is More Effective Than Selective Publishing of Statistically Significant Results." *PLOS ONE*, vol. 9, no. 1, 2014. doi:10.1371/journal.pone.0084896. Accessed 10 Apr. 2017.

Wong, K. C. "Null Hypothesis Testing (I): 5% Significance Level." *East Asian Archives of Psychiatry*, vol. 26, no. 3, 2016, pp. 112–13. *MEDLINE Complete*, search.ebscohost.com/login.aspx?direct=true&db=mdc&AN=27703100&site=eds-live. Accessed 10 Apr. 2017.

Zhang, Pan, and Cristopher Moore. "Scalable Detection of Statistically Significant Communities and Hierarchies, Using Message-Passing for Modularity." *PNAS*, vol. 111, no. 51, 2014, pp. 18144–49, doi:10.1073/pnas.1409770111. Accessed 10 Apr. 2017.

SOLOMON FOUR-GROUP DESIGN

FIELDS OF STUDY

Experimental Design; Research Design; Statistical Analysis

ABSTRACT

The Solomon four-group design is a hybrid research design that aims to minimize testing threat and confounding variables. It has offers greater internal and external validity than traditional pretest-posttest designs do. However, it is complex and tedious to design.

PRINCIPAL TERMS

- **causation:** a relationship between two events, actions, or other phenomena in which one event is determined to have directly caused or otherwise affected the other.
- **external validity:** the extent to which the results of a study can be generalized to apply to other populations and settings.
- **internal validity:** the extent to which the results of a study can confidently be attributed to the quality being tested, as determined by how well the study was designed to minimize systematic errors.
- **pretest-posttest design:** a research design in which subjects are tested both before and after the experimental treatment or intervention.

MINIMIZING CONFOUNDING VARIABLES

The Solomon four-group design is a highly rigorous research design seen as a benchmark in some types of research. It is especially valued in educational and sociological studies. It combines features of two-group pretest-posttest design and posttest-only design in a hybrid form. As the name implies, the design includes four randomized groups. Group A receives the pretest, treatment, and posttest. Group B is a control group that receives the pretest and posttest but no treatment. Group C receives no pretest but is treated and posttested. Group D is a control that receives no pretest or treatment, only the posttest.

The Solomon four-group design is intended to assess pretest sensitization effects, or testing threat. This is the extent to which a pretest influences and potentially skews experimental results. By including control groups for both pretested and posttest-only subjects, the researcher accounts for this factor. The

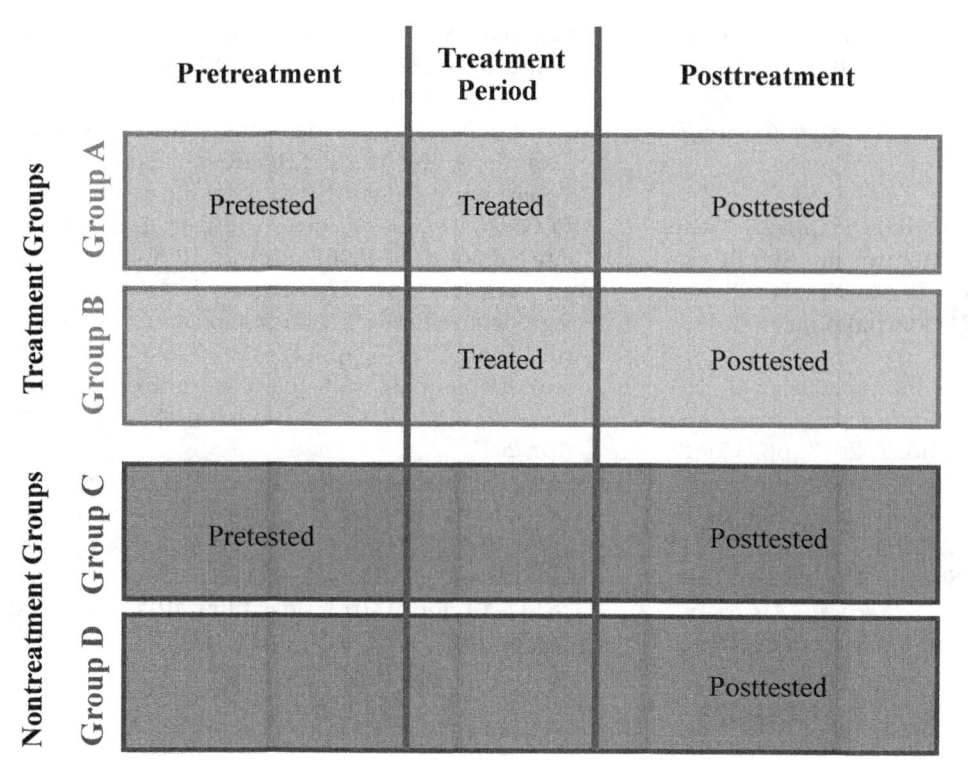

	Pretreatment	Treatment Period	Posttreatment
Treatment Groups — Group A	Pretested	Treated	Posttested
Treatment Groups — Group B		Treated	Posttested
Nontreatment Groups — Group C	Pretested		Posttested
Nontreatment Groups — Group D			Posttested

Solomon four-group design therefore minimizes the effect of confounding variables on the study outcome. The method is effective for analyzing the differences, if any, between various permutations of the pretest, treatment, and posttest outcomes of a dependent variable.

Once the experiment has been conducted, the structure of the Solomon four-group design allows for sophisticated statistical analysis. Methods such as analysis of variance (ANOVA) can be used to determine if pretesting influenced the results. If the pretest does have an effect, it can be evaluated whether it influences the treatment or if the effect is independent of treatment. These questions are answered by directing the comparison between pretest, treatment, and posttest effects.

Increased Validity, Increased Costs

The chief advantage of the Solomon four-group design is its high level of rigor. Unlike simpler experimental or quasi-experimental designs, it ensures high levels of both internal validity and external validity. By comparing the results of each group, researchers can eliminate confounding variables. For example, any temporal distortion due to external factors is accounted for. This rigorous design helps prevent researchers from mistakenly assuming causation. The ability to generalize results and the level of statistical power make it a favored design in many situations.

However, the Solomon four-group design is not always necessarily the best choice for an experiment. This is largely because it is a very complex design to organize and execute. For many researchers, the time and resources required for this design are rarely available. Even if the design is feasible, the complexity of the comparisons and statistical analysis involved can be daunting. Furthermore, some research is not overly concerned with pretest sensitization, making the extra level of rigor unnecessary. If researchers are willing to accept the issues with internal and external validity, a simpler pretest-posttest design can be used, if it is sufficiently powered.

The Solomon Four-Group Design in Practice

The Solomon four-group design is often used in educational research. This is because pretesting of educational interventions is known to have a potential effect on study participants. For example, imagine researchers want to know the impact of a tutoring program on student test scores. A pretest-posttest design might be used. However, students taking the pretest might apply that experience to the tutoring and/or the posttest. This could skew results compared to students who were not pretested.

Using the Solomon four-group design would allow researchers to account for this additional variable. One group of students would take a pretest, receive tutoring, and take a posttest. Another group would be pretested and posttested but not receive tutoring. A third group would not be pretested but would be tutored and take a posttest. The final group would only take a posttest, with no pretest or tutoring. Comparing the test results of each group would reveal whether the pretest influenced the experimental outcomes.

APPLICATIONS

Though cumbersome to use, the Solomon four-group pretest-posttest design is a powerful tool for researchers. It is an invaluable design when the study faces significant threats to internal and external validity in measuring the outcome of interest. It provides critical information about the study itself that can be used to improve the research process. Even to researchers who may rarely use the Solomon four-group design, understanding its advantages and drawbacks is helpful in order to fully grasp the principles of good research design. Knowing the tradeoffs between rigor and complexity is important not only to conducting research but also to evaluating research by others.

Mandy McBroom-Zahn

FURTHER READING

Bernard, Harvey Russell. *Social Research Methods: Qualitative and Quantitative Approaches.* SAGE Publications, 2000.

Bonate, Peter L. *Analysis of Pretest-Posttest Designs.* Chapman & Hall/CRC, 2000.

Bordens, Kenneth S., and Bruce B. Abbott. *Research Design and Methods: A Process Approach.* 9th ed., McGraw-Hill Education, 2014.

Chang, Todd P., et al. "Pediatric Emergency Medicine Asynchronous E-learning: A Multicenter Randomized Controlled Solomon Four-Group Study." *Academic Emergency Medicine*, vol. 21, no. 8, 2014, pp. 912–919, doi:10.1111/acem.12434.

Marczyk, Geoffrey R., et al. *Essentials of Research Design and Methodology.* John Wiley & Sons, 2005. Essentials of Behavioral Science Series. *eBook Collection (EBSCOhost)*, search.ebscohost.com/login.aspx?direct=true&db=nlebk&AN=128595&site=ehost-live. Accessed 20 Mar. 2017.

Sahai, Hardeo, and Mohammed I. Ageel. *Analysis of Variance: Fixed, Random and Mixed Models.* Birkhäuser, 2012.

SAMPLE PROBLEM

Medical researchers are interested in the effect of online learning modules on emergency medicine trainees. Trainees at four large hospitals are included in a Solomon four-group experimental design. A written pretest is used to measure trainees' baseline knowledge. The treatment is a series of web-based, interactive modules. Exam scores are used to quantify posttest results. Define the four study groups (A, B, C, and D) based on the tests and/or treatment they receive.

Answer:

Group A receives the written pretest, exposure to the learning modules, and the posttest. Group B receives the written pretest and the posttest. Group C is exposed to the learning modules and receives the posttest. Group D only receives the posttest.

SPLIT-PLOT TYPE DESIGNS

FIELDS OF STUDY

Research Design; Experimental Design

ABSTRACT

A split-plot research design offers an alternative to the use of a completely randomized design in situations where the possible value (level) of one of the factors being studied cannot easily be changed. As such a situation occurs frequently, the split-plot design is an effective choice in many experimental scenarios.

PRINCIPAL TERMS

- **block:** a group of relatively homogenous experimental units.
- **completely randomized design:** an experimental design in which experimental units are randomly

[Figure: Split-plot design diagram showing Block 1 and Block 2, each divided into Main Treatment and No Main Treatment regions, with Subtreatment A and Subtreatment B areas containing Sample 1, Sample 2, and Sample 3.]

assigned to treatments so that each has the same chance of receiving a treatment.
- **factor:** an independent variable or a confounding variable that affect the outcome, or dependent, variable.
- **split-plot factor:** a factor that is easy to change.
- **whole-plot factor:** a factor that is difficult to change.

What Is a Split-Plot Design?

A split-plot research design offers an alternative to the use of a completely randomized design in situations where the possible value (level) of one of the factors being studied cannot easily be changed. Factors are independent variables that are manipulated during the experiment. They can have multiple levels. For example, a factor might be type of saddle, and it might have two levels, English and Western. When using a split-plot design, at least two factors are included. For example, an experiment might be designed to study how different types of saddle (factor 1) affect the time in which different types of horses (factor 2) complete a race course.

When a completely randomized design is used to conduct an experiment, all possible combinations of levels, consisting of one level from each factor, are randomly applied. Each combination is called a treatment. For example, there may be two levels of saddle—English (E) and Western (W)—and three levels of horse—Thoroughbred (T), Arabian (A), and quarter (Q)—being studied. The following unique treatments can be created: E/T, E/A, E/Q, W/T, W/A, W/Q. These treatments would then be applied in random order. Because of this,, the levels of any factor may have to be changed at any point in the sequence. There is no way to control how many times a factor will have to be changed.

A split-plot design may be used instead of a completely randomized design when one of the factors being studied is difficult to change. This factor is called the whole-plot factor. The number of levels in the whole-plot factor equals the number of whole plots used in the design. In many cases, the experimenter will replicate the testing of each level of whole-plot factor to ensure accurate results. This would mean that the number of whole plots required would be doubled. All of the whole plots, or largest experimental units, serve as blocks. Factors that are easy to change are called split-plot factors.

When using a split-plot design, the experiment is conducted in a complex sequence that minimizes the number of times that the whole-plot factor needs to be changed. This is accomplished by first randomly applying the levels of the whole-plot factor to each of the whole plots. Then each of the levels of the split-plot factor is randomly applied to the subplots within each whole plot.

Advantages and Disadvantages

A split-plot design should be used when one of the factors being studied is difficult to change. These types of factors can add significant additional time and other costs to the experimental process when using a completely randomized design. For example, temperature is a difficult, timely factor to change. By using a split-plot design instead, these costs can be minimized. This design also allows for the study of multiple factors using fewer resources.

One downside of a split-plot design is that the statistical analysis that needs to be performed can be slightly more complicated than with a completely randomized design. Thus, if there are no difficult-to-change factors, a completely randomized design might be the most efficient alternative. However, ever-improving statistical software is reducing this difference. In any case, it is important that the statistical calculations that are performed are appropriate to the type of design selected. Using calculations intended for a different design model will lead to inaccurate results.

Real-World Applications of Split-Plot Design

A split-plot design is often used for studies conducted in the agricultural field. The word "plot," in this context, stems from the use of plots of land in agricultural studies to test factors such as fertilizer types and irrigation methods. This design has also been employed in industrial experiments to test the strength of a certain procedure or product.

Improving Research Design with Split-Plot Designs

Many real-world scenarios in governmental, business, and scientific environments contain factors that are difficult to change. Examples of these factors include chemicals that are difficult to spray in small areas, kilns that take a long time to change temperature, and machine parts that are difficult to change. Researchers are studying countless numbers of such factors on an everyday basis. This makes the split-plot design a common and important experimental design.

Maura Valentino, MSLIS

Further Reading

Christensen, Larry B., et al. *Research Methods, Design, and Analysis.* 12th ed., Pearson, 2014.

Creswell, John W. *Research Design: Qualitative, Quantitative, and Mixed Methods Approaches.* 4th ed., Sage Publications, 2014.

Federer, Walter T., and Freedom King. *Variations on Split Plot and Split Block Experiment Designs.* Wiley-Interscience, 2007.

Goos, Peter. *The Optimal Design of Blocked and Split-Plot Experiments.* Springer, 2002.

Jones, Bradley, and Christopher J. Nachtsheim. "Split-Plot Designs: What, Why, and How." *Journal of Quality Technology*, vol. 41, no. 4, 2009, pp. 340–61. *Business Source Complete*, search.ebscohost.com/login.aspx?direct=true&db=bth&AN=44704221&site=ehost-live. Accessed 31 May 2017.

Leedy, Paul D., and Jeanne Ellis Ormrod. *Practical Research: Planning and Design.* 11th ed., Pearson, 2016.

> **SAMPLE PROBLEM**
>
> An agribusiness grows two different types of crops in two greenhouses. The business wants to study the effect of using two different types of honeybees to pollinate two different crops, which are watered at two different levels. While the crop factor is easy to change (crops can be easily planted in a small section of a greenhouse), as is the water factor (the irrigation system used allows small sections of each greenhouse to be watered at a different rate), the honeybee factor is difficult to change because there is no way to restrict honeybees from visiting all areas of a greenhouse. Because one of the factors is hard to change, a split-plot design will be used. In this design, instead of having both types of honeybee in both greenhouses, one type will be introduced into the first greenhouse, and the other will be introduced into the second greenhouse.
>
> Define the groups that will be tested. Specifically, identify which factor is the whole-plot factor, which factors are the split-plot factors, and which treatments (combinations of factors) will be tested.
>
> **Answer:**
>
> The whole-plot factor is the honeybee factor (HB), and the split-plot factors are the crop factor (C) and the water factor (W). As there are two levels of the whole-plot factor (HB1, HB2), two whole plots will be created to divide the treatments into two groups. Each group of treatments to be tested will contain one level of the whole-plot factor combined with each of the levels of the crop factor (C1, C2) and the water factor (W1, W2). This design will result in four unique treatments being tested in each greenhouse:
>
> - Greenhouse 1: HB1/C1/W1, HB1/C1/W2, HB1/C2/W1, HB1/C2/W2
> - Greenhouse 2: HB2/C1/W1, HB2/C1/W2, HB2/C2/W1, HB2/C2/W2

STRATIFIED RANDOM SAMPLING / RANDOMIZED BLOCK DESIGN

FIELDS OF STUDY

Sampling Design; Sampling Techniques; Experimental Design; Statistical Analysis

ABSTRACT

Stratified random sampling divides a population into relatively homogenous subgroups and then draws samples from each group in order to study the entire population. Randomized block design, a form of experimental design, is similar to stratified random sampling. In randomized block design, researchers divide subjects into homogenous subgroups and then conduct the same experiment on each subgroup.

PRINCIPAL TERMS

- **experimental group:** the subjects in an experimental research study who receive the intervention or treatment being tested.
- **interaction:** in statistics, an effect by which the impact of an independent variable on the dependent variable is altered, in a nonadditive way, by the influence of another independent variable or variables.
- **strata:** homogenous subgroups that are drawn from a larger population and do not overlap with one another.
- **variability:** the extent to which outcomes in a data set differ from one another.

OVERVIEW

Stratified random sampling divides a large population into homogenous subgroups called strata. Researchers then draw samples from each of the strata. This allows researchers to be sure that their sample is representative of the larger population. In particular, stratified random sampling is better than simple random sampling at ensuring minority groups are represented in the sample. Stratified random sampling can be used to ensure that the size of various groups in the final

POPULATION

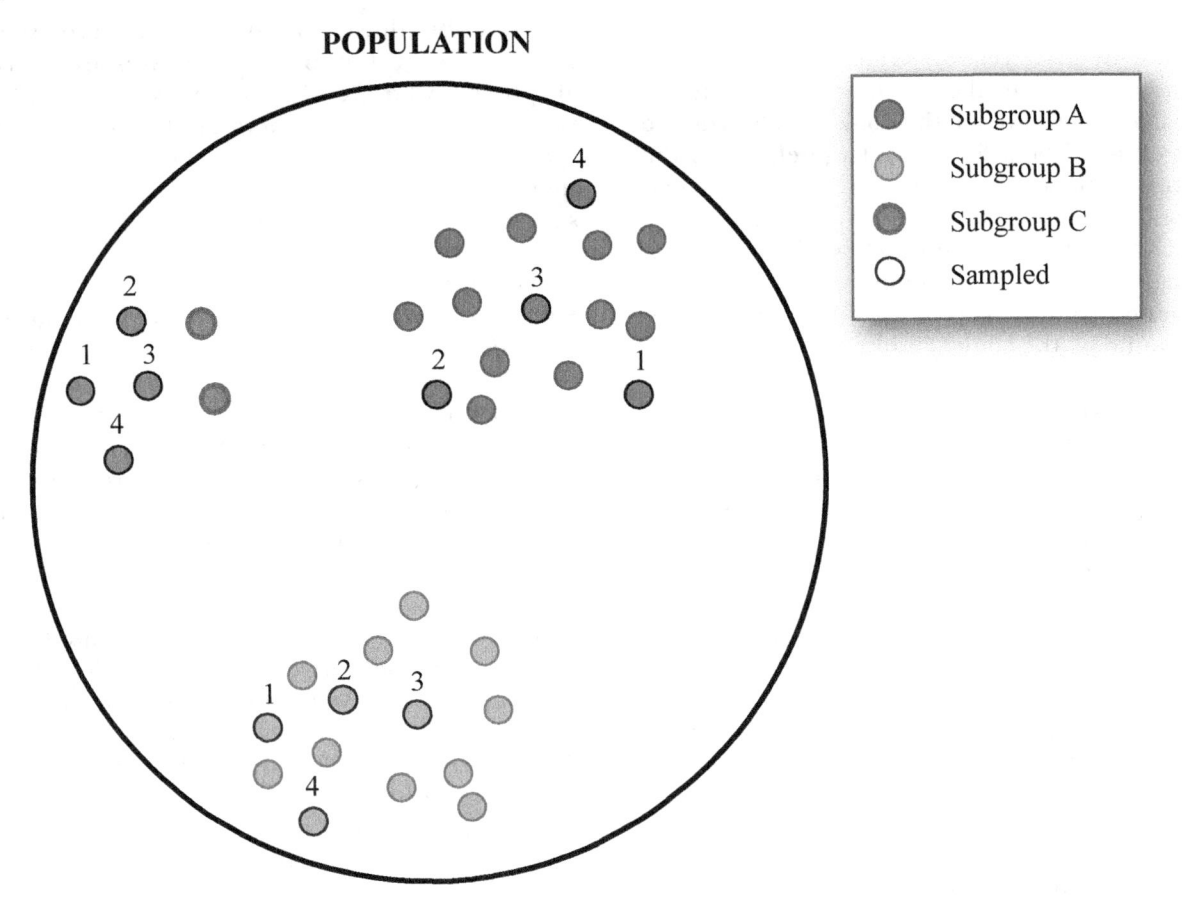

sample are proportionate to their size in the larger population. Furthermore, the strata are more homogenous than the total population so there is less variability in the data. Thus, stratified random sampling offers greater statistical precision than simple random sampling.

Randomized block design is based on a similar concept. Randomized block design ensures that the effect being studied truly caused the result, rather than another variable or chance occurrence. Researchers divide the subjects into subgroups called blocks, similar to strata. The researcher then conducts the same experiment on each block. By reducing the variability within each block, this design enables researchers to estimate the treatment effect with more certainty.

Stratified random sampling begins by dividing the population into homogeneous subgroups that do not overlap with one another. These divisions are made by identifying a variable, such as gender. For example, a teacher might divide a class of students into a group of boys and a group of girls. These groups are the two strata. The teacher then applies an intervention, such as reading a book about manners to the class. After reading the book, the teacher analyzes the effect of the book on student behavior. Because the teacher divided the class into strata, it is possible to judge the effect of the book on the individual strata as well as the entire class. The strata can help the teacher determine if the book affects girls differently than boys.

Stratified random sampling can help to identify an interaction between the independent variable being studied (the book) and a separate independent variable (gender) on the dependent variable (student behavior). If the entire class is studied without being divided into strata, the teacher might find a large amount of variability between students, from no effect to a radical change in behavior. By first creating strata, the teacher can reduce variability in the outcomes. For example, the teacher may find that the book had a negligible effect on girls' behavior but a large effect on the behavior of boys. Furthermore, the teacher could also read the book to one class and then read a different book to a separate class. The class that was read the book on manners would be the experimental group. The class that was read a normal book would be the control group.

REPRESENTATION

Stratified random sampling allows researchers to study both a population as a whole and smaller subgroups. A researcher may want to study the residents of New York City, where there is a wide variety of ethnicities, language groups, and religions. A randomized sample could result in a sampling error. For example, a researcher might intend to survey a diverse sample of residents. However, if the researchers use a purely random sample, the project might end up with a sample of sixty New Yorkers who are Catholic and forty who are Protestant. Due to pure chance, the random sampling failed to include any residents of other faiths.

The research team can improve their study by using stratified random sampling. In doing so, researchers can ensure they include twenty residents who are Buddhist, twenty who are Protestant, twenty who are Catholic, twenty who are Jewish, and twenty who are Muslim. By using a stratified random sample, researchers can test a hypothesis within each subgroup (strata). The researcher can then compare the strata to make more accurate statements about the entire population.

USE IN MANY FIELDS

Stratified random sampling and randomized block design are used in many fields. For example, sociologists use stratified random sampling to ensure that workers from all levels of a corporation, from the facilities staff to the executives, are surveyed. City planners use it to ensure that they survey residents from all parts of a town, rather than just those who live in the most populated areas. Geographers also use this method to ensure that they have examined an entire topographical area.

Stratified random sampling and randomized block design both reduce variability within samples. This allows researchers to draw more definitive conclusions from their studies by reducing variance in the data they generate. Using these methods, researchers can demonstrate with greater clarity that an effect is due to the variable under study rather than an unknown factor or an interaction effect.

Allison Hahn, PhD

SAMPLE PROBLEM

Assume that a nutrition company wants to market a new sports beverage to athletes. The company claims that the beverage will improve athletic performance by 25 percent. The company wants to test this product with the help of a university sports program. The program includes novice, junior varsity, and varsity athletes. Furthermore, each athlete plays one of three sports: track, gymnastics, or basketball. How could the researchers improve the study by using stratified random sampling?

Answer:

First, determine which characteristic will be used to determine strata. Then define the strata. Next, identify an independent variable other than the sports beverage that could influence the results.

The characteristic used to define the strata is the sport: track, gymnastics, and basketball. In some sports, such as track, it can be easy to calculate a 25 percent improvement based on an athlete's speed. However, in gymnastics it might be more difficult to calculate performance improvement due to the different skills involved. Once the strata are defined, researchers can further divide each individual stratum based on their team level: novice, junior varsity, and varsity. In this way, the researchers could determine with more certainty if the team level is a variable influencing the results.

FURTHER READING

Daniel, Johnnie. *Sampling Essentials: Practical Guidelines for Making Sampling Choices.* Sage Publications, 2012.

Levy, Paul S., and Stanley Lemeshow. *Sampling of Populations: Methods and Applications.* John Wiley & Sons, 2013.

Kaminska, Olena, and Peter Lynn. "Survey-Based Cross-Country Comparisons Where Countries Vary in Sample Design: Issues and Solutions." *Journal of Official Statistics*, vol. 33, no. 1, 2017, pp. 123–36.

Robinson, Oliver C. "Sampling in Interview-Based Qualitative Research: A Theoretical and Practical Guide." *Qualitative Research in Psychology*, vol. 11, no. 1, 2014, pp. 25–41. *Academic Search Complete*, search.ebscohost.com/login.aspx?direct=true&db=a9h&AN=92017423&site=ehost-live. Accessed 19 Apr. 2017.

Thompson, Steven K. *Sampling.* 3rd ed., John Wiley & Sons, 2012.

Tillé, Yves, and Alina Matei. "Basics of Sampling for Survey Research." *The SAGE Handbook of Survey Methodology*, edited by Christof Wolf et al., Sage Publications, 2016, pp. 311–28.

STUDENT'S *t*-TEST

FIELDS OF STUDY

Statistical Analysis; Variance Analysis; Hypothesis Testing

ABSTRACT

Student's *t*-test is a method of statistical analysis that enables researchers to prove or disprove a hypothesis. It can be used to compare the means of two sets of data to check whether the difference between them is statistically significant. It can also be used to compare the mean of a single data set against the mean estimated for the null hypothesis, the prediction that no significant differences will be found between the control and experimental groups.

PRINCIPAL TERMS

- **independent samples:** two different sample populations or randomized members of the same population measured at separate points in time.
- **nonrandom sampling:** one of several non-probabilistic sampling methods in which individuals within the population being studied are selected for examination in a manner that does not involve random chance, so that not all members of the population have a chance of being included, and the probability of selecting a particular sample cannot be calculated.
- **paired *t*-test:** a method of statistical analysis used to test a hypothesis by comparing two measures taken from the same population, usually in a before-and-after scenario.
- **sample size:** the number of subjects included in a survey or experiment.
- **standard deviation:** a measurement of the degree of dispersion present in a distribution; greater dispersion results in a larger standard deviation.

A Method of Comparing Data

Comparing two data sets requires the selection of a metric common to both, so that the values of the metric for each set can be determined to be equal or unequal. If the values are not equal, then the next question to be answered is whether the difference between them is statistically significant or more likely the result of chance. Often, the metric that is used for comparison in data analysis is the standard deviation (SD) of each data set. Yet, in some cases, such as when the sets of data have a small sample size, the SD may be too small to be useful. In other cases, the exact SD is not or cannot be known with precision. In those situations, the SD is estimated and the data sets can be compared using the mean of each distribution instead. This is what Student's *t*-test does.

Developed in 1908 by chemist and statistician W. S. Gosset, who published his work under the pseudonym Student, this test assumes that the populations being compared follow a normal distribution and their variances are the same. It also typically assumes that random sampling was used to select study subjects and assign them to groups.

Experiments that use Student's *t*-test for data analysis are usually designed to test whether a treatment or intervention, when applied to a group of subjects, produces a change. The test identifies whether a change has occurred and thus helps prove or disprove the null hypothesis (the prediction that no significant differences will be found between the control and experimental groups). This then provides direction for further research into the phenomenon, as the *t*-test alone does not explain how or why a change occurred.

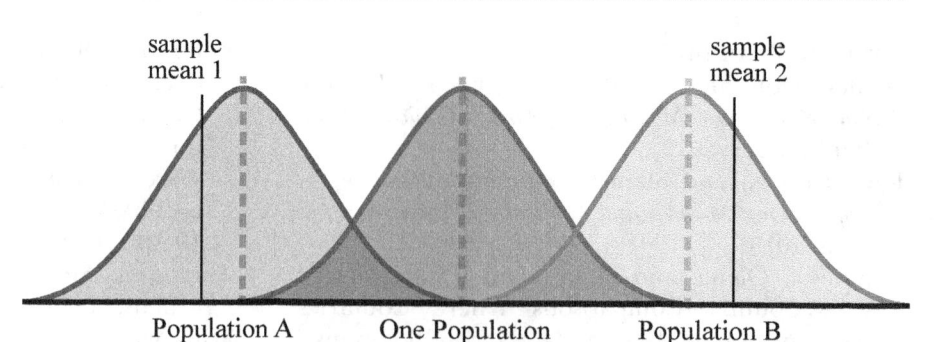

To Pair or Not to Pair

There are actually several types of *t*-tests: one-sample, two-sample, and paired. A simple one-sample *t*-test calculates a test statistic known as the *t*-value. The *t* value is the ratio of the effect size (the difference between the sample mean and the null hypothesis mean) and the standard error of variability. The

---

SAMPLE PROBLEM

Researchers are interested in comparing the health of the average person in two different countries, A and B. One of the factors they wish to assess is body weight. To perform their study, the researchers will recruit a random sample of seventeen participants (seven in country A and ten in country B) from a population of twenty and record each subject's weight in pounds. These figures are then used to calculate the mean for each group, and the means compared. The researchers decided to use a confidence level of 95 percent.

The hypotheses to be tested are as follows:

H_0: There is no significant difference in the average weight of Country A and Country B.

H_a: There is a significant difference in the average weights of Country A and Country B.

Because the variation in weights is unknown and may be either higher or lower, a two-tailed test is appropriate. This test evaluates both the higher and lower tails of the distribution for significance. To accomplish this, an overall significance level of $p < 0.05$ will be used. Thus, the average weights will be considered significantly different if the test statistic (*t*-value) is in the top or bottom 2.5 percent ($p < 0.025$) of its distribution.

Based on this information and the following test results, can the null hypothesis be rejected?

Group	Two-Tailed P value	Mean	Confidence Interval at 95%
Country A	0.60	145	14
Country B	0	138	6.1

Answer:

According to the *t*-test table, the mean for County Group A is 145 pounds, and the mean for Country Group B is 138 pounds. In this sample problem, there is no difference statistically between Country A and Country B weights because the overall *P* value is higher than $p < 0.05$. Therefore, the null hypothesis is accepted.

Additionally, the confidence interval provides a range of values likely containing the true mean for the overall population; the level of confidence indicates that if repeated samples were taken, 95 percent of them would contain the true mean. The table shows that Country A has a confidence interval of 14, giving it a range of 131–159 pounds, and Country B has a confidence interval of 6.1, giving it a range of 131.9–144.1 pounds.

	Independent Variable			
Replications	1	2	3	4
1	Treatment			
2		Treatment		
3			Treatment	
4				Treatment
5				

posttest. The posttest may also serve as the second pretest, or a separate pretest may be given. In the second phase, the groups switch roles. The experimental group from the first phase becomes the control group, and vice versa. At the end of the second phase, both groups are given a second posttest.

This design is quasi-experimental because participants are not randomly assigned to groups, as they would be in a true experimental design. Quasi-experimental designs are used for research studies that cannot use random assignment. Usually this is because participants are already sorted into groups. The lack of random assignment poses a threat to internal validity, or the extent to which the study outcome can be attributed to changes in the independent variable, rather than some other, confounding variable. However, the nature of the switching replications design—specifically, the fact that all participants will be part of the experimental group—somewhat offsets this threat. In addition, it is possible to use random assignment in a switching replications design if circumstances permit, creating a hybrid design. This would make the study a true experiment.

When to Use Switching Replications Design

The switching replications design is best used in a setting where treatments like the one being tested are regularly repeated, such as a school where classes are divided into semesters. Because of this, the design is often used by organizations that are deciding whether or not to implement or continue a certain program. The different groups are typically preselected due to the nature of the organization. For example, the two groups could be the students in two different classes or employees on two different shifts.

The switching process strengthens the experiment's internal validity in multiple ways. First, it mitigates the problems caused by lack of random assignment by ensuring that both control and experimental data will be collected from every single participant, regardless of which group they are in. Second, it removes social threats, such as competition and resentment, that can be caused by one group not receiving the treatment. Finally, it addresses ethical concerns that arise in true experiments when participants who truly need the program or treatment are assigned to the control group and therefore do not receive it. In a switching replications design, all participants will eventually receive the treatment.

External validity is also enhanced because replication is inherent in the design. However, researchers using this design must keep in mind that the replication in the second phase is not completely independent from the first phase. The repeated pre- and post-testing could cause a priming effect in the second phase. This could skew results and thus threaten the validity of the experiment. When priming effects are a concern, the Solomon four-group design may be a better choice.

Switching Replications Design in Action

In addition to its advantages with respect to internal validity, the switching replications design increases efficiency in allocating resources. For example, a school may want to investigate whether or not to adopt an interactive online textbook. Instead of purchasing access to the textbook for every single student, the school can purchase access for just one class and then conduct a switching replications study.

Two classes studying the same subject at the same grade level would be selected for this study. All

standard error of variability is a measure of how precisely the sample approximates the mean for the overall population.

In a two-sample *t*-test, the *t*-value is the ratio of the difference between two sample means to the variability. For the test to provide meaningful results, both data sets must be from different independent samples.

A paired *t*-test may be used when there is only one group of subjects and nonrandom sampling is used, as in the single-group pretest-posttest design. In that case, each subject is essentially matched with itself, and this *t*-test checks for differences between each subject's before-and-after measurements. The *t*-value here is calculated just as in the one-sample *t*-test.

Importance of Student's *t*-Test

The purpose of Student's *t*-test is to test a hypothesis that compares two small sets of data. Without Student's *t*-test, a researcher could not discover whether to reject the null hypothesis or accept the hypothesis. Student's *t*-test is used in a variety of fields of study, including zoology, education, and medicine.

Scott Zimmer, JD

Further Reading

Adams, Kathrynn A., and Eva K. Lawrence. *Research Methods, Statistics, and Applications.* Sage Publications, 2015.

Buglear, John. *Practical Statistics: A Handbook for Business Projects.* Kogan Page, 2014.

Efron, Bradley, and Trevor Hastie. *Computer Age Statistical Inference: Algorithms, Evidence, and Data Science.* Cambridge UP, 2016.

Haslam, S. Alexander, and Craig McGarty. *Research Methods and Statistics in Psychology.* 2nd ed., Sage Publications, 2014.

Levine, David M., and David Stephan. *Even You Can Learn Statistics and Analytics: An Easy to Understand Guide to Statistics and Analytics.* 3rd ed., Pearson Education, 2015.

Vogt, W. Paul, and R. Burke Johnson. *The Sage Dictionary of Statistics & Methodology: A Nontechnical Guide for the Social Sciences.* 5th ed., Sage Publications, 2016.

Switching Replications Design

Fields of Study

Research Design; Experimental Design; Quasi-experiment

Abstract

The switching replications design enables researchers to pretest and posttest two groups by switching their roles. It is a robust design often used to address whether an organization needs to implement a program.

Principal Terms

- **control group:** the subjects in an experiment who do not receive any intervention or who receive a placebo or sham intervention.
- **experimental group:** the subjects in an experimental research study who receive the intervention or treatment being tested.
- **pretest-posttest design:** a research design in which subjects are tested both before and after the experimental treatment or intervention.
- **priming effect:** a psychological effect in which exposure to a stimulus influences an individual's response to a subsequent stimulus.
- **replication:** the repetition of an experiment or study with the aim of achieving the same results in order to confirm findings or reduce error.

Strong Experimental Design

The switching replications design is a robust quasi-experimental design that administers a treatment twice, switching the roles of the experimental group and the control group between treatments. It is a pretest-posttest design, meaning that the groups are tested both before the study begins and after each treatment phase. In the first phase, both experimental and control groups are given a pretest. Next, the treatment is provided to the experimental group, and then both groups are given a

students in both classes would be pretested at the beginning of the first semester of the school year, to evaluate their baseline knowledge of the subject. Then, throughout the first semester, class 1 would use the interactive online textbook and class 2 would use the regular textbook. At the end of the semester, both classes would be given a posttest to evaluate what and how well they have learned. Then, for the second semester, the groups would switch, so that class 2 would use the new textbook and class 1 would use the old one. At the end of the second semester, both classes would be given a second posttest. By this time, all participants would have spent one semester using the old textbook and one semester using the new one. The school can then compare how the students performed with each textbook to determine if the interactive textbook produced better results and, if so, whether access should be purchased for the entire school.

Real-World Value

The switching replications design is widely used in organizational studies as a method of evaluating new programs. It can be used by schools and corporations seeking to understand the effects of potential programs. By switching group roles to give the treatment to both groups, this design addresses both internal and external threats to validity.

Pamelyn Witteman, PhD

Further Reading

Creswell, John W. *Research Design: Qualitative, Quantitative, and Mixed Methods Approaches.* 4th ed., Sage Publications, 2014.

Field, Andy. *Discovering Statistics Using SPSS.* 3rd ed., Sage Publications, 2009.

Orcher, Lawrence T. *Conducting Research: Social and Behavioral Science Methods.* 2nd ed., Taylor & Francis, 2014.

SAMPLE PROBLEM

A researcher decides to study the effectiveness of a ten-week training program designed to help machinists reduce the number of defective items that they produce. The factory where the study is being conducted employs 500 machinists. However, there are only enough instructors available to teach 250 machinists at a time. Because of this, the researcher decides to use a switching replications design. Create a timeline for measurement and treatment of the machinists using this design.

Answer:

The treatment (independent variable) is the ten-week training program. The dependent variable is the number of defective items produced by the machinists. The timeline is as follows:

1. The machinists are divided into two groups of 250, group A and group B. If the machinists work two different shifts, the easiest way to assign groups would be to make the first shift group A and the second shift group B. However, if they all work the same shift, the machinists may be assigned to groups using random assignment, which would make the design a hybrid switching replications design.
2. Both groups are measured to determine the number of defective items they are producing. This is the pretest.
3. Group A is given the ten-week training program. Group B is assigned as the control group, so members of the group do not receive the training.
4. After group A completes the training program, the number of defective items produced by both groups is measured again. This is the first posttest.
5. Next, group B is given the ten-week training program, and group A functions as the control group.
6. After group B completes the training program, the number of defective items being produced by both groups is once again measured. This is the second posttest.
7. Analysis can then be conducted to determine if the program had the desired effect of reducing the number of defective items produced by the machinists.

Patten, Mildred L., and Michelle Newhart. *Understanding Research Methods: An Overview of the Essentials*. 10th ed., Routledge, 2017.

Swanson, Richard A., and Elwood F. Holton III, editors. *Research in Organizations: Foundations and Methods of Inquiry*. Berrett-Koehler Publishers, 2005.

Trochim, William M. K. "Hybrid Experimental Designs." *Research Methods Knowledge Base*, 2006, www.socialresearchmethods.net/kb/exphybrd.php. Accessed 3 May 2017.

Trochim, William M. K. "Other Quasi-Experimental Designs." *Research Methods Knowledge Base*, 2006, www.socialresearchmethods.net/kb/quasioth.php. Accessed 3 May 2017.

Vogt, W. Paul. *Quantitative Research Methods for Professionals*. Pearson, 2007.

Systematic Sampling

FIELDS OF STUDY

Sampling Techniques; Sampling Design

ABSTRACT

Systematic sampling is a sampling method that uses randomization to select a sample from a population and then draws additional samples at a predetermined and fixed interval. Systematic sampling is more representative than other randomized sampling systems, such as simple random sampling. This decreases the risk of bias in the sampling process.

PRINCIPAL TERMS

- **cluster sampling:** a probabilistic sampling method in which the population being studied is divided into subsets called "clusters" based on a shared characteristic unrelated to the research question, such as geographic location, and then multiple clusters are selected at random for comprehensive examination.
- **simple random sampling:** a probabilistic sampling method in which members of the sample group are drawn from the population at random, so that each member of the population has an equal chance of being included.
- **uniform sampling:** a sampling method in which samples are drawn from a population in a consistent pattern with a predetermined population, sample size, and starting point.

What Is Systematic Sampling?

Systematic sampling is a sampling technique that uses simple random sampling to select a starting point and then draws samples from a population at a fixed interval. This method is used for populations that are relatively uniform. Like uniform sampling, the sample population and the desired sample size are determined prior to the sampling process. However, the starting point in systematic sampling is selected using simple random sampling. Uniform sampling, on the other hand, starts at a predetermined (nonrandom) point. Samples are then chosen at fixed sampling intervals. Therefore, each element in the population has an equal probability of selection. For these reasons, systematic sampling is considered random.

Researchers must consider traits, patterns, or cycles that could affect the randomness and representativeness of a systematic sample. Systematic sampling provides results that are generally representative of the total population. Systematic sampling can also be faster and easier to perform than other sampling methods.

Using Systematic Sampling

To conduct a systematic sample, researchers first identify the population to be sampled (N). They then determine the desired sample size (n) for their study. The researchers then calculate the sampling interval (k), also known as the skip. To do this, researchers divide the total population by the desired sample size ($N / n = k$). For example, if researchers wanted a sample size of 4 from a population of 24 units, then the sampling interval would be equal to 6 (24 / 4 = 6). In this example, every 6th unit after a randomly selected starting point would be sampled.

In many cases, the total population is not evenly divisible by the sample size. In these instances, the sampling interval will not be a whole number. If

POPULATION

SAMPLE POPULATION

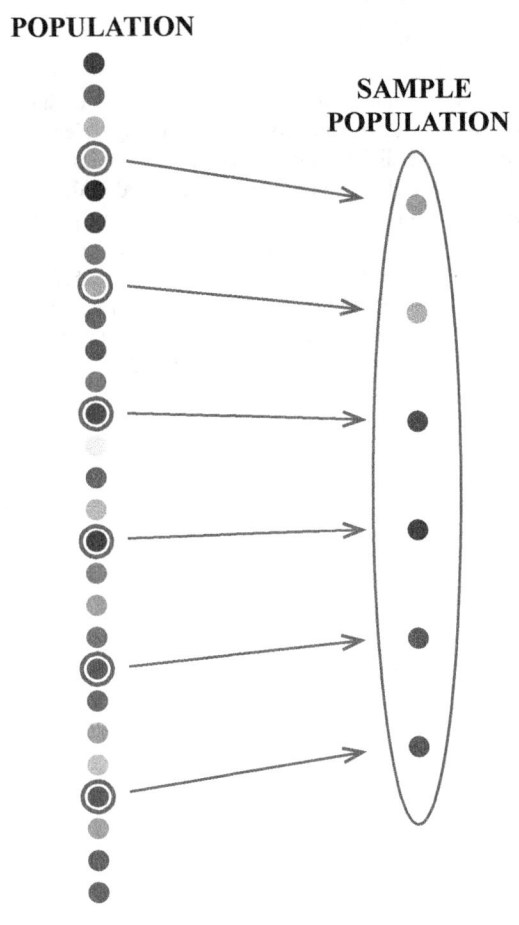

sampling interval k is not a whole number, it cannot be rounded up or down. If k is rounded up or down, it will affect the equal probability with which each sample can be selected. To overcome this problem, the starting point must be selected at random from all integers between 0 and N. The researchers then draw samples from every kth unit. Each noninteger selected is then rounded down. For example, if researchers wanted to draw 4 samples from a population of 25 units, then the sampling interval would equal to 6.25. If the randomly selected starting point is 2, then the numbers 2, 8.25, 14.5, and 20.75 would be calculated. In this example, the 2nd, 8th, 14th, and 20th units would be selected. If the researchers reach the end of the sample list without attaining the desired sample size, researchers return to the beginning of the list and start again in a continuum in which 1 follows N.

Systematic sampling is typically used for populations in which the sequencing is done ahead of time without the influence of the researchers. For example, systematic sampling could be used to select every kth widget that comes off of an assembly line or every kth student who enrolls in an after-school program.

Context and Considerations

Systematic sampling is often compared with simple random sampling. Among the advantages of systematic sampling is its simplicity and comprehensiveness. It allows a researcher to systematize the random selection of subjects and ensure its representativeness. It extends the sampling process to the whole population without the need for complex mathematical formulas.

Compared to simple random sampling, systematic sampling is less likely to leave large groups unrepresented. However, researchers must carefully take into account any patterns or cycles within the ordered population that could affect the randomness and representativeness of the sample. For example, in the sample problem above, if every sixth home is at a corner and therefore worth more than other homes on the block, the selected sample would no longer be random. Any findings would be flawed. Therefore, whenever possible, researchers should first examine all members of the sample population to ensure that the units are relatively homogenous. Researchers must ensure that all units have the same chance of selection.

Systematic random sampling is similar to cluster sampling. In this method, the population is divided into nonoverlapping primary units called clusters. One or more primary units are then randomly selected and all the secondary units within are sampled. In systematic random sampling, the population is divided into primary units once the sampling interval has been calculated (every kth unit). When the starting point is randomly selected, every secondary unit within the selected primary unit is sampled.

Trudy Mercadal, PhD

> **SAMPLE PROBLEM**
>
> A city government wants to establish a series of bus stops along Broadway Boulevard, a street with 300 houses. Researchers want to determine how residents perceive this initiative.
>
> The research team chooses to survey homeowners living on Broadway Boulevard. Given time constraints, sampling all 300 homeowners is impossible. The team decides to sample 48 homeowners. How can the researchers use systematic sampling to draw 48 samples?
>
> **Answer:**
>
> To calculate the sampling interval, the researchers divide the total population (300) by the desired sample size (48) in the following manner: 300 / 48 = 6.25. The sampling interval is 6.25. The researchers then randomly select a starting point from all 300 households and sample units at intervals of 6.25. Selected nonintegers are rounded down. The researchers randomly select 10 as the starting point. Thus, the homeowners of the houses at 10, 16 (10 + 6.25 = 16.25), 22 (10 + 6.25 x 2 = 22.5), 28 (10 + 6.25 x 3 = 28.75), 35 (10 + 6.25 x 4 = 35), and so on are selected. For the final unit (10 + 6.25 x 47 = 303.75), the researchers start from the beginning of the list and select the 4th household.

Further Reading

Arnab, Raghunath. *Survey Sampling Theory and Applications.* Academic Press, 2017.

Blair, Edward, and Blair, Johnny. *Applied Survey Sampling.* Sage Publications, 2014.

Eldar, Yonina C. *Sampling Theory: Beyond Bandlimited Systems.* Cambridge UP, 2015.

Elsayir, Habib Ahmed. "Comparison of Precision Systematic Sampling with Some Other Probability Samplings." *American Journal of Theoretical and Applied Statistics,* vol. 3, no. 4, 2014, pp. 111–16.

Ott, R. Lyman, and Longnecker, Michael. *An Introduction to Statistical Methods and Data Analysis.* 7th ed., Cengage Learning, 2016.

Rees, Colin. *Rapid Research Methods for Nurses, Midwives and Health Professionals.* Wiley Blackwell, 2016.

Rose, Angela M. C., et al. "A Comparison of Cluster and Systematic Sampling Methods for Measuring Crude Mortality." *Bulletin of the World Health Organization,* vol. 84, 2006, pp. 290–96.

Time-Series Designs

FIELDS OF STUDY

Research Design

ABSTRACT

In a time-series research design, multiple measurements are made prior to and after the treatment or intervention. Time-series experiments are considered quasi-experiments because they do not use an experimental group and a control group.

PRINCIPAL TERMS

- **longitudinal sampling:** an experiment that examines a single group of subjects over an extended period of time.
- **quasi-experiment:** a study in which subjects are not randomly assigned to control and experimental groups or the experimenter cannot control which subjects receive the experimental condition.
- **single case study:** a research method in which a selected individual or small group is observed and examined thoroughly.

OVERVIEW

Time-series designs represent an approach to research generally considered less rigorous than the practice of randomly assigning subjects of a study to experimental and control groups. For this reason, studies using time series are sometimes referred to as quasi-experiments. Instead of using the two-group setup, a time-series design examines one or more groups of subjects multiple times throughout the duration of the research. In this respect, time series are like the single case study, a design in which a selected individual or small group is observed and examined thoroughly.

In a comparative study using a pretest and a posttest, subjects are assessed once before the treatment or intervention and then once afterward. Longitudinal sampling relies on the time-series design to examine subjects over an extended period of time. Taking multiple measurements throughout the duration of the study shows whether and how trends develop over time, particularly social trends. Taking a single measurement, as in a cross-sectional study, is more akin to a snapshot of a particular moment in time. Both designs are useful in some circumstances, but they are not interchangeable.

DATA POINTS

In a time-series study, the same subjects function as both the experimental and the control group. A time-series study may use one or more groups of subjects, but all groups receive the experimental condition. Because the researcher must record and organize numerous measurements (pretests and posttests), time-series studies may use an alphabetic notation to distinguish where in the series a given measurement falls. The first measurement is a baseline that establishes the starting value for the trait being studied. After this baseline has been recorded, an interval of time passes. The length of interval varies, but in most research, interval length is consistent throughout the study, so that measurements are taken daily, weekly, monthly, yearly, and so on.

At the end of the interval, another measurement (pretest) is taken, and another interval then passes, after which more pretests are done. This process is repeated for however many pretests are deemed necessary. At a given point, the subjects undergo the experimental condition. Further measurements (posttests) are taken at intervals afterward. The time-series design thereby allows researchers to see how the independent variable in the experiment affects the dependent variable over time.

In a reversal time-series study, the subjects are exposed to the experimental condition multiple times, with a withdrawal period in between. Measurements are taken throughout. Such studies only work for certain interventions or treatments. This type of research helps demonstrate the connection or lack thereof between the independent and dependent variables.

INVESTIGATING EFFICACY

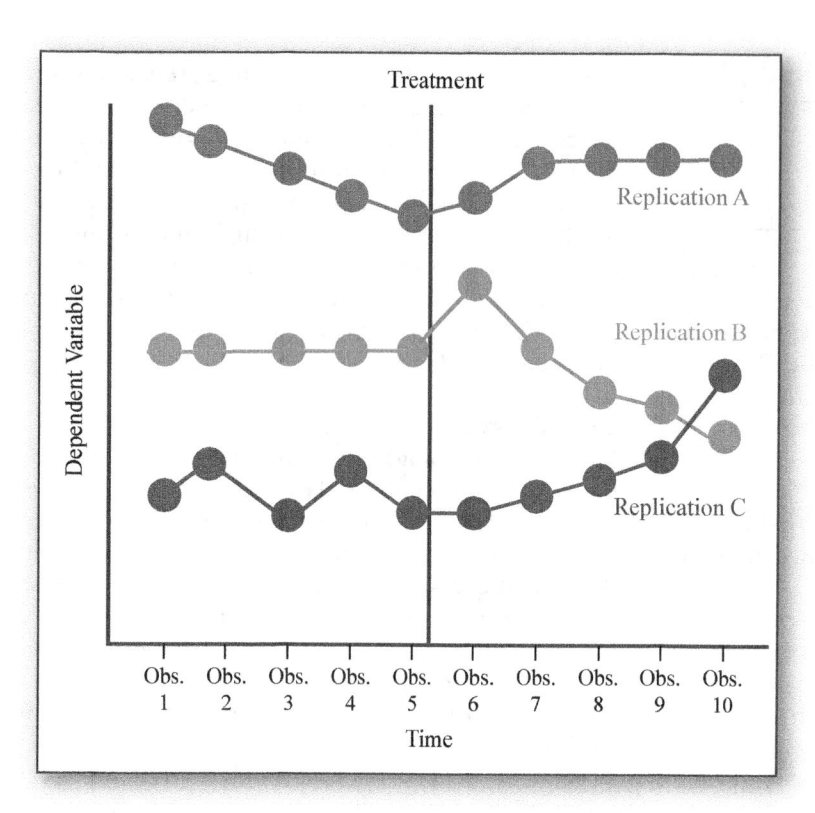

A study examining the efficacy of a weight loss support group might utilize the reversal time-series design. The subjects would be people who wish to lose excess weight. At the outset, the subjects would be weighed in order to have a baseline of data against which later measurements could be compared. Next, the intervention would be applied. In this case, the subjects would have a support group meeting. After this, an interval of time (a month, perhaps) would pass, and then the subjects would be measured by recording their weight and the date. After another interval, another support group meeting would be held and measurements would be taken again the following month.

Over time, this process would create a series of data points that could be graphed and analyzed. The researchers could then observe any patterns or trends in the data. Because of the reversal, they could see whether there were group-wide changes in weight in direct response to the presence or absence of the meetings. In order to make this a true experiment and determine causality, however, the researchers would need to assign participants randomly to the support group and to a control group.

TIME SERIES IN HEALTH AND MEDICINE

Time-series design is useful to researchers in a wide range of scientific fields, particularly those in medicine, pharmacology, psychology, and psychiatry. Those professions specialize in developing various treatments for physical and mental conditions. Time-series research can help them to demonstrate how effective a particular approach is. Time-series data can be viewed as preferable to the more limited information that may be derived from a pretest/posttest study, since it serves as an extended series of pretests, treatments, and posttests. The presence of repeated measurements tends to make the information collected more reliable, increasing its persuasive power. When the stakes are as high as public health and well-being, the additional time and effort needed to conduct a time-series study are not difficult to justify.

Scott Zimmer, JD

FURTHER READING

Goodwin, C. James. *Research in Psychology.* John Wiley & Sons, 2010.

Hanbury, Andria, et al. "Immediate versus Sustained Effects: Interrupted Time Series Analysis of a Tailored Intervention." *Implementation Science,* vol. 8, no. 130, 2013, doi:10.1186/1748-5908-8-130. Accessed 20 Mar. 2017.

Marczyk, Geoffrey R., et al. *Essentials of Research Design and Methodology.* John Wiley & Sons, 2005. Essentials of Behavioral Science Series.

Sexton-Radek, Kathy. "Single Case Designs in Psychology Practice." *Health Psychology Research,* vol. 2, no. 3, 2014, pp. 98–99, doi:10.4081/

hpr.2014.1551. Accessed 20 Mar. 2017.

St. Clair, Travis, et al. "The Validity and Precision of the Comparative Interrupted Time-Series Design: Three Within-Study Comparisons." *Journal of Educational and Behavioral Statistics*, vol. 41, no. 3, 2016, pp. 269–99, doi:10.3102/1076998616636854. Accessed 20 Mar. 2016.

Svoronos, Theodore, et al. "Clarifying the Interrupted Time Series Study Design." *BMJ Quality & Safety*, vol. 24, no. 7, 2015, pp. 475–76, doi:10.1136/bmjqs-2015-004122. Accessed 20 Mar. 2017.

Type I and Type II Errors

FIELDS OF STUDY

Hypothesis Testing; Sampling Techniques; Statistical Analysis

ABSTRACT

Type I and type II errors are the two main types of error that can be made in the interpretation of research results. Type I error is committed when the null hypothesis is rejected even though it is true. Type II error occurs when the null hypothesis is accepted even though it is false. Researchers must avoid committing either type of error as they analyze the data they have collected.

PRINCIPAL TERMS

- **critical region:** the set of values for which the null hypothesis is rejected.
- **null hypothesis:** a prediction that no significant differences will be found between the control and experimental groups.
- **P value:** the probability of the test result occurring by chance when assuming that the null hypothesis is true.
- **random sampling:** one of several probabilistic sampling methods in which individuals within the population being studied are selected for examination at random, so that each member of the population has a chance (often an equal chance) of being included.

Testing a Hypothesis

The purpose of many types of research is to test a hypothesis in the hope of explaining an observed phenomenon. A hypothesis is a prediction of what will happen when an intervention is applied to an experimental group. A simple hypothesis might be that a tutoring program will cause students' test scores to increase. To test this hypothesis, researchers would need to collect data by selecting an experimental and control group using random sampling to avoid bias. Then they would give both groups a baseline test before providing tutoring to the experimental group. Finally, they would retest both groups.

With every hypothesis that is tested, the opposite of that hypothesis is implicitly present. This is called the null hypothesis, which researchers hope to disprove. The null hypothesis predicts that there will be no significant difference observed between the groups under study. Once researchers arrive at a result, they conclude based on the data whether the null hypothesis should be rejected and the alternative hypothesis can be proven true.

In all kinds of experiments, there is the potential for error in determining whether to accept or reject a hypothesis. Type I error occurs when researchers reject the null hypothesis even though it is really true. In some cases, the researchers mistakenly think that the intervention caused a change when, in reality, it did not. Instead, there was random variation in results caused by chance. In other cases, type I error is caused by either conscious or unconscious biases that lead to tainted results. Type II error occurs if the researchers find that the intervention caused no statistically significant changes when, in reality, it did. This leads them to accept the null hypothesis despite it really being false. Inadequate sample size can lead to this error. The establishment of a P value, using random sampling, and controlling for extraneous variables and confounding can help to protect against these errors. Researchers should calculate the likelihood of committing these errors at the outset of the

Analyzing Results Using the *P* Value

The statistical analysis of results requires calculating the probability that, assuming the null hypothesis were true, the results achieved could have been

	H_0 is actually:	
	True	**False**
Reject H_0	Type I error (α)	Correct Decision
Fail to reject H_0	Correct Decision	Type II error (β)

due to random chance. This probability is called the *P* value. When used to locate the results on a coordinate plane, it gives a visual demonstration of where the results fall in relation to the critical region of the graph. If the *P* value of the results is outside of the critical region, then the hypothesis is rejected and the null hypothesis is accepted. In this case, the intervention had no statistically significant effect. If the *P* value of the results falls inside the critical region, then the hypothesis is accepted and the null hypothesis is rejected. In general, a smaller *P* value means that there is a greater likelihood of rejecting the null hypothesis. In this case, the results give enough evidence that there is a real difference between the groups in support of the alternative hypothesis.

Importance of Sample Size

In discussions of type I and type II errors, the size of the population under study is also highly relevant. Depending on the difference being observed, a larger sample can usually make detecting a true difference between groups easier. In the tutoring example, a smaller, random sample of students may be suitable for obtaining a fairly representative result to allow the researcher to disprove the null hypothesis. However, maybe the researchers wanted to study a smaller difference, such as the difference in the effect of tutoring on test scores for boys and girls. In that case, a larger sample size would increase the test's power to detect this difference and avoid a type II error. The researchers should try to set sample size based on the difference they hope to observe before beginning the experiment. This should help minimize error and cost.

False Positives and False Negatives

Type I error is sometimes referred to as a "false positive." This is because it shows an effect where one does not really exist. This could happen if scientists determined through testing that a new vaccine had an effect in fighting a certain virus when, in reality, it does not. Similarly, type II error is known as a "false negative." In the example, it would involve the researchers analyzing the test results and interpreting that the vaccine does not have an effect against the virus when it in fact does. Depending on the circumstances, either type of error can be serious or even catastrophic. Therefore, it is important to keep these errors in mind when analyzing test results.

Scott Zimmer, JD

Further Reading

Akobeng, Anthony K. "Understanding Type I and Type II Errors, Statistical Power and Sample Size." *Acta Paediatrica*, vol. 105, no. 6, 2016, pp. 605–9.

Bishara, Anthony J., and James B. Hittner. "Reducing Bias and Error in the Correlation Coefficient Due to Nonnormality." *Educational and Psychological Measurement*, vol. 75, no. 5, 2015, pp. 785–804.

Drummond, Gordon B, and Sarah L. Vowler. "Type I: Families, Planning and Errors." *British Journal of Pharmacology*, vol. 168, no. 1, 2013, pp. 2–6.

Green, Samuel B., et al. "Type I and Type II Error Rates and Overall Accuracy of the Revised Parallel

Analysis Method for Determining the Number of Factors." *Educational and Psychological Measurement*, vol. 75, no. 3, 2015, pp. 428–57.

Ioannidis, John P. A., et al. "Optimal Type I and Type II Error Pairs When the Available Sample Size Is Fixed." *Journal of Clinical Epidemiology*, vol. 66, no. 8, 2013, pp. 903–10.

Lawson, John. *Design and Analysis of Experiments with R*. CRC Press, 2015.

Turing, Alan

British Mathematician

Twentieth-century British mathematician Alan Turing is regarded as one of the founders of modern computing. His research into machines and human thought helped shape the field of artificial intelligence.

- **Born:** June 23, 1912; London, England
- **Died:** June 7, 1954; Wilmslow, England
- **Primary field:** Mathematics
- **Specialties:** Computability theory; logic; statistics

Early Life

Alan Mathison Turing was born in London, England, on June 23, 1912. As a student at the Sherborne School, he excelled at science and math, and he won a number of mathematics prizes despite his unconventional solutions. Turing went on to attend King's College, Cambridge, where he continued to pursue mathematics at his own pace and using his own methods, with growing success. By 1933, Turing had begun to explore mathematical logic, which focuses on proofs and computation.

In 1936, Turing completed his first major paper, "On Computable Numbers, with an Application to the Entscheidungsproblem." In the paper, published in the *Proceedings of the London Mathematical Society*, Turing describes a machine that could perform simple, carefully defined operations on paper tape. Turing modeled his machine on the action of a human following explicit instructions. This theoretical "Turing machine" was essentially a computer, although the technology did not yet exist to build the machine as described.

Also in 1936, Turing traveled to the United States and began his graduate studies at Princeton University. He earned his PhD from the university in 1938. He soon returned to Cambridge, where he began work on an analog mechanical device.

Life's Work

World War II began soon after Turing's return to England. The British Government Code and Cypher School (GCCS) asked Turing to use his mathematical skills to help decipher the codes being used by the Germans. German scientists had developed a device known as the Enigma cipher machine, which generated constantly changing codes that were nearly impossible to break. Turing helped design the Bombe, a machine that successfully deciphered the Enigma code, while working for the British government. Turing was also responsible for monitoring the communications of German submarines so that ships could be safely rerouted to avoid them.

Building on Turing's work, other members of the GCCS designed Colossus, the first electronic, programmable, and digital computer. Unlike later computers, Colossus required that those operating it change some of the machine's wiring manually to set it up for a new job. Although its functions were limited, Colossus proved that digital electronic computing machinery was feasible. For his part in the war effort, Turing was granted the title of Officer of the Order of the British Empire.

After the war, Turing went to work for Britain's National Physical Laboratory (NPL), where he designed a computer that was based on programs rather than on the rearrangement of electronic parts. Turing's proposed computer could handle many different types of tasks ranging from numerical work to algebra, file management, and code breaking. However, this computer was never built.

In 1947, Turing left the NPL for the University of Manchester, where he worked on the development of the Manchester Automatic Digital Machine, another early computer. In 1950, he published his paper "Computing Machinery and Intelligence," introducing the concept of artificial intelligence. Turing believed that it was possible to create a

Passport photo of Alan Turing at age 16. [Public domain], via Wikimedia Commons

machine that would imitate the processes of the human brain. Turing also proposed the "Turing test," which could be used to explore the question of whether a computer can think for itself.

The Turing test challenges a machine's capability to perform humanlike conversation. During the test, a human judge conducts a conversation with a human and a machine. If the judge cannot tell which is which, then the machine passes the test. Turing originally proposed the test to address objectively the question of whether machines can think. Turing believed that if the judge cannot tell the machine from the human after a reasonable amount of time, the machine is somewhat intelligent.

During the final years of his life, Turing worked on the concept that later came to be called artificial life. Turing's main focus in biology was the physical structure of living things. He was interested in how and why organisms develop particular shapes, and he wondered how simple cells know how to grow into complex forms. Turing saw the development of natural forms such as plants and animals as nothing more than a simple set of steps, or an algorithm. He used a computer to simulate a chemical mechanism that the genes of a zygote, or egg, use to determine the anatomical structure of an animal or plant. At the same time, he was experimenting with neural networks and brain structure. Turing's ultimate goal was to merge already established biological theory with mathematics and computers.

Turing was elected a fellow of the Royal Society of London in 1951, primarily in recognition of his work on Turing machines in the 1930s. In 1952, he published a theoretical paper on morphogenesis, the formation of living organisms. He also researched such topics as quantum theory and relativity theory.

In 1952, Turing was arrested and tried for engaging in a homosexual relationship, which was considered a crime in Britain. To avoid prison, he agreed to receive estrogen injections for one year in order to "neutralize" his homosexuality. Additionally, he was perceived as a security risk and subsequently lost his security clearance and ability to work for the government. Turing died on June 7, 1954. At the time, it was ruled a suicide.

Impact

Turing made many contributions to mathematics, logic, and statistics. He is best remembered, however, for his contributions to computability, machine design, and artificial intelligence. His work on computability, especially the universal Turing machine concept, was the first modern work on the theory of computation and became a central idea in recursive function theory (an active area of research in mathematical logic) and in automata theory (an important theoretical discipline within computer science). The value of Turing's efforts in the design of code-breaking equipment to the war effort was also significant. His work at the NPL resulted in the creation of one of the first operating modern computers, which was used for important scientific and engineering applications in the 1950s. Turing's work also influenced the design of later computers, though his design ideas largely fell outside the mainstream of computer design developments.

Turing was especially influential as the foremost champion of artificial intelligence research in the first decade of modern computing. He introduced the distinction between robotics and artificial intelligence research, arguing that the future of artificial intelligence lay in the use of the stored-program computer, not in the construction of special-purpose robots that could mimic vision or other human attributes. The Turing test has endured as the principal test of success in artificial intelligence research. In the decades since Turing's death, the field of artificial intelligence has advanced to the point at which computers can be programmed to read stories and answer questions about them, assist medical doctors in diagnoses, play chess at the expert level, and assist humans in numerous information-processing tasks that have become commonplace in society.

—*Bob Crepeau*

FURTHER READING

Copeland, B. Jack, ed. *Alan Turing's Automatic Computing Engine: The Master Codebreaker's Struggle to Build the Modern Computer*. New York: Oxford UP, 2005. Print. Provides a detailed history of Turing's contributions to computer science and presents diagrams and illustrations explaining the hardware, software, and other features of Turing's computers.

Graham-Cumming, John. "Alan Turing: Computation." *New Scientist* 214.2867 (2012): 2. Print. Describes Turing's scientific achievements and discusses his efforts as a code breaker during World War II and his foundational work in computer science.

Hodges, Andrew. *Alan Turing: The Enigma—The Centenary Edition*. Princeton: Princeton UP,

---

TURING DESCRIBES THE TURING MACHINE

At one level, mathematics can be viewed as a method of rearranging symbols according to a set of rules to obtain true statements about the things represented by the symbols. British mathematician George Boole showed in An Investigation of the Laws of Thought (1854) that conclusions about the truth or falsity of a combination of statements could be arrived at by using similar methods. In 1928, German mathematician David Hilbert posed what most mathematicians considered to be the major questions about such a symbolic system for mathematics. One of these was the issue of decidability: Can the truth of any statement be determined by the mechanical manipulation of symbols in a finite number of steps?

Alan Turing, then a researcher at King's College, Cambridge, focused on the question of exactly what is meant by a mechanical procedure as performed by a human mathematician. In 1937, he published the groundbreaking paper "On Computable Numbers, with an Application to the Entscheidungsproblem" in the Proceedings of the London Mathematical Society. (The German word Entscheidungsproblem, literally "decision problem," is the term commonly used in reference to Hilbert's question.) In his paper, Turing describes the simplest type of machine that could perform the same manipulation of symbols that mathematicians perform.

Turing's theoretical device, which came to be referred to as a "Turing machine," is similar to a typewriter in that it can print out a string of symbols but different in that it can also read symbols and modify them. In addition, its behavior is controlled by the symbols read by it rather than by a human operator. For simplicity, Turing describes the machine as acting not on a sheet of paper but on an endless paper tape that can be read only one symbol at a time. The internal workings of the Turing machine involve a memory register that can exist in only a limited number of states and a set of rules that determine, on the basis of any symbol that can be read and the state of the internal memory, what symbol will be printed to replace it on the tape, what the new state of the internal memory will be, and which direction the tape will move, if at all.

In this important paper, Turing also describes a programmable Turing machine that can read, from the tape, a description of the rules of any other Turing machine and then behave as if it is that machine. Using the properties of such a programmable, or universal, Turing machine, Turing demonstrates that the answer to Hilbert's "decision problem" is a definite no; there is no mechanical procedure that can directly demonstrate the truth or falsity of a mathematical statement.

Turing's paper played an influential role in the development of computer technology and artificial intelligence. By reducing the manipulations and mental processes of a human mathematician to operations that could be performed by a machine that could be built, Turing suggested that the thinking involved in solving mathematical problems is not very different from that used by humans in planning and in the other types of problem-solving behavior that constitute human intelligence. If this is true, then there is no fundamental reason a suitable machine could not be programmed to display intelligence.

2012. Print. Offers a biography of Turing, covering his influence on modern computer sciences, artificial intelligence, and the gay rights movement.

Petzold, Charles. *The Annotated Turing: A Guided Tour through Alan Turing's Historic Paper on Computability and the Turing Machine.* Hoboken: Wiley, 2008. Print. Guides readers through Turing's landmark paper, offering detailed explanation and examples.

Turing, Alan. "Computing Machinery and Intelligence." *Mind* 59 (1950): 433–60. Print. Provides Turing's counterarguments to common objections against artificial intelligence and introduces the Turing test, which decides when a computer has achieved intelligence.

Wu, Chien-Shiung

CHINESE AMERICAN PHYSICIST

Nuclear physicist Chien-Shiung Wu disproved the principle of conservation of parity and proved the law of vector current in beta decay. Wu was the first woman elected president of the American Physical Society and honored with Israel's Wolf Prize.

- **Born:** May 31, 1912; Liu Ho, China
- **Died:** February 16, 1997; New York, New York
- **Primary field:** Physics
- **Specialties:** Nuclear physics; atomic and molecular physics; quantum mechanics

EARLY LIFE

Chien-Shiung Wu was born in Liu Ho, China, near Shanghai, on May 31, 1912. At this time, Chinese girls were educated at home, if at all, but Wu's father, Wu Zhong-Yi, was progressive. With Wu's mother, he operated a local school for girls, the first school of its kind in China. Wu attended this institution until she was nine years old. When Wu had completed her studies at the girls' primary school, her father encouraged her to go to Suzhou to attend a private high school. Studying under leading scholar Hu Shi, Wu learned English and became politically active. She graduated as valedictorian of her class.

Wu decided she wanted to become a physicist, and with the encouragement of Hu Shi, she entered the National Central University in Nanjing in 1930. By this time, it was not unusual for women to study science. Although Wu was studying during the early years of the Sino-Japanese War, and Japan had invaded China, she managed to complete her degree in physics in four years.

In 1936, Wu traveled to the United States, where she pursued her graduate work in physics at the University of California at Berkeley. One of her most influential professors was Ernest O. Lawrence, winner of the Nobel Prize in Physics in 1939 for inventing the particle accelerator known as the cyclotron. Wu was inducted into the Phi Beta Kappa academic honor society in recognition of her outstanding work as a graduate student.

Wu received her PhD in 1940, and accepted a job as a research assistant at Berkeley. During the early years of World War II, she taught at Smith College in Northampton, Massachusetts, and then Princeton University in New Jersey, where she was the first woman to teach in the Physics Department. In 1942, she married Luke Chia-Liu Yuan, also a physicist.

LIFE'S WORK

In the early 1940s, Wu began conducting experiments to test Enrico Fermi's 1934 theory of beta decay, the radioactive transformation of an atom from one atomic number to another, caused by the emission of beta particles from the nucleus. In March 1944, Wu joined Columbia University in New York City, where she would remain until her retirement. Her first position was on the scientific staff of the Division of War Research. This group was working on the Manhattan Project to develop the atomic bomb. Specifically, Wu worked on radiation detection. In this capacity, she developed a gaseous diffusion process for separating radioactive uranium-235 from common uranium-238. She also helped to develop a more sensitive Geiger counter. After the war, Wu stayed on at Columbia as a research associate and teacher. Her son, Vincent, was born in 1948. In 1952, she was made associate professor of physics.

In the 1950s, theoretical physicists Chen Ning Yang and Tsung-Dao Lee were looking for an experimental physicist to conduct research that would disprove the principle of conservation of parity. Lee was a colleague of Wu's on the Columbia faculty, so he knew of her expertise in beta decay, and asked for her help.

It had long been known that physical properties of particles—such as mass, energy, momentum, and electrical charge—remain unchanged after a nuclear reaction. *Parity* refers to the property of symmetry of the physical laws. Parity, like other physical properties, had been believed since 1924 to be conserved in nuclear reactions. In other words, particles emitted during a nuclear reaction should be emitted in all directions equally. However, in 1956, Yang and Lee found theoretical evidence that parity is not always conserved in nuclear reactions, specifically in certain weak interactions, such as those occurring in beta

Chien-Shiung Wu (1912-1997), professor of physics at Columbia University, shown with a "Dr. Brode" (probably Wallace Brode) and a group of Science Talent Search winners, 1958. By Smithsonian Institution from United States [No restrictions], via Wikimedia Commons

decay. But they needed experimental results to prove their hypothesis.

Wu's team consisted of a group of top scientists from the National Bureau of Standards. In 1957, the team tested Yang and Lee's hypothesis with the radioactive material cobalt-60. First, they cooled the cobalt to 0.01 degree above absolute zero, approximately -459 degrees Fahrenheit. This was to minimize any random thermal movements of the nuclei, so that the scientists could record the disintegration of the radioactive atoms without the interference of other effects. When the electrons emitted during the cobalt's decay moved in the direction opposite that of the magnetic field that Wu had set up, they contravened the law of parity.

Worldwide recognition followed for Wu, Lee, and Yang. Scientists at Columbia and the University of Chicago conducted similar experiments utilizing other weak reactions, and their results confirmed Wu's research. This disproved forever the law of conservation of parity. When Yang and Lee received the Nobel Prize in Physics in 1957 for disproving the conservation of parity, many scientists were disappointed that Wu was not included in the award.

Wu did, however, receive numerous other awards and honors. For instance, in 1957, Wu was named a full professor at Columbia University. She was made a member of the National Academy of Sciences in 1958; at that time, she was only the seventh woman so honored and was also the first Chinese American to be made a member. Also in 1958, Wu became the first woman to receive the Research Corporation Award, given annually to outstanding scientists. In her acceptance speech, Wu remarked on the uniqueness of winning the award for destroying a law rather than for establishing one. Also that year, Wu received an honorary doctorate of science degree from Princeton University. She was the first woman to receive this honor from Princeton.

While continuing to research and teach, Wu and her husband edited the book *Nuclear Physics* (1961). In 1963, Wu's experiments provided evidence to confirm the theoretical law of vector current in beta decay. This theory had been proposed in 1958 by physicists Murray Gell-Mann and Richard P. Feynman. Wu's research culminated in her publication of *Beta Decay* in 1965. The text became a standard reference for physicists. That same year, she was awarded the Chi-Tsin Achievement Award from Taiwan's Chi-Tsin Culture Foundation.

Wu also performed experiments to confirm the theory that electromagnetic radiation is emitted when an electron and a positron collide. Her success in measuring low-energy electrons emitted by beta decay supported Fermi's theory of weak interactions. Later, Wu conducted research in ultra-low-temperature physics, muonic and pionic X-rays, and spectroscopic examinations of hemoglobin. She coedited, with Vernon W. Hughes, a three-volume text called *Muon Physics* (1975–1977).

The honors that Wu received for her research included a number of firsts. In 1975, the American Physical Society honored Wu with the National Medal of Science, the Tom Bonner Prize, and election to the presidency of the society. Wu was the first woman appointed to this role and the first woman honored with Israel's Wolf Prize in 1978. She was also the first scientist to receive the Wolf Prize in the category of physics. An asteroid was named for Wu in 1990; she was the first living scientist to receive that honor.

Wu retired from her position at Columbia University in 1981, after thirty-seven years, and was named professor emeritus. She died of a stroke on February 16, 1997, in New York City.

IMPACT

Though Wu never won a Nobel Prize, she was considered one of the foremost female physicists during her lifetime. She is remembered primarily for her research in nuclear forces and structure, which helped disprove the principle of conservation of parity. Until 1957, the conservation of parity, related to symmetry, was considered a basic law of nuclear physics. Wu showed that although parity might be conserved in strong, electromagnetic interactions, the same does not necessarily hold true in weak interactions of subatomic particles. The discovery of this lack of symmetry in parity was hugely influential to the scientific community, since the idea of symmetry has been tied to the laws of physics for centuries.

Wu's findings in nuclear physics led to further research by other influential physicists, including Feynman, Gell-Mann, Robert Marshak, and George Sudarshan. While Wu's work disproved the law of conservation of parity, it did not explain why this was so. Feynman and Gell-Mann were later able to explain it with the V-A theory of weak interactions.

—*Ellen Bailey*

FURTHER READING

Bertulani, Carlos A. *Nuclear Physics in a Nutshell*. Princeton: Princeton UP, 2007. Print. Discusses the atomic nucleus and explains the theories of

---

WU PROVES THE LAW OF CONSERVATION OF VECTOR CURRENT IN BETA DECAY

The theory of conservation of vector current in beta decay was first proposed by Murray Gell-Mann and Richard P. Feynman in 1958. The theory originated from Enrico Fermi's experiments in beta decay in the 1930s, in which he initiated a theory of weak interactions. He described the weak interaction of beta decay as an electron and a neutrino interacting with a current operator that converts a neutron into a proton or a proton into a neutron. Analogous to electromagnetism, Fermi described this current as a vector current, like an electric current. Though Fermi's theory accounted for the spectrum shape of the beta decay in the fastest beta decays, it did not fit particularly well with all known beta decays. When this became clear to scientists, it was decided that something other than vector currents could be at work. For a long time, physicists favored the theory that scalar and tensor currents are responsible for the interaction; but this did not support Fermi's analogy to electromagnetism, which has a vector current.

Despite intermittent experiments, the scalar and tensor theory, though flawed, was widely accepted and mostly overlooked as an area of study. Chien-Shiung Wu's discovery disproving the law of conservation of parity, however, renewed interest in the subject of beta decay. In 1958, George Sudarshan and Robert Marshak proposed, in the name of universality, that the currents are indeed vector and axial vector. Experiments proved this theory correct. The importance of this discovery was the restored analogy to electromagnetism. Feynman and Gell-Mann further suggested that the vector current in beta decay is conserved in the manner of an electromagnetic current; this means that the rate of change of its time component, plus the divergence of its space, equals zero. It was a good theory, but once again, as with the law of parity, Wu decided to see if she could prove law quantitatively.

To test the conservation of vector current, Wu and her collaborators performed an experiment that had been done earlier but with inconclusive results. Wu and her team compared symmetric and asymmetric states of protons and neutrons; isospin is the term for their relating symmetry. An isospin singlet is in an asymmetric state and an isospin triplet is in a symmetric state. The experiment was completed in 1963 and successfully demonstrated the conservation of weak vector current. Not only had Wu proved a new law of nature, but her discovery also marked the crucial first step toward the unification of weak and electromagnetic interaction.

the physicists who have studied it. Includes Wu's disproving of the principle of conservation of parity.

Cooperman, Stephanie H. *Chien-Shiung Wu: Pioneering Physicist and Atomic Researcher.* New York: Rosen, 2004. Print. Describes Wu's life, career, and legacy.

Garwin, Richard L., and Tsung-Dao Lee. "Chien-Shiung Wu." *Physics Today* 50.10 (1997): 120–21. Print. An obituary for Chien-Shiung Wu, with biographical data and information regarding her scientific career.

TIME LINE OF INVENTIONS AND SCIENTIFIC ADVANCEMENTS

Date	Invention or discovery
4–5 billion years ago	Sun starts to produce energy.
10 million years ago	Humans make the first tools from stone, wood, antlers, and bones.
1–2 million years ago	Humans discover fire.
25,000–50,000 bce	Humans first wear clothes.
10,000 bce	Earliest boats are constructed.
8000–9000 bce	Beginnings of human settlements and agriculture.
6000–7000 bce	Hand-made bricks first used for construction in the Middle East.
4000 bce	Iron used for the first time in decorative ornaments.
3500 bce	Humans invent the wheel.
3000–600 bce	Bronze Age: Widespread use of copper and its important alloy bronze.
2000 bce	Water-raising and irrigation devices like the shaduf (shadoof) introduce the idea of lifting things using counterweights.
c1700 bce	Semites of the Mediterranean develop the alphabet.
0–1500 bce	Ancient societies invent some of the first machines for moving water and agriculture.
1000 bce	Iron Age begins: iron is widely used for making tools and weapons in many parts of the world.
600 bce	Thales of Miletus discovers static electricity.
c.150–100 bce	Gear-driven, precision clockwork machines (such as the Antikythera mechanism) are in existence.
c.50 bce	Roman engineer Vitruvius perfects the modern, vertical water wheel.
62 ce	
	Hero of Alexandria, a Greek scientist, pioneers steam power.
105 ce	Ts'ai Lun makes the first paper in China.
27 bce–395 ce	Romans develop the first, basic concrete called pozzolana.
~600 ce	Windmills are invented in the Middle East.
700–900 ce	Chinese invent gunpowder and fireworks.
800–1300 ce	Islamic "Golden Age" sees the development of a wide range of technologies, including ingenious clocks and feedback mechanisms that are the ancestors of modern automated factory machines.
1000 ce ??	Chinese develop eyeglasses by fixing lenses to frames that fit onto people's faces.
1450	Johannes Gutenberg pioneers the modern printing press, using rearrangeable metal letters called movable type.
1470s	The first parachute is sketched on paper by an unknown inventor.

Time Line of Inventions and Scientific Advancements

1530s	Gerardus Mercator helps to revolutionize navigation with better mapmaking.
1590	A Dutch spectacle maker named Zacharias Janssen makes the first compound microscope.
~1600	Galileo Galilei designs a basic thermometer.
16th century	Antoni van Leeuwenhoek and Robert Hooke independently develop microscopes.
1600	William Gilbert publishes his great book De Magnete describing how Earth behaves like a giant magnet. It's the beginning of the scientific study of magnetism.
1609	Galileo Galilei builds a practical telescope and makes new astronomical discoveries.
1643	Galileo's pupil Evangelista Torricelli builds the first mercury barometer for measuring air pressure.
1650s	Christiaan Huygens develops the pendulum clock (using Galileo's earlier discovery that a swinging pendulum can be used to keep time).
1687	Isaac Newton formulates his three laws of motion.
1700s	Bartolomeo Cristofori invents the piano.
1701	English farmer Jethro Tull begins the mechanization of agriculture by inventing the horse-drawn seed drill.
1703	Gottfried Leibniz pioneers the binary number system now used in virtually all computers.
1712	Thomas Newcomen builds the first practical (but stationary) steam engine.
1700s	Christiaan Huygens conceives the internal combustion engine, but never actually builds one.
1737	William Champion develops a commercially viable process for extracting zinc on a large scale.
1757	John Campbell invents the sextant, an improved navigational device that enables sailors to measure latitude.
1730s–1770s	John Harrison develops reliable chronometers (seafaring clocks) that allow sailors to measure longitude accurately for the first time.
1751	Axel Cronstedt isolates nickel.
1756	Axel Cronstedt notices steam when he boils a rock—and discovers zeolites.
1769	Wolfgang von Kempelen develops a mechanical speaking machine: the world's first speech synthesizer.
1770s	Abraham Darby III builds a pioneering iron bridge at a place now called Ironbridge in England.
~1780	Josiah Wedgwood (or Thomas Massey) invents the pyrometer.
1783	French Brothers Joseph-Michel Montgolfier and Jacques-Étienne Montgolfier make the first practical hot-air balloon.

1791	Reverend William Gregor, a British clergyman and amateur geologist, discovers a mysterious mineral that he calls menachite. Four years later, Martin Klaproth gives it its modern name, titanium.
1800	Italian Alessandro Volta makes the first battery (known as a Voltaic pile).
1801	Joseph-Marie Jacquard invents the automated cloth-weaving loom. The punched cards it uses to store patterns help to inspire programmable computers.
1803	Henry and Sealy Fourdrinier develop the papermaking machine.
1806	Humphry Davy develops electrolysis into an important chemical technique and uses it to identify a number of new elements.
1807	Humphry Davy develops the electric arc lamp.
1814	George Stephenson builds the first practical steam locomotive.
1816	Robert Stirling invents the efficient Stirling engine.
1820s–1830s	
	Michael Faraday builds primitive electric generators and motors.
1827	Joseph Niepce makes the first modern photograph.
1830s	William Sturgeon develops the first practical electric motor.
1830s	Louis Daguerre invents a practical method of taking pin-sharp photographs called
1830s	William Henry Fox Talbot develops a way of making and printing photographs using reverse images called negatives.
1830s–1840s	Charles Wheatstone and William Cooke, in England, and Samuel Morse, in the United States, develop the electric telegraph (a forerunner of the telephone).
1836	Englishman Francis Petit-Smith and Swedish-American John Ericsson independently develop propellers with blades for ships.
1839	Charles Goodyear finally perfects a durable form of rubber (vulcanized rubber) after many years of unsuccessful experimenting.
1840s	Scottish physicist James Prescott Joule outlines the theory of the conservation of energy.
1840s	Scotsman Alexander Bain invents a primitive fax machine based on chemical technology.
1849	James Francis invents a water turbine now used in many of the world's hydropower plants.
1850s	Henry Bessemer pioneers a new method of making steel in large quantities.
1850s	Louis Pasteur develops pasteurization: a way of preserving food by heating it to kill off bacteria.
1850s	Italian Giovanni Caselli develops a mechanical fax machine called the pantelegraph.
1860s	Frenchman Étienne Lenoir and German Nikolaus Otto pioneer the internal combustion engine.

Time Line of Inventions and Scientific Advancements

1860s	James Clerk Maxwell figures out that radio waves must exist and sets out basic laws of electromagnetism.
1860s	Fire extinguishers are invented.
1861	Elisha Graves Otis invents the elevator with built-in safety brake.
1867	Joseph Monier invents reinforced concrete.
1868	Christopher Latham Sholes invents the modern typewriter and QWERTY keyboard.
1876	Alexander Graham Bell patents the telephone, though the true ownership of the invention remains controversial even today.
1870s	Thomas Edison develops the phonograph, the first practical method of recording and playing back sound on metal foil.
1870s	Lester Pelton invents a useful new kind of water turbine known as a Pelton wheel.
1877	Thomas Edison invents his sound-recording machine or phonograph—a forerunner of the record player and CD player.
1877	Edward Very invents the flare gun (Very pistol) for sending distress flares at sea.
1880	Thomas Edison patents the modern incandescent electric lamp.
1880	Pierre and Paul-Jacques Curie discover the piezoelectric effect.
1880s	Thomas Edison opens the world's first power plants.
1880s	Charles Chamberland invents the autoclave (steam sterilizing machine).
1880s	Charles and Julia Hall and Paul Heroult independently develop an affordable way of making aluminum.
1880s	Carrie Everson invents new ways of mining silver, gold, and copper.
1881	Jacques d'Arsonval suggests heat energy could be extracted from the oceans.
1883	Charles Eastman invents plastic photographic film.
1884	Charles Parsons develops the steam turbine.
1885	Karl Benz builds a gasoline-engined car.
1886	Josephine Cochran invents the dishwasher.
1888	Friedrich Reinitzer discovers liquid crystals.
1888	Nikola Tesla patents the alternating current (AC) electric induction motor and, in opposition to Thomas Edison, becomes a staunch advocate of AC power.
1899	Everett F. Morse invents the optical pyrometer for measuring temperatures at a safe distance.
1890s	French brothers Joseph and Louis Lumiere invent movie projectors and open the first movie theater.
1890s	German engineer Rudolf Diesel develops his diesel engine—a more efficient internal combustion engine without a sparking plug.
1895	German physicist Wilhelm Röntgen discovers X rays.
1895	American Ogden Bolton, Jr. invents the electric bicycle.

1901	Guglielmo Marconi sends radio-wave signals across the Atlantic Ocean from England to Canada
1901	The first electric vacuum cleaner is developed.
1903	Brothers Wilbur and Orville Wright build the first engine-powered airplane.
1905	Albert Einstein explains the photoelectric effect.
1905	Samuel J. Bens invents the chainsaw.
1906	Willis Carrier pioneers the air conditioner.
1906	Mikhail Tswett discovers chromatography.
1907	Leo Baekeland develops Bakelite, the first popular synthetic plastic.
1907	Alva Fisher invents the electric clothes washer.
1906-8	Frederick Gardner Cottrell develops the electrostatic smoke precipitator (smoke-stack pollution scrubber).
1908	American industrialist and engineer Henry Ford launches the Ford Model T, the world's first truly affordable car.
1909	German chemists Fritz Haber and Zygmunt Klemensiewicz develop the glass electrode, enabling very precise measurements of acidity.
1912	American chemist Gilbert Lewis describes the basic chemistry that leads to practical, lithium-ion rechargeable batteries (though they don't appear in a practical, commercial form until the 1990s).
1912	Hans Geiger develops the Geiger counter, a detector for radioactivity.
1919	Francis Aston pioneers the mass spectrometer and uses it to discover many isotopes.
1920s	John Logie Baird develops mechanical television.
1920s	Philo T. Farnsworth invents modern electronic television.
1920s	Robert H. Goddard develops the principle of the modern, liquid-fueled space rocket.
1920s	German engineer Gustav Tauschek and American Paul Handel independently develop primitive optical character recognition (OCR) scanning systems.
1920s	Albert W. Hull invents the magnetron, a device that can generate microwaves from electricity.
1921	Karel Capek and his brother coin the word "robot" in a play about artificial humans.
1921	John Larson develops the polygraph ("lie detector") machine.
1928	Thomas Midgley, Jr. invents coolant chemicals for air conditioners and refrigerators.
1928	The electric refrigerator is invented.
1930s	Peter Goldmark pioneers color television.

Time Line of Inventions and Scientific Advancements

1930s	Laszlo and Georg Biro pioneer the modern ballpoint pen.
1930s	Maria Telkes creates the first solar-powered house.
1930s	Wallace Carothers develops neoprene (synthetic rubber used in wetsuits) and nylon, the first popular synthetic clothing material.
1930s	Robert Watson Watt oversees the development of radar.
1930s	Arnold Beckman develops the electronic pH meter.
1931	Harold E. Edgerton invents the xenon flash lamp for high-speed photography.
1932	Arne Olander discovers the shape memory effect in a gold-cadmium alloy.
1936	W.B. Elwood invents the magnetic reed switch.
1938	Chester Carlson invents the principle of photocopying (xerography).
1938	Roy Plunkett accidentally invents a nonstick plastic coating called Teflon.
1939	Igor Sikorsky builds the first truly practical helicopter.
1940s	English physicists John Randall and Harry Boot develop a compact magnetron for use in airplane radar navigation systems.
1942	Enrico Fermi builds the first nuclear chain reactor at the University of Chicago.
1945	US government scientist Vannevar Bush proposes a kind of desk-sized memory store called Memex, which has some of the features later incorporated into electronic books and the World Wide Web (WWW).
1947	John Bardeen, Walter Brattain, and William Shockley invent the transistor, which allows electronic equipment to made much smaller and leads to the modern computer revolution.
1949	Bernard Silver and N. Joseph Woodland patent barcodes—striped patterns that are initially developed for marking products in grocery stores.
1950s	Charles Townes and Arthur Schawlow invent the maser (microwave laser). Gordon Gould coins the word "laser" and builds the first optical laser in 1958.
1950s	Stanford Ovshinksy develops various technologies that make renewable energy more practical, including practical solar cells and improved rechargeable batteries.
1950s	European bus companies experiment with using flywheels as regenerative brakes
1950s	Percy Spencer accidentally discovers how to cook with microwaves, inadvertently inventing the microwave oven.
1954	Indian physicist Narinder Kapany pioneers fiber optics.
1956	First commercial nuclear power is produced at Calder Hall, Cumbria, England.
1957	Soviet Union (Russia and her allies) launch the Sputnik space satellite.
1957	Lawrence Curtiss, Basil Hirschowitz, and Wilbur Peters build the first fiber-optic gastroscope.
1958	Jack Kilby and Robert Noyce, working independently, develop the integrated circuit.
integrated circuits	Transistors

1959	IBM and General Motors develop Design Augmented by Computers-1 (DAC-1), the first computer-aided design (CAD) system.
1960s	Joseph-Armand Bombardier perfects his Ski-Doo® snowmobile.
1960	Theodore Maiman invents the ruby laser.
1962	William Armistead and S. Donald Stookey of Corning Glass Works invent light-sensitive (photochromic) glass.
1963	Ivan Sutherland develops Sketchpad, one of the first computer-aided design programs.
1964	IBM helps to pioneer e-commerce with an airline ticket reservation system called SABRE.
1965	Frank Pantridge develops the portable defibrillator for treating cardiac arrest patients.
1966	Stephanie Kwolek patents a super-strong plastic called Kevlar.
1967	Japanese company Noritake invents the vacuum fluorescent display (VFD).
1968	Alfred Y. Cho and John R. Arthur, Jr invent a precise way of making single crystals called molecular beam epitaxy (MBE).
1969	World's first solar power station opened in France.
1969	Long before computers become portable, Alan Kay imagines building an electronic book, which he nicknames the Dynabook.
1969	Willard S. Boyle and George E. Smith invent the CCD (charge-coupled device): the light-sensitive chip used in digital cameras, webcams, and other modern optical equipment.
1969	Astronauts walk on the Moon.
1960s	Douglas Engelbart develops the computer mouse.
1960s	James Russell invents compact discs.
1971	Electronic ink is pioneered by Nick Sheridon at Xerox PARC.
1971	Ted Hoff builds the first single-chip computer or microprocessor.
1973	Martin Cooper develops the first handheld cellphone (mobile phone).
1973	Robert Metcalfe figures out a simple way of linking computers together that he names Ethernet. Most computers hooked up to the Internet now use it.
1974	First grocery-store purchase of an item coded with a barcode.
1975	Whitfield Diffie and Martin Hellman invent public-key cryptography.
1975	Pico Electronics develops X-10 home automation system.
1976	Steve Wozniak and Steve Jobs launch the Apple I: one of the world's first personal home computers
1970s–1980s	James Dyson invents the bagless, cyclonic vacuum cleaner.

1970s–1980s	Scientists including Charles Bennett, Paul Benioff, Richard Feynman, and David Deutsch sketch out how quantum computers might work.
1980s	Japanese electrical pioneer Akio Morita develops the Sony Walkman, the first truly portable player for recorded music.
1981	Stung by Apple's success, IBM releases its own affordable personal computer (PC).
1981	The Space Shuttle makes its maiden voyage.
1981	Patricia Bath develops laser eye surgery for removing cataracts.
1981–1982	Alexei Ekimov and Louis E. Brus (independently) discover quantum dots.
1983	Compact discs (CDs) are launched as a new way to store music by the Sony and Philips corporations.
1987	Larry Hornbeck, working at Texas Instruments, develops DLP® projection—now used in many projection TV systems.
1989	Tim Berners-Lee invents the World Wide Web.
1990	German watchmaking company Junghans introduces the MEGA 1, believed to be the world's first radio-controlled wristwatch.
1991	Linus Torvalds creates the first version of Linux, a collaboratively written computer operating system.
1994	American-born mathematician John Daugman perfects the mathematics that make iris scanning systems possible.
1994	Israeli computer scientists Alon Cohen and Lior Haramaty invent VoIP for sending telephone calls over the Internet.
1995	Broadcast.com becomes one of the world's first online radio stations.
1995	Pierre Omidyar launches the eBay auction website.
1996	WRAL-HD broadcasts the first high-definition television (HDTV) signal in the United States.
1997	Electronics companies agree to make Wi-Fi a worldwide standard for wireless Internet.
2001	Apple revolutionizes music listening by unveiling its iPod MP3 music player.
2001	Richard Palmer develops energy-absorbing D3O plastic.
2001	The Wikipedia online encyclopedia is founded by Larry Sanger and Jimmy Wales.
2001	Bram Cohen develops BitTorrent file-sharing.
2001	Scott White, Nancy Sottos, and colleagues develop self-healing materials.
2002	iRobot Corporation releases the first version of its Roomba® vacuum cleaning robot.
2004	Electronic voting plays a major part in a controversial US Presidential Election.
2004	Andre Geim and Konstantin Novoselov discover graphene.

2005	A pioneering low-cost laptop for developing countries called OLPC is announced by MIT computing pioneer Nicholas Negroponte.
2007	Amazon.com launches its Kindle electronic book (e-book) reader.
2007	Apple introduces a touchscreen cellphone called the iPhone.
2010	Apple releases its touchscreen tablet computer, the iPad.
2010	3D TV starts to become more widely available.
2013	Elon Musk announces "hyperloop"—a giant, pneumatic tube transport system.
2015	Supercomputers (the world's fastest computers) are now a mere 30 times less powerful than human brains.
2016	Three nanotechnologists win the Nobel Prize in Chemistry for building miniature machines out of molecules.

GLOSSARY

active factor: the independent variable that has the most profound effect on the dependent variable or outcome.

alpha (α): variable representing the likelihood of a type I error, which occurs when the null hypothesis is rejected despite being true.

before-and-after graph: a graph that depicts the relationship between two variables, typically either independent and dependent variables or pretest and posttest scores, before and after a treatment for both the experimental group and the control group.

block: a group of relatively homogenous experimental units.

causation: a relationship between two events, actions, or other phenomena in which one event is determined to have directly caused or otherwise affected the other.

cluster sampling: a probabilistic sampling method in which the population being studied is divided into subsets called "clusters" based on a shared characteristic unrelated to the research question, such as geographic location, and then multiple clusters are selected at random for comprehensive examination.

completely randomized design: an experimental design in which experimental units are randomly assigned to treatments so that each has the same chance of receiving a treatment.

confidence interval: a range of values within which the value of an unknown population parameter has a given probability of falling.

control group: the subjects in an experiment who do not receive any intervention or who receive a placebo or sham intervention.

critical region: the set of values for which the null hypothesis is rejected.

cutoff point: a designated value used to divide a range of numbers into two relevant categories; in a regression discontinuity design, the numbers are the participants' pretest scores, and the categories are the experimental group and the control group.

error: in statistics, a situation in which the null hypothesis is found true despite being false, or found false despite being true.

experimental group: the subjects in an experimental research study who receive the intervention or treatment being tested.

experimental group: the subjects in an experimental research study who receive the intervention or treatment being tested.

external validity: the extent to which the results of a study can be generalized to apply to other populations and settings.

factor: an independent variable or a confounding variable that affect the outcome, or dependent, variable.

full factorial design: an experimental research design that analyzes all possible combinations of factors in an experiment.

independent samples: two different sample populations or randomized members of the same population measured at separate points in time.

interaction: in statistics, an effect by which the impact of an independent variable on the dependent variable is altered, in a nonadditive way, by the influence of another independent variable or variables.

internal validity: the extent to which the results of a study can confidently be attributed to the quality being tested, as determined by how well the study was designed to minimize systematic errors.

longitudinal sampling: an experiment that examines a single group of subjects over an extended period of time.

nonrandom sampling: one of several non-probabilistic sampling methods in which individuals within the population being studied are selected for examination in a manner that does not involve random chance, so that not all members of the population have a chance of being included, and the probability of selecting a particular sample cannot be calculated.

P value: the probability of a test result occurring by chance when assuming that the null hypothesis is true.

paired *t*-test: a method of statistical analysis used to test a hypothesis by comparing two measures taken from the same population, usually in a before-and-after scenario.

pretest-posttest design: a research design in which subjects are tested both before and after the experimental treatment or intervention.

pretest-posttest design: a research design in which subjects are tested both before and after the experimental treatment or intervention.

priming effect: a psychological effect in which exposure to a stimulus influences an individual's response to a subsequent stimulus.

quasi-experiment: a study in which subjects are not randomly assigned to control and experimental groups or the experimenter cannot control which subjects receive the experimental condition.

replicate: an experimental run that tests the same set of treatment levels as another run within the same experiment, or an experimental subject that receives the same set of treatment levels as another subject in the experiment.

replication: the repetition of an experiment or study with the aim of achieving the same results in order to confirm findings or reduce error.

sample size: the number of subjects included in a survey or experiment.

sample size: the number of subjects included in a survey or experiment.

simple random sampling: a probabilistic sampling method in which members of the sample group are drawn from the population at random, so that each member of the population has an equal chance of being included.

single case study: a research method in which a selected individual or small group is observed and examined thoroughly.

spatial effects: environmental elements, such as sunlight, that affect the area under study and could therefore confound experimental results.

split-plot factor: a factor that is easy to change.

standard deviation: a measurement of the degree of dispersion present in a distribution; greater dispersion results in a larger standard deviation.

statistical analysis: the collection, examination, description, manipulation, and interpretation of quantitative data.

strata: homogenous subgroups that are drawn from a larger population and do not overlap with one another.

subject: the variable that is to be studied to determine the effect of a treatment; could be a person, an animal, or an object, depending on the experiment.

treatment: the process or intervention applied or administered to members of an experimental group.

treatment: the process or intervention applied or administered to members of an experimental group.

treatment: the process or intervention applied or administered to members of an experimental group.

two-way interaction: in statistics, an effect by which the impact of one independent variable on the dependent variable is altered, in a nonadditive way, by the influence of a second independent variable.

uniform sampling: a sampling method in which samples are drawn from a population in a consistent pattern with a predetermined population, sample size, and starting point.

variability: the extent to which outcomes in a data set differ from one another.

variable: in research or mathematics, an element that may take on different values under different conditions.

whole-plot factor: a factor that is difficult to change.

Bibliography

Adams, Kathrynn A., and Eva K. Lawrence. *Research Methods, Statistics, and Applications.* Sage Publications, 2015.

Akroyd, Wallace Ruddell. *Three Philosophers: Lavoisier, Priestley, and Cavendish.* Westport: Greenwood, 1970. Print. An account of the chemical revolution in relatively nontechnical terms, focusing mainly on Lavoisier and Priestley, with a short chapter on Cavendish.

Alferes, Valentim R. *Methods of Randomization in Experimental Design.* Sage Publications, 2012.

Almeder, Robert F. *The Philosophy of Charles S. Peirce: A Critical Introduction.* Totowa, N.J.: Rowman & Littlefield, 1980. An analysis of Peirce's philosophy, stressing his epistemological realism, which contains a perceptive and detailed discussion of his theory of knowledge.

Arnab, Raghunath. *Survey Sampling Theory and Applications.* Academic Press, 2017.

Barrett, Peter. *Science and Theology Since Copernicus: The Search for Understanding.* London: Continuum International, 2004. Print. Traces the legacy of Copernicus over four hundred years. Examines the history of the debate between science and Christianity, attempting to fashion a philosophical basis for the simultaneous embrace of scientific method and religious faith in the modern world.

Benford, Gregory. "Leaping the Abyss." *Reason* 4 (2002): 24–31. Print.

Bernard, Harvey Russell. *Social Research Methods: Qualitative and Quantitative Approaches.* SAGE Publications, 2000.

Bernstein, Jeremy. *Oppenheimer: Portrait of an Enigma.* Chicago: Dee, 2005. Print. Biography of Oppenheimer by a physicist who studied under him and was a staff writer at the *New Yorker*.

Berra, Tim M. *Charles Darwin: The Concise Story of an Extraordinary Man.* Baltimore: Johns Hopkins UP: 2009. Print. A biography of Darwin that includes information on his family life, his travels on the *Beagle*, and his work and publications related to evolution.

Bertulani, Carlos A. *Nuclear Physics in a Nutshell.* Princeton: Princeton UP, 2007. Print. Discusses the atomic nucleus and explains the theories of the physicists who have studied it. Includes Wu's disproving of the principle of conservation of parity.

Biemer, Paul P., et al., editors. *Total Survey Error in Practice.* John Wiley & Sons, 2016.

Bird, Kai, and Martin J. Sherwin. *American Prometheus: The Triumph and Tragedy of J. Robert Oppenheimer.* New York: Vintage, 2006. Print. A Pulitzer Prize–winning biography copiously footnoted and containing information drawn from a multitude of interviews.

Blackmore, John T. "Three Autobiographical Manuscripts by Ernst Mach." *Annals of Science* 35.4 (July 1978): 401–19. Print. Offers a collection of original source material as an aid to understanding Mach's role in the history of physics, psychology, and science in general.

Blair, Edward, and Johnny Blair. *Applied Survey Sampling.* Sage Publications, 2015.

Bonate, Peter L. *Analysis of Pretest-Posttest Designs.* Chapman & Hall/CRC, 2000.

Bordens, Kenneth S., and Bruce B. Abbott. *Research Design and Methods: A Process Approach.* 9th ed., McGraw-Hill Education, 2014.

Bowman-Kruhm, Mary. *The Leakeys: A Biography.* Westport, CT: Greenwood, 2005. Print. Intended for general and high school-level readers and written in a style that moves between the serious and the breezy. Outline structure makes the complexities of the Leakeys' interactions with one another and of the fossil finds easy to follow. Bibliography.

Box, George E. P., et al. *Statistics for Experimenters: Design, Innovation, and Discovery.* 2nd ed., Wiley-Interscience, 2005.

Buglear, John. *Practical Statistics: A Handbook for Business Projects.* Kogan Page, 2014.

Cappelleri, Joseph C., and William M. K. Trochim. "An Illustrative Statistical Analysis of Cutoff-Based Randomized Clinical Trials." *Journal of Clinical Epidemiology*, vol. 47, no. 3, 1994, pp. 261–70.

Chang, Todd P., et al. "Pediatric Emergency Medicine Asynchronous E-learning: A Multicenter Randomized Controlled Solomon Four-Group Study." *Academic Emergency Medicine*, vol. 21, no. 8, 2014, pp. 912–919, doi:10.1111/acem.12434.

Christensen, Larry B., et al. *Research Methods, Design, and Analysis.* 12th ed., Pearson, 2014.

Clewer, Alan G., and David H. Scarisbrick. *Practical Statistics and Experimental Design for Plant and Crop Science.* John Wiley & Sons, 2001.

Close, Frank. *The Infinity Puzzle: Quantum Field Theory and the Hunt for an Orderly Universe*. Basic, 2011. An account that does not rely on mathematics to give the reader an appreciation of the intricacies of quantum field theory and its development to the present day.

Conkin, Paul K. *Puritans and Pragmatists: Eight Eminent American Thinkers*. Bloomington: Indiana University Press, 1968. One of the finest overviews of American intellectual history. Places Peirce within the context of the development of American thought between Jonathan Edwards and George Santayana.

Connor, James A. *Kepler's Witch: An Astronomer's Discovery of Cosmic Order amid Religious War, Political Intrigue, and the Heresy Trial of His Mother*. New York: Harper, 2005. Print. A biography of Kepler that focuses in particular on his experiences regarding religious and political conflict. Also includes translations of some of Kepler's personal writings.

Cooper, Michael. *"A More Beautiful City": Robert Hooke and the Rebuilding of London after the Great Fire*. Stroud: Sutton, 2003. Print. Offers a brief sketch of Hooke's life and work in science, while discussing in detail his architectural work and role in rebuilding London.

Cooperman, Stephanie H. *Chien-Shiung Wu: Pioneering Physicist and Atomic Researcher*. New York: Rosen, 2004. Print. Describes Wu's life, career, and legacy.

Copeland, B. Jack, ed. *Alan Turing's Automatic Computing Engine: The Master Codebreaker's Struggle to Build the Modern Computer*. New York: Oxford UP, 2005. Print. Provides a detailed history of Turing's contributions to computer science and presents diagrams and illustrations explaining the hardware, software, and other features of Turing's computers.

Creswell, John W. *Research Design: Qualitative, Quantitative, and Mixed Methods Approaches*. 4th ed., Sage Publications, 2014.

Criss, Doug. "Stephen Hawking Says We've Got About 1,000 Years to Find a New Place to Live." *CNN*, 18 Nov. 2016, www.cnn.com/2016/11/17/health/hawking-humanity-trnd. Accessed 6 Jan. 2017.

Daniel, Johnnie. *Sampling Essentials: Practical Guidelines for Making Sampling Choices*. Sage Publications, 2012.

Darwin, Charles. *On the Origin of the Species and the Voyage of the Beagle*. London: Everyman's Lib., 2003. Print. Two of Darwin's most celebrated and influential books on evolution and natural selection. Includes an introductory essay by evolutionary biologist Richard Dawkins.

De Waal, Cornelis. *On Peirce*. Belmont, Calif.: Wadsworth/Thomson Learning, 2001. Brief (91-page) overview of Peirce's philosophy, designed to introduce his ideas to students.

———. *On Pragmatism*. Belmont, Calif.: Wadsworth/Thomson Learning, 2005. Describes the main figures and the central issues of pragmatism, with individual chapters devoted to Peirce, Dewey, James, and other philosophers.

De Winter, Joost, and Riender Happee. "Why Selective Publication Of Statistically Significant Results Can Be Effective." *PLOS ONE*, vol. 8, no. 6, 2013, doi:10.1371/journal.pone.0066463. Accessed 10 Apr. 2017.

Desmond, Adrian, and James Moore. *Darwin: The Life of a Tormented Evolutionist*. New York: Norton, 1994. Print. A detailed biography of Darwin that explores his life's work and its historical and social context, as well as his character, theology, and views on evolution.

Dooley, Larry M., and James R. Lindner. "The Handling of Nonresponse Error." *Human Resources Development Quarterly*, vol. 14, no. 1, 2003, pp. 99–110. *Business Source Complete*, search.ebscohost.com/login.aspx?direct=true&db=bth&AN=11578762&site=ehost-live. Accessed 31 May 2017.

Efron, Bradley, and Trevor Hastie. *Computer Age Statistical Inference: Algorithms, Evidence, and Data Science*. Cambridge UP, 2016.

Einstein, Albert. *Relativity: The Special and General Theory*. Trans. Robert W. Lawson. New York: Penguin, 2006. Print. Explains Einstein's theories of special and general relativity in an accessible way for a general audience.

———. *The World As I See It*. New York: Open Road, 2010. Print. A collection of Einstein's writings on religion, politics, ethics, and other topics.

Eldar, Yonina C. *Sampling Theory: Beyond Bandlimited Systems*. Cambridge UP, 2015.

Elsayir, Habib Ahmed. "Comparison of Precision Systematic Sampling with Some Other Probability Samplings." *American Journal of Theoretical and Applied Statistics*, vol. 3, no. 4, 2014, pp. 111–16.

'Espinasse, Margaret. *Robert Hooke*. Berkeley: U of California P, 1956. Print. Proposes that Hooke's conflict with Newton changed how science was pursued in England, from broad, practical empirical studies to narrower mathematical induction.

Feder, Kenneth L., and Michael Alan Park. *Human Antiquity: An Introduction to Physical Anthropology and Archaeology*. 5th ed. New York: McGraw, 2007. Print. Excellent updated account of human origins and early hominid forms, written in a readable style. Photographs, illustrations, bibliography.

Federer, Walter T., and Freedom King. *Variations on Split Plot and Split Block Experiment Designs*. Wiley-Interscience, 2007.

Ferguson, Kitty. *Stephen Hawking: An Unfettered Mind*. New York: Macmillan, 2012. Print.

Festing, Michael F. W. "Randomized Block Experimental Designs Can Increase the Power and Reproducibility of Laboratory Animal Experiments." *ILAR Journal*, vol. 55, no. 3, 2014, pp. 472–76. *MEDLINE Complete*, search.ebscohost.com/login.aspx?direct=true&db=mdc&AN=25541548&site=ehost-live. Accessed 31 May 2017.

Feynman, Richard Phillips, and Laurie M. Brown. *Feynman's Thesis: A New Approach to Quantum Theory*. Hackensack, NJ: World Scientific, 2010. Print. Describes the development of Feynman's thesis and explains the thought process that led to his groundbreaking theories.

Feynman, Richard Phillips, Ralph Leighton, and Edward Hutchings. *"Surely You're Joking, Mr. Feynman!" Adventures of a Curious Character*. New York: Bantum, 1989. Print. A series of anecdotes relating Feynman's life in his own words. His views on learning through understanding and questioning everything come through the humor.

Field, Andy. *Discovering Statistics Using SPSS*. 3rd ed., Sage Publications, 2009.

Garwin, Richard L., and Tsung-Dao Lee. "Chien-Shiung Wu." *Physics Today* 50.10 (1997): 120–21. Print. An obituary for Chien-Shiung Wu, with biographical data and information regarding her scientific career.

Gibbs, N. M. "Errors in the Interpretation of 'No Statistically Significant Difference.'" *Anaesthesia and Intensive Care*, vol. 31, no. 2, 2013, pp. 151–3. *MEDLINE Complete*, search.ebscohost.com/login.aspx?direct=true&db=mdc&AN=23577371&site=eds-live. Accessed 10 Apr. 2017.

Gingerich, Owen. *The Book Nobody Read: Chasing the Revolutions of Nicolaus Copernicus*. New York: Walker, 2004. Print. An examination of every copy of the original printing of Copernicus's *De revolutionibus* in existence, demonstrating who read the work, what they thought of it, and how exactly Copernicus's ideas spread throughout Europe.

Gleick, James. *Genius: The Life and Science of Richard Feynman*. New York: Pantheon, 1992. Print. Biography of Feynman covering his scientific achievements and his personality.

Goldsmith, Barbara. *Obsessive Genius: The Inner World of Marie Curie*. New York: Norton, 2005. Print. Biography of both the public and private figure of Curie, compiled through family interviews and Curie's letters, diaries, and workbooks. Illustrations, notes, bibliography.

Goodwin, C. James. *Research in Psychology*. John Wiley & Sons, 2010.

Goos, Peter. *The Optimal Design of Blocked and Split-Plot Experiments*. Springer, 2002.

Goudge, Thomas A. *The Thought of C. S. Peirce*. Toronto: University of Toronto Press, 1950. One of the most perceptive studies of Peirce's thought. Sees Peirce's philosophy as resting on a conflict within his personality that produced tendencies toward both naturalism and Transcendentalism.

Graetzer, Hans G., and David L. Anderson. *The Discovery of Nuclear Fission: A Documentary History*. New York: Reinhold, 1970. Print. Reprints the original papers and reports by scientists who first uncovered the problem and meaning of nuclear fission.

Graham-Cumming, John. "Alan Turing: Computation." *New Scientist* 214.2867 (2012): 2. Print. Describes Turing's scientific achievements and discusses his efforts as a code breaker during World War II and his foundational work in computer science.

Griffiths, Anthony J. F., et al. *An Introduction to Genetic Analysis*. 10th ed. New York: Freeman, 2012. Print. A chapter on Mendelian genetics provides a clear explanation of Mendel's experiments, using contemporary terminology. Includes problems with answers, a glossary, and an index.

Groves, Robert M. "Nonresponse Rates and Nonresponse Bias in Household Surveys." *Public Opinion Quarterly*, vol. 70, no. 5, 2006, pp. 646–75. *Academic Search Complete*, search.ebscohost.com/login.aspx?direct=true&db=a9h&AN=23631890&site=ehost-live. Accessed 31 May 2017.

Hanbury, Andria, et al. "Immediate versus Sustained Effects: Interrupted Time Series Analysis of a

Tailored Intervention." *Implementation Science*, vol. 8, no. 130, 2013, doi:10.1186/1748-5908-8-130. Accessed 20 Mar. 2017.

Haslam, S. Alexander, and Craig McGarty. *Research Methods and Statistics in Psychology*. 2nd ed., Sage Publications, 2014.

Hawking, Jane. *Music to Move the Stars: A Life with Stephen Hawking*. New York: Macmillan, 1999. Print.

Hawking, Stephen, and Roger Penrose. *The Nature of Space and Time*. Princeton: Princeton UP, 1996. Print.

Hawking, Stephen. *A Brief History of Time*. New York: Bantam, 1988. Print.

———. *A Briefer History of Time*. Rev. ed. New York: Bantam, 2005. Print.

———. *Black Holes and Baby Universes*. New York: Bantam, 1993. Print.

———. *My Brief History*. New York: Bantam, 2013, Print.

———. *The Universe in a Nutshell*. New York: Bantam, 2001. Print.

Hawking, Stephen, ed. *The Dreams That Stuff Is Made Of: The Most Astounding Papers of Quantum Physics and How They Shook the Scientific World*. Philadelphia: Running, 2011. Print.

Hedges, Larry V., and Jennifer Hanis-Martin. "Can Non-randomized Studies Provide Evidence of Causal Effects? A Case Study Using the Regression Discontinuity Design." *Education Research on Trial: Policy Reform and the Call for Scientific Rigor*, edited by Pamela Barnhouse Walters et al., Routledge, 2009, pp. 105–24.

Henig, Robin Marantz. *The Monk in the Garden: The Lost and Found Genius of Gregor Mendel, the Father of Genetics*. Boston: Houghton Mifflin, 2001. Print. A biography of Mendel, recounting his life, his experiments, and the importance of his genetic discoveries.

Henry, John. *Moving Heaven and Earth: Copernicus and the Solar System*. Cambridge: Icon, 2001. Print. Argues that Copernicus's discovery had revolutionary effects for the cultural status afforded to theoretical science and mathematics in Western culture.

Hern, Alex. "Stephen Hawking: AI Will Be 'Either Best or Worst Thing' for Humanity." *The Guardian*, 19 Oct. 2016, www.theguardian.com/science/2016/oct/19/stephen-hawking-ai-best-or-worst-thing-for-humanity-cambridge.

Hirshfeld, Alan. *The Electric Life of Michael Faraday*. New York: Walker, 2006. Print. A biography of Michael Faraday and descriptions of his discoveries.

Hocking, Ronald R. *Methods and Applications of Linear Models: Regression and the Analysis of Variance*. 3rd ed., John Wiley & Sons, 2013.

Hodges, Andrew. *Alan Turing: The Enigma—The Centenary Edition*. Princeton: Princeton UP, 2012. Print. Offers a biography of Turing, covering his influence on modern computer sciences, artificial intelligence, and the gay rights movement.

Hoffmann, Christoph. "Representing Difference: Ernst Mach and Peter Salcher's Ballistic-Photographic Experiments." *Endeavour* 33.1 (Mar. 2009): 18–23. Print. Presents photographs depicting the bullet experiment conducted by Mach.

Hooke, Robert. *The Diary of Robert Hooke, 1672–1680*. Ed. Henry W. Robinson and Walter Adams. London: Wykeham, 1968. Print. A detailed record of eight years in Hooke's life that testifies to the variety, intensity, and burden of his workload.

Inwood, Stephen. *The Forgotten Genius: The Biography of Robert Hooke, 1635–1703*. San Francisco: MacAdam, 2005. Print. Explains Hooke's varied scientific achievements and discusses the attendant controversies, suggesting that a tendency in Hooke to overstate his claims led him into conflicts.

Isaacson, Walter. *Einstein: His Life and Universe*. New York: Simon, 2007. Print. A full biography of Einstein's life and work.

Jacob, Robin Tepper, et al. *A Practical Guide to Regression Discontinuity*. MDRC, Aug. 2012. *MDRC*, www.mdrc.org/publication/practical-guide-regression-discontinuity. Accessed 8 May 2017.

James, Frank A. J. L. *Michael Faraday: A Very Short Introduction*. New York: Oxford UP, 2010. Print. A brief introduction to Faraday and his work. References and further reading included.

Jardine, Lisa. *The Curious Life of Robert Hooke: The Man Who Measured London*. New York: Harper, 2005. Print. Discusses Hooke's grueling schedule of work for the Royal Society, his partnership with Wren, and his official duties for London and how these affected his research, professional standing, and health.

Jones, Bradley, and Christopher J. Nachtsheim. "Split-Plot Designs: What, Why, and How." *Journal of Quality Technology*, vol. 41, no. 4, 2009, pp. 340–61. *Business Source Complete*, search.ebscohost.

com/login.aspx?direct=true&db=bth&AN=44704221&site=ehost-live. Accessed 31 May 2017.

Jungnickel, Christa, and Russell McCormmach. *Cavendish: The Experimental Life*. Lewisburg: Bucknell UP, 1999. Print. Examines Cavendish's discoveries within the context of the elite society in which he and his father developed their scientific interests.

Kaminska, Olena, and Peter Lynn. "Survey-Based Cross-Country Comparisons Where Countries Vary in Sample Design: Issues and Solutions." *Journal of Official Statistics*, vol. 33, no. 1, 2017, pp. 123–36.

Kanipe, Jeff. *Chasing Hubble's Shadows: The Search for Galaxies at the Edge of Time*. New York: Hill, 2007. Print. Chronicles astronomers' efforts to map the universe farther and farther away from Earth. Covers such topics as dark matter, redshifts, the Milky Way Galaxy, and other clusters of galaxies. Illustrations, bibliography, index.

Karwatka, Dennis. "Ernst Mach and the Mach Number." *Tech Directions* 69.5 (Dec. 2009): 14. Print. Recounts Mach's life and career, noting his contributions to the development of several branches of physics, including optics and mechanics.

Klein, Maury. *The Power Makers: Steam, Electricity, and the Men Who Invented Modern America*. New York: Bloomsbury, 2008. Print. Narrative history describing the introduction of steam and electric power to the United States. Faraday's contributions to the study of electricity are detailed.

Kolbert, Elizabeth. "'I Think We Have It': Is the Higgs Boson a Disapppointment?" *New Yorker*. Condé Nast, 5 July 2012. Web. 24 Dec. 2013.

Krauss, Lawrence M. *Quantum Man: Richard Feynman's Life in Science*. New York: Norton, 2011. Print. A biography of Feynman that discusses the physicist's life and personality, as well as his scientific legacy. Includes information from Feynman's lectures.

Kumar, Manjit. *Quantum: Einstein, Bohr, and the Great Debate about the Nature of Reality*. New York: Norton, 2011. Print. Detailed examination of Brownian Motion and the clashing theories of two twentieth-century scientists about the character and behavior of matter on the molecular and subatomic scale.

Leakey, Mary. *Disclosing the Past: An Autobiography*. New York: Doubleday, 1984. Print. Autobiography that examines some of the painful difficulties Leakey experienced with her marriage and with her scientific competitors. Index, bibliography.

Leedy, Paul D., and Jeanne Ellis Ormrod. *Practical Research: Planning and Design*. 11th ed., Pearson, 2016.

Levine, David M., and David Stephan. *Even You Can Learn Statistics and Analytics: An Easy to Understand Guide to Statistics and Analytics*. 3rd ed., Pearson Education, 2015.

Levy, Paul S., and Stanley Lemeshow. *Sampling of Populations: Methods and Applications*. John Wiley & Sons, 2013.

Linden, Ariel, et al. "Evaluating Program Effectiveness Using the Regression Point Displacement Design." *Evaluation & the Health Professions*, vol. 29, no. 4, 2006, pp. 407–23, doi:10.1177/0163278706293402. Accessed 19 May 2017.

Lindner, James R., et al. "Handling Nonresponse in Social Sciences Research." *Journal of Agricultural Education*, vol. 42, no. 4, 2001, pp. 43–53.

Magee, Judith. *Art and Nature: Three Centuries of Natural History Art from Around the World*. Vancouver: Greystone, 2010. Print. Presents an illustrated history of natural science and geographical explorations since 1700, encompassing the results of Brown's Australian expedition and the accompanying botanical drawings of Ferdinand Bauer.

Mailet, Hélène. *Hawking Incorporated: Stephen Hawking and the Anthropology of the Knowing Subject*. Chicago: U of Chicago P, 2012. Print.

Marczyk, Geoffrey R., et al. *Essentials of Research Design and Methodology*. John Wiley & Sons, 2005. Essentials of Behavioral Science Series.

Martin, Victoria. "A Layperson's Guide to the Higgs Boson." *The University of Edinburgh*. School of Physics and Astronomy / University of Edinburgh, 2 July 2012, www.ph.ed.ac.uk/higgs/laypersons-guide. Accessed 24 July 2012.

McCormmach, Russell. *Speculative Truth: Henry Cavendish, Natural Philosophy, and the Rise of Modern Theoretical Science*. New York: Oxford UP, 2004. Print. Explores the new theories of natural philosophy that emerged in the second half of the eighteenth century, including Cavendish's mechanical theory of heat. Includes an edition of Cavendish's manuscript about the mechanical theory of heat.

McCoy, Terrence. "How Stephen Hawking, Diagnosed with ALS Decades Ago, Is Still Alive." *The Washington Post*, 24 Feb. 2015, www.washingtonpost.

com/news/morning-mix/wp/2015/02/24/how-stephen-hawking-survived-longer-than-possibly-any-other-als-patient. Accessed 6 Jan. 2017.

McEvoy, J. P., and Oscar Zarate. *Introducing Stephen Hawking.* New York: Totem, 1997. Print.

Meitner, Lise. "Looking Back." *Bulletin of the Atomic Scientists* 20.9 (1964): 2–7. Print. An autobiographical account of Meitner's life, covering her youth through the discovery of atomic fission.

Mendel, Gregor. *Experiments in Plant-Hybridisation.* Trans. William Bateson. Charleston: Nabu, 2011. Print. Reprints Bateson's 1909 translation of Mendel's lectures on his groundbreaking experiments.

Merali, Zeeya. "Stephen Hawking: 'There Are No Black Holes.'" *Scientific American.* Scientific American, 27 Jan. 2014. Web. 15 Aug. 2014.

Miller, David Philip. *Discovering Water: James Watt, Henry Cavendish, and the Nineteenth Century "Water Controversy."* Burlington: Ashgate, 2004. Print. Describes how Cavendish, Watt, and Lavoisier's independent discoveries that water is a compound of airs and not a combination of elements became an issue of controversy among nineteenth-century scientists.

Misak, Cheryl, ed. *The Cambridge Companion to Peirce.* New York: Cambridge University Press, 2004. Collection of essays discussing Peirce's philosophy and his place within the pragmatist tradition. The essays include examinations of Peirce and medieval thought, his account of perception, and his theory of signs.

Moore, Edward C. *American Pragmatism: Peirce, James, and Dewey.* New York: Columbia University Press, 1961. An analysis of American pragmatism based on its three primary figures. Provides an excellent comparison of their different positions.

Morell, Virginia. *Ancestral Passions: The Leakey Family and the Quest for Humankind's Beginnings.* New York: Simon, 1995. Print. Detailed work that treats all members of the complicated Leakey family; shows how the family came to dominate the field of anthropology, and discusses the animosities within the family as well as among scientists. Bibliography, index.

Ogilvie, Marilyn Bailey. *Marie Curie: A Biography.* Amherst, NY: Prometheus, 2010. Print. A biography of Curie's life, including a discussion of her discovery of radium and her Nobel Prize awards. Illustrations, bibliography, index.

Orcher, Lawrence T. *Conducting Research: Social and Behavioral Science Methods.* 2nd ed., Taylor & Francis, 2014.

Ott, R. Lyman, and Longnecker, Michael. *An Introduction to Statistical Methods and Data Analysis.* 7th ed., Cengage Learning, 2016.

Ottaviani, Jim, and Leland Myrick. *Feynman.* New York: First Second, 2011. Print. Biography in graphic novel format provides insight into Feynman's life from childhood through his work on the *Challenger* disaster.

Overbye, Dennis. "A Pioneer as Elusive as His Particle." *The New York Times,* 15 Mar. 2014, www.nytimes.com/2014/09/16/science/a-discoverer-as-elusive-as-his-particle-.html. Accessed 20 Mar. 2017.

Pais, Abraham, and Robert Crease. *J. Robert Oppenheimer: A Life.* New York: Oxford UP, 2007. Print. Biography cowritten by a physicist who knew Oppenheimer and a historian at the Brookhaven National Laboratory.

Patten, Mildred L., and Michelle Newhart. *Understanding Research Methods: An Overview of the Essentials.* 10th ed., Routledge, 2017.

Petzold, Charles. *The Annotated Turing: A Guided Tour through Alan Turing's Historic Paper on Computability and the Turing Machine.* Hoboken: Wiley, 2008. Print. Guides readers through Turing's landmark paper, offering detailed explanation and examples.

Pike, Gary R. "Adjusting for Nonresponse in Surveys." *Higher Education: Handbook of Theory and Research,* edited by John C. Smart, vol. 22, Springer, 2007, pp. 411–49.

Potter, Vincent G. *Charles S. Peirce: On Norms and Ideals.* Amherst: University of Massachusetts Press, 1967. An analysis of Peirce's attempt to establish aesthetics, ethics, and logic as the three normative sciences. The author places particular emphasis on the role of "habit" in the universe.

Rees, Colin. *Rapid Research Methods for Nurses, Midwives and Health Professionals.* Wiley Blackwell, 2016.

Reilly, Francis E. *Charles Peirce's Theory of Scientific Method.* New York: Fordham University Press, 1970. A discussion of Peirce's ideas concerning the method and the philosophy of science.

Rife, Patricia. *Lise Meitner and the Dawn of the Nuclear Age.* Boston: Birkhäuser, 1999. Print. A biography interpreting Meitner's life and describing her work leading to the discovery of fission.

Rose, Angela M. C., et al. "A Comparison of Cluster and Systematic Sampling Methods for Measuring Crude Mortality." *Bulletin of the World Health Organization*, vol. 84, 2006, pp. 290–96.

Rossiter, Margaret W. *Women Scientists in America: Forging a New World Since 1972.* Vol. 3. Baltimore: Johns Hopkins UP, 2012. Print. Biographies of American women scientists and their significant research from 1972 onward. Includes Margaret Geller. Illustrations, bibliography, index.

Ruse, Michael. *Defining Darwin: Essays on the History and Philosophy of Evolutionary Biology.* New York: Prometheus, 2009. Print. Contains essays on subject including Darwin's life, pre-Darwinian theories, evolutionary ethics, evolutionary development, and scientists such as Alfred Russel Wallace, Herbert Spencer, Baptiste de Lamarck, and Gregor Mendel.

Sahai, Hardeo, and Mohammed I. Ageel. *Analysis of Variance: Fixed, Random and Mixed Models.* Birkhäuser, 2012.

Sample, Ian. *Massive: The Missing Particle That Sparked the Greatest Hunt in Science.* Basic, 2010. Presents the history of the search for the Higgs boson in a highly readable format that describes the development of Higgs's theories in the context of the history of subatomic and elementary particle physics.

Segrè, Emilio. *From X-rays to Quarks: Modern Physicist and Their Discoveries.* 1980. Mineola, NY: Dover, 2007. Print. Includes a detailed chapter on the discoveries made by Becquerel and the Curies. Illustrations, bibliography, indexes.

Sexton-Radek, Kathy. "Single Case Designs in Psychology Practice." *Health Psychology Research*, vol. 2, no. 3, 2014, pp. 98–99, doi:10.4081/hpr.2014.1551. Accessed 20 Mar. 2017.

Sime, Ruth Lewin. *Lise Meitner: A Life in Physics.* Berkeley: U of California P, 1996. Print. Provides a comprehensive chronicle of Meitner's life, career, and contributions to atomic and nuclear physics.

Singh, Simon. *Big Bang: The Origin of the Universe.* New York: Fourth Estate, 2004. Print.

Skagestad, Peter. *The Road of Inquiry: Charles Peirce's Realism.* New York: Columbia University Press, 1981. Focuses on Peirce's theory of scientific method but also contains an introduction with considerable biographical information.

Sparberg, Esther B. "A Study of the Discovery of Fission." *American Journal of Physics* 32.1 (1964): 2–8. Print. Reviews the history of the discovery of fission and discusses Meitner's place in that history.

St. Clair, Travis, et al. "The Validity and Precision of the Comparative Interrupted Time-Series Design: Three Within-Study Comparisons." *Journal of Educational and Behavioral Statistics*, vol. 41, no. 3, 2016, pp. 269–99, doi:10.3102/1076998616636854. Accessed 20 Mar. 2016.

Stephenson, Bruce. *Kepler's Physical Astronomy.* Princeton: Princeton UP, 1994. Print. Explains Kepler's astronomical theories and discusses the development of his laws of planetary motion.

Svoronos, Theodore, et al. "Clarifying the Interrupted Time Series Study Design." *BMJ Quality & Safety*, vol. 24, no. 7, 2015, pp. 475–76, doi:10.1136/bmjqs-2015-004122. Accessed 20 Mar. 2017.

Swanson, Richard A., and Elwood F. Holton III, editors. *Research in Organizations: Foundations and Methods of Inquiry.* Berrett-Koehler Publishers, 2005.

Thompson, Steven K. *Sampling.* 3rd ed., John Wiley & Sons, 2012.

Tillé, Yves, and Alina Matei. "Basics of Sampling for Survey Research." *The SAGE Handbook of Survey Methodology*, edited by Christof Wolf et al., Sage Publications, 2016, pp. 311–28.

Trochim, William M. K., and Joseph C. Cappelleri. "Cutoff Assignment Strategies for Enhancing Randomized Clinical Trials." *Controlled Clinical Trials*, vol. 13, no. 3, 1992, pp. 190–212.

Trochim, William M. K. *The Regression Point Displacement Design for Evaluating Community-Based Pilot Programs and Demonstration Projects.* Unpublished manuscript, 1996. *Research Methods Knowledge Base*, www.socialresearchmethods.net/research/RPD/RPD.pdf. Accessed 19 May 2017.

_____. "Hybrid Experimental Designs." *Research Methods Knowledge Base*, 2006, www.socialresearchmethods.net/kb/exphybrd.php. Accessed 3 May 2017.

_____. "Other Quasi-Experimental Designs." *Research Methods Knowledge Base*, www.socialresearchmethods.net/kb/quasioth.htm. Accessed 19 May 2017.

_____. "Regression Point Displacement Analysis." *Research Methods Knowledge Base*, www.socialresearchmethods.net/kb/statrpd.php. Accessed 19 May 2017.

Tully, Christopher G. *Elementary Particle Physics in a Nutshell.* Princeton UP, 2011. Explains the concepts of elementary particle physics and Higgs interactions in an understandable way, but requires familiarity with the mathematics of quantum mechanics.

Turing, Alan. "Computing Machinery and Intelligence." *Mind* 59 (1950): 433–60. Print. Provides Turing's counterarguments to common objections against artificial intelligence and introduces the Turing test, which decides when a computer has achieved intelligence.

Van Assen, Marcel A. L. M., et al. "Why Publishing Everything Is More Effective Than Selective Publishing of Statistically Significant Results." *PLOS ONE*, vol. 9, no. 1, 2014. doi:10.1371/journal.pone.0084896. Accessed 10 Apr. 2017.

Van Leeuwen, Nikki, et al. "Regression Discontinuity Design: Simulation and Application in Two Cardiovascular Trials with Continuous Outcomes." *Epidemiology*, vol. 27, no. 4, 2016, pp. 503–11.

Voelkel, James Robert. *The Composition of Kepler's Astronomia nova*. Princeton: Princeton UP, 2001. Print. Explores Kepler's process in writing *Astronomia nova* and the development of his theories about the movement of the planets.

Vogt, W. Paul, and R. Burke Johnson. *The Sage Dictionary of Statistics & Methodology: A Nontechnical Guide for the Social Sciences*. 5th ed., Sage Publications, 2016.

Vogt, W. Paul. *Quantitative Research Methods for Professionals*. Pearson, 2007.

Walter, Alan E., and Hélène Langevin-Joliot. *Radiation and Modern Life: Fulfilling Marie Curie's Dream*. Amherst, NY: Prometheus, 2004. Print. Covers the science behind radiation and the numerous applications of radiation since its discovery. Illustrations, glossary, index.

Weinberg, Steven. *Cosmology*. New York: Oxford UP, 2008. Print. An in-depth introduction to modern cosmological research. Discusses numerous topics, including those that Geller has worked on. Illustrations, glossary, index.

Westman, Robert S. *The Copernican Question: Prognostication, Skepticism, and Celestial Order*. Berkeley: U of California P, 2011. Print. Chronicles the development of the Copernican model of the universe and places Kepler's work within a larger context.

White, Michael, and John Gribbin. *Stephen Hawking: A Life in Science*. New York: Penguin, 1993. Print.

Wong, K. C. "Null Hypothesis Testing (I): 5% Significance Level." *East Asian Archives of Psychiatry*, vol. 26, no. 3, 2016, pp. 112–13. *MEDLINE Complete*, search.ebscohost.com/login.aspx?direct=true&db=mdc&AN=27703100&site=eds-live. Accessed 10 Apr. 2017.

Wyman, Peter A., et al. "Designs for Testing Group-Based Interventions with Limited Numbers of Social Units: The Dynamic Wait-Listed and Regression Point Displacement Designs." *Prevention Science*, vol. 16. no. 7, 2015, pp. 956–66, doi:10.1007/s11121-014-0535-6. Accessed 19 May 2017.

Yoon, Carol Kaesuk. *Naming Nature: The Clash between Instinct and Science*. New York: Norton, 2009. Print. Presents an overview of the history and controversies involved in taxonomy (the science of organizing and classifying living organisms) since Swedish botanist Carl Linnaeus founded the specialty in the eighteenth century.

Zhang, Pan, and Cristopher Moore. "Scalable Detection of Statistically Significant Communities and Hierarchies, Using Message-Passing for Modularity." *PNAS*, vol. 111, no. 51, 2014, pp. 18144–49, doi:10.1073/pnas.1409770111. Accessed 10 Apr. 2017.

INDEX

A

A Large Ion Collider Experiment (ALICE) 134
abductive logic programming (ALP) 1, 2
abductive reasoning 1, 58, 59, 121
  abduction
    deduction and induction 1
    practice 2
    research 2
  abductive logic programming 1
  artificial intelligence 2
  deductive reasoning 1
  inductive reasoning 1
active factor 193
alpha (α) 251, 252
alternative hypothesis 183
ambulatory organisms 210
amyotrophic lateral sclerosis (ALS) 100–103
analysis of covariance (ANCOVA)
  advantages and disadvantages 4
  covariate 2
  defined 3–4
  interaction 2
  posttest 2
  practice 4
  pretest 3
  sample problem 5
  significance 4
  treatment 3
analysis of variance (ANOVA) 3–4, 6, 48
  advantages and disadvantages 6–7
  defined 6
  Latin square designs 135
  nested designs 170
  normal distribution 6
  practice 7
  sample problem 8
  significance 7
  sum of squares between 6
  sum of squares within 6
  variance 6
animal models 33
Arisbe 191
artificial intelligence (AI) 2, 116, 273
artificial radioactivity 154
atomic bomb 187, 188
attribute
  pretest-posttest experimental research design 199, 200
  sampling framework 246
automata theory 274

B

B^2FH theory 15
backward reasoning 121
Banks, Joseph 12, 13
Bayesian network 22
Beagle 56, 57
Becquerel, Antoine Henri 52
before-after-control-impact (BACI) 9
  advantages and disadvantages 10
  control group 9
  design 9–10
  impact group 9
  interaction 9
  sample problem 11
before-and-after graph 224
bell curve 6, 65
best-fitting model, stepwise regression for 228
beta decay transition 278, 279
bias
  internal validity 123
  nonresponse error 177
  qualitative research 210
  quantitative research 213
big bang theory 100
big data 44
black hole radiation 100–104
blending 157
blocked randomization 220
Born-Oppenheimer approximation 186
Boyle's gas law 19
Brown, Robert
  early life 12
  history 11
  impact 13–14
  invisible atomic universe, physical evidence 11–12
  life's work 12–13
Burbidge, E. Margaret
  cosmic evolution 16
  early life 14
  history 14
  impact 15–16
  life's work 15

C

carryover effect 47
Cartesian coordinate system 141
case-control study 49, 50
case study 20
  defined 20–21
  disadvantage 21
  findings 21
  holistic study 110
  longitudinal sampling 143
  methodology 20
  qualitative data 20
  qualitative research 211
  quantitative data 20
  use 21–22
categorical data 25
categorical variable 27
category analysis 125
causal networks 22
  applications 23–24
  Bayesian network 22
  dependency 22
  latent variable 22
  learning causality 23
  parent variable 22
  relationships 23
  research, basis of 23
causal reasoning 159
  agreement method 159
  concomitant variation 159
  determining cause 159–160
  difference method 159
  joint method 159
  positives and negatives 160–161
  practice 161
  residues, method 159
causation 40
  correlation and 44
    caution 46
    controlled experiments 45–46
    negative correlation 44, 45
    positive correlation 44, 45
    relationships 45
  correlational research 42–43
  explanatory research 74
  external validity 78
  internal validity 123
  Solomon four-group design 253
Cavendish, Henry
  early life 17
  history 17
  hydrogen, discovery of 20
  impact 19
  life's work 18
celestial spheres 39–40
census, sampling vs. 248
  advantages and disadvantages 249–250
  convenience sampling 248
  data collection 249
  methods 249
  sample problem 250
  simple random sampling 248
  surveying 250
  utilization 250
  voluntary response sampling 249
center of distribution. 107
central tendency 63–64
chance-based methods 219
Chi-square goodness
  fit test 24
    categorical data 25
    degrees of freedom 25
    distribution of values 26
    expected data 25
    goodness of fit 25
    independence 25
    limitations 25
    observed data 25
    sample problem 26
    testing 25
  independence 27
    categorical variable 27
    contingency table 27
    degrees of freedom 27
    goodness of fit 27
    limitations 28
    P value 27
    sample problem 28–29
    testing 27–28
class frequency 107
class interval 107
classic field experiment 87
classical theory 101
cluster sampling 30, 164, 220
  benefits and drawbacks 31
  multistage sampling 164
  power of 31–32
  probabilistic sampling 201, 202

randomness and interspersion 244
sample problem 32
sampling design *vs.* experimental design 241–242
simple random sampling 30
stratified random sampling 30
systematic random sampling 30
systematic sampling 266
use 31
coefficient of determination 41
cognitive bias 181
cognitive process 116
cohort study
  longitudinal sampling 143
  prospective cohort design 206
  retrospective cohort design 234
commentariolus 38, 39
comparative research 32, 33
  animal models 33
  cross-cultural study 33
  design 33
  generalization 33
  in vivo testing 33
  lab rats 34
  principle 33
  quantitative *vs.* qualitative 33–34
  variation 33–34
completely randomized designs 35
  control group 35
  designs 35–36
  experimental unit 35
  randomization 35
  reliability and generalizability 36
  sample problem 37
  treatment 35
  unintended consequences 36
completely randomized design 255, 256
concomitant variation 159
confidence intervals (CIs) 183–184
  one-tailed and two-tailed t-tests 183
  significance levels 251
confounding variable 87, 123
construct validity 123
content analysis 125
contingency table 27
continuous variable 65, 66, 107
control group 9, 35
  experimental research 72
  field experiment 87
  hypothesis testing 115

hypothesis-based study 117
laboratory experiment 133
nonequivalent dependent variables design 172
pretest-posttest experimental research design 199
replication, manipulation and randomization 232
sampling design *vs.* experimental design 242
switching replications design 263
convenience sampling 175, 248
Cook, James 12
Copernican theory 127
Copernicus, Nicolaus
  celestial spheres 39–40
  early life 37–38
  history 37
  impact 39
  life's work 38
correlation
  causations and 44
    caution 46
    controlled experiments 45–46
    negative correlation 44, 45
    positive correlation 44, 45
    relationships 45
  descriptive research 60
  modeling 40
    causation 40
    hidden variables 40
    limitations 41
    sample problem 42
    trend line 40
    uses 40–41
    value 41
    variable 40
  research 42
    causation 42–43
    correlation *vs.* causation 43
    digital 44
    large experimental studies 43
    negative correlation 43
    positive correlation 43
    problems 43–44
    relationships 43
covariate 2
credibility 211
critical region
  significance levels 251
  type I and type II errors 271
cross-cultural study 33
crossed designs 170–171

crossover repeated measures design 47
  advantages and limitations 48
  application 48
  carryover effect 47
  crossover designs and clinical trials 48
  experimental unit 47
  repeated measures 47
  sample problem 49
  testing 47–48
  treatment 47
  washout period 47
cross-sectional sampling 49
  case-control study 49, 50
  dependent variable 49
  independent variable 49–50
  limits and benefits 50
  mortality 50
  using 50
cultural bias 181
Curie, Marie
  early life 51–52
  history 51
  impact 53
  life's work 52–53
  radium and uses 52–54
cutoff point 224

D

Darwin, Charles 157
  early life 55–56
  history 55
  impact 57
  life's work 56–57
  natural selection 57–58
data collection 21
deductive reasoning 1, 58, 121
  abductive reasoning 58, 59
  certainty 59
  fallacies 60
  inductive reasoning 58, 59
  limitations 59
  mathematical proof 59
  proofs 204
  testing theories 59
deep-learning models 116
degrees of freedom 25, 27
dependency 22
dependent variable 48, 49
  explanatory research 74

field experiment 87
  laboratory experiment 133
  nonequivalent dependent variables design 172
  pretest-posttest experimental research design 199
descriptive research 213
directed acyclic graphs (DAGs) 23
discrete variable 65, 107
descriptive research 60, 74–75
  applications 61
  census data 61
  correlation 60
  exploratory research 76
  qualitative research 60–61
  quantitative data 60
  quantitative research 60–61
  sample size 60
  strengths and limitations 61
descriptive statistics 62, 63
  central tendency 62–64
  dispersion 62, 64
  distribution 62
  mean, median, and mode 63–64
  quantitative data 62
  sample problem 64
  standard deviation 62
  uses 62
dispersion 63, 64
distribution 63–65
  continuous variable 65
  discrete variable 65
  histogram 65
  predictive value 66
  probability/frequency 65
  randomness and interspersion 244
  sample problem 67
  skewed distribution 65, 66
  variables, types of 66
diversity sampling 175, 176
double-blind studies 134

E

Einstein, Albert
  early life 69–70
  history 69
  impact 71
  life's work 70–71
  relativity theory 71–72
electromagnetic induction 83–84
electromagnetic rotation 82

electromagnetism 70, 82
US Elementary and Secondary Education Act 224
environmental bias 133
ethnography 110
Euler's identity 205
expected data 25
experimental design, sampling design *vs.* 241
  cluster sampling 241–242
  control group 242
  design considerations 242–243
  experimental group 242
  gender bias 243
  importance 243
  random sampling 242
  variable 242
experimental group
  field experiment 87
  independent variable 87
  laboratory experiment 133
  nonequivalent dependent variables design 172
  pretest-posttest experimental research design 199, 200
  sampling design *vs.* experimental design 242
  stratified random sampling/randomized block design 258
  switching replications design 263
experimental mortality 124
experimental research 72
  applications 73
  control group 72
  maintaining control 73
  null hypothesis 72
  quantitative research 213
  randomization 72
  replication, manipulation and randomization 232
  sample group 72
  uses 72–73
  variable 72
explanatory research 74
  causation 74
  control 75
  dependent variable 74
  independent variable 74
  pet ownership and longevity 75
  quantitative data 74
  relationship 74–75
exploratory research 76
  applications 77
  benefits and drawbacks 77
  conduct 76–77
  descriptive research 76
  intermediate step 77
  observational research 76
experimental unit 35, 47
external validity 78
  causation 78
  generalizability
    populations 78, 79
    situations 78, 79
  importance 79
  internal validity 78–79, 123
  posttest-only design 197, 198, 200
  sampling bias 78
  Solomon four-group design 253

F
factorial design 171
fallibilism 192
false negative 272
false positive 272
Faraday effect 83
Faraday, Michael
  early life 81
  electromagnetic induction 83–84
  history 81
  impact 83
  life's work 81–83
Feynman diagrams 85, 86
field experiment 87, 133
  benefits and drawbacks 88
  confounding variable 87
  control group 87
  dependent variable 87
  experimental blocks 88
  experimental group 87
  harm reduction 88
  necessity of 88
  observational research 87
  testing 87–88
fixed effect 170
Feynman, Richard 84
  early life 84–85
  Feynman diagrams 85, 86
  history 84
  impact 85–86
  life's work 85
fit test, Chi-square goodness 24
  categorical data 25

degrees of freedom 25
distribution of values 26
expected data 25
goodness of fit 25
independence 25
limitations 25
observed data 25
sample problem 26
testing 25
fractional factorial designs 89
confounding 89–90
experimental condition 90
experimental run 90
factor 90
fractionalizing 90
information with the fewest runs 90–91
interaction 90
Plackett-Burman design 90–91
practice 91
sample problem 91
significance 91
F-statistic 167
full factorial design 92, 193
applications 93–94
combinations 93
experimental condition 92
experimental run 92
factor 92
interaction 93
sample problem 94
setting up independent variables 93
types of 93
funding effect bias 182

G
Galilei, Galileo 99
Geller, Margaret 95
early life 95
Great Wall 96, 97
history 95
impact 96
life's work 95–96
gender bias 243
generalizability 36
populations 78
replication, manipulation and randomization 232
situations 78
goodness of fit 24–26, 28

grand unified theory 101
Great Wall 96

H
Hahn, Otto 153
haphazard sampling 219
Hawking, Stephen
black hole radiation 100–104
early life 99–100
history 99
life's work 100–102
significance 102–103
heredity 158
hidden variables 40, 41
Higgs boson 101, 105, 106
Higgs, Peter 104
early life 104
Higgs boson 105, 106
history 104
impact 105–106
life's work 104–105
histogram 65, 66, 107
advantages and disadvantages 108
class frequency 107
class interval 107
continuous variable 107
control measures 108
discrete variable 107
frequency 107
importance 108
qualitative data 107
sample problem 109
simple visual aid 107–108
history 124
holism 110
holistic study 109, 110
case study 110
considerations of 110–111
holism and ethnography 110
importance 111
participant observation 110
practice 111
realist ethnography 110
Hooke's law 113
Hooke, Robert
early life 112
first scientific best seller 113, 114
history 112
impact 113–114

life's work 112–113
hormone therapy (HT) 43
hydrogen 20
hypothesis-based study 117
   biomedicine 118
   control group 117
   independent variable 117
   null hypothesis 117
   parameter 117
   research purpose 117
   significance 118
   variable 117
hypothesis testing 115
   computers reading comics 116
   control group 115
   importance 116
   null hypothesis 115
   parameter 115
   scientific method 115
   testing and retesting 115–116
   variable 115

I

impact group 9
in vivo testing 33
independence 25, 27
   categorical variable 27
   contingency table 27
   degrees of freedom 27
   goodness of fit 27
   limitations 28
   P value 27
   sample problem 28–29
   testing 27–28
independent rariable manipulation 119
   covariate 119
   elimination 119
   extraneous variable 119
   inclusion 119
   manipulating variables 119
   manipulation 120
   statistical control 119–120
   street lighting experiment 120
independent samples 261
independent variable 49–50
   experimental group 87
   explanatory research 74
   hypothesis-based study 117
   impacts 117–118

laboratory experiment 133
Mann-Whitney U test 150
nonequivalent dependent variables design 172
pretest-posttest experimental research design 199
inductive reasoning 1, 59, 59, 121
   backward reasoning 121
   deductive reasoning 121
   foundational methodology 122
   inferential extrapolation 122
   proofs 204
   primary applications 122
   theory and practice 121
inferential extrapolation 122
instrumentation 124
inference 239
internal validity 78, 123
   bias 123
   causation 123
   confounding 123
   external validity 123
   importance 124
   lab 124
   posttest-only design 197
   Solomon four-group design 253
   threats to 123–124
   validity and minimizing confounding 123
interpretivism 195, 196
interpretive methods 125
   category analysis 125
   content analysis 125
   entrepreneurial research 126
   importance 126
   interpretive research 125–126
   interpretive rigor 126
   narrative analysis 125
   qualitative data 125
   subjectivity 125
interspersion
   animal behavior experiments 245
   cluster sampling 244
   design strategy 245
   distribution 244
   stratified random sampling 244
   strengths and difficulties 245
   testing 244–245
inverse relationship 43
inverse variation 141
Investigator 13

J
James, William 191
joint method 159

K
Kepler, Johannes 127
  early life 127
  history 127
  impact 128–129
  life's work 127–128
  planetary motion laws 128, 129
Kruskal-Wallis test 130
  comparing tests 131
  limitation 131
  Mann-Whitney U test 130
  nominal variable 130
  one-way analysis of variance 130
  ordinal variable 130
  sample problem 132
  studying differences among groups 131
  testing 130
Kruskal-Wallis H test 150

L
laboratory experiment 133
  benefits and limitations 134
  control group 133
  dependent variable 133
  environmental bias 133
  experimental group 133
  importance 134–135
  independent variable 133
  Large Hadron Collider 134
  A Large Ion Collider Experiment 134
  manipulation 133
  testing 133–134
Latin square designs 135
  action 136–137
  advantages and disadvantages 136
  ANOVA 135
  application 137
  block designs and variability 135–136
  nuisance factor 135
  residual 135
  sample problem 137
  treatment 135
  variance 135
Lamarck, Jean-Baptiste 57
Large Hadron Collider (LHC) 134

latent variable 22
Leakey, Louis Seymour Bazett 127
Leakey, Mary
  early life 138
  history 138
  impact 139
  life's work 138–139
least squares 226
linear relationships 140
  applications 142
  determination 141
  direct relationship 140
  inverse relationship 140
  logarithmic 140
  polynomial 140
  sample problem 142
  statistical analysis 141–142
  straight/curved lines 140–141
  variable 140
London study 120
longitudinal sampling 143
  advantages and disadvantages 144
  case study 143
  cohort study 143
  correlation 143
  foundations 143–144
  method selection 144
  observational research 143
  practical applications 144
  time-series design 143, 269

M
Mach bands 149
Mach, Ernst
  early life 147
  history 147
  impact 148–149
  life's work 147–148
  sound, speed 148–149
Mach number 148
Manchester Automatic Digital Machine 273
manipulation, randomization and replication 232
  control group 232
  experimental research 232
  generalization 232
  quantitative data 232
  sampling and experimental design 232
  strategies 234
  strengths and drawbacks 232–233

survey sample example 234
Mann-Whitney *U* Test 130, 150
  applications 151
  comparing statistical tests 150–151
  independent variable 150
  Kruskal-Wallis *H* test 150
  nonparametric 150
  normal distribution 150
  ordinal variable 150
  sample problem 152
  testing nonparametric data 150
mathematical proof 59, 204
maturation 124
mean 63–64
mean square 167
Measurement, Learning & Evaluation (MLE) Project 144
median 63–64
Meitner, Lise 153
  early life 153
  history 153
  impact 154–155
  life's work 153–154
  nuclear fission 154–155
Mendel, Gregor
  early life 156
  heredity 158
  history 156
  impact 158
  life's work 156–157
Metaphysical Club 190
Mill, John Stuart 159–161
Mill's methods 121–122
mode 63–64
multiple case study 161, 162
  considerations 162–163
  disadvantages 163
  multiple case study 162
  pilot case 162
  robustness 162
  sample size 162
  testing 162
multiple linear regression model 228
multistage sampling 164
  advantages and disadvantages 164
  cluster sampling 164
  primary units 164
  random sampling 164
  sample problem 166
  secondary units 164
  use in 165
Muon Physics 279

N

narrative analysis 125
natural experiments 233
natural selection 57
negative correlation 41, 43, 44, 45
nested analysis of variance
  applications 168
  benefits and limitations 167–168
  F-statistic 167
  mean square 167
  nested variable 167
  nominal variable 167
  practice 168
  sample problem 169
  testing 167
  variance 167
nested designs
  ANOVA 170
  applications 171
  limitations, benefits and comparison 170–171
  nesting factors 170
  practice 171
  random effect 170
  sample problem 172
nested variable 167
Newton, Isaac 19
Newtonian mechanics 70
nonambulatory organisms 210
nonequivalent dependent variables (NEDV) design 172
  challenges and ensuring effectiveness 173
  control group 172
  dependent variable 172
  design option 174
  education, research in 174
  experimental group 172
  independent variable 172
  nonrandom assignment 172
  research studies 172–173
  testing 172–173
nonlinear relationships 140
  applications 142
  determination 141
  direct relationship 140
  inverse relationship 140

logarithmic 140
polynomial 140
sample problem 142
statistical analysis 141–142
straight/curved lines 140–141
variable 140
non-probabilistic sampling
  action 176
  convenience sampling 175
  diversity sampling 175
  implications 176–177
  methodologies, definition 175
  quota sampling 175
  random sampling 175
  techniques 175–176
nonrandom assignment 172
nonrandom sampling 261
nonresponse error 177
  bias 177
  causes 178
  importance 178
  overcoverage 177
  responsiveness 177–178
  sample problem 179
  telephone surveys 178
  undercoverage 177
normal science 115
nominal variable 130
nuclear fission 154–155
Nuclear Physics 279
null hypothesis 72
  hypothesis-based study 117
  hypothesis testing 115
  one-tailed and two-tailed *t*-tests 183
  type I and type II errors 271

O

objectivity 181
  cognitive bias 181
  cultural bias 181
  double-blind 181
  funding effect bias 182
  maintaining 181, 182
  positivist methods 195
  qualitative data 181
  quantitative data 181
  research ethics 182
  scientific method 181
  subjectivity 181
  systematic bias 181
observational research
  exploratory research 76
  field experiment 87
  longitudinal sampling 143
observed data 25
one-tailed and two-tailed *t*-tests 183
  action 185
  alternative hypothesis 183
  confidence interval 183–184
  data sets 183
  importance 185
  limitations and considerations 184
  null hypothesis 183
  sample distribution 183
  sample problem 185
  standard deviation 183
one-way analysis of variance 130, 131
Oppenheimer, J. Robert
  atomic bomb 187, 188
  early life 185–186
  history 185
  impact 187
  life's work 186–187

P

P value 27
  significance levels 251
  type I and type II errors 271, 272
paired *t*-test 261–263
parent variable 22
parity 277–279
partial variation 141
particle physics 105
Penicillium fungus 122
perchloroethene 82
phaneroscopy 191
photoelectric effect 71
Peirce, Charles Sanders
  early life 189
  history 189
  life's work 189–192
  significance 192
pilot case 162
placebo-controlled studies 134
Plackett-Burman design 90, 193
  active factor 193
  benefits 194
  full factorial design 193

sample problem 195
  testing 193–194
  two-way interaction 193
  use 194
Ponelis's research 126
population
  generalizability 78, 79
  prospective cohort design 206
  retrospective cohort design 235
  sample frame error 239
positive correlation 43, 44, 45
positivism 196
positivist methods 195, 197
  benefits and burdens 196
  interpretivism 195
  objectivity 195
  positivism vs interpretivism 196
  qualitative data 195
  quantitative data 195
  quantitative social research 196–197
  replication 196
  subjectivity 196
positivistic version of reality 191
posttest 2, 198
posttest-only design 197
  effects, validity 198
  external validity 197
  internal validity 197
  limited utility 198–199
  using posttests 198
  pretest 197
  research
    prospective cohort design 206
    retrospective cohort design 235
  sample size 197
  tool 199
post-test design 143
power 131
pragmatic maxim 191
pragmaticism 190
pragmatism 190
predictive value 66
pretest 3, 197
pretest-posttest design
  experimental research 199
    advantages and disadvantages 200
    attribute 199
    control group 199
    dependent variable 199

education 200–201
experimental group 199
improving research design 200
independent variable 199
useful 201
Solomon four-group design 253
switching replications design 263
primary units 164
priming effect 263
probabilistic sampling 30, 201
  action 203
  cluster sampling 201
  comparing sampling methodologies 203
  power 203
  random sampling 201–202
  studying groups with 202–203
probability distribution 66
proofs 204
  challenge of 204–205
  deductive reasoning 204
  elegance of 205
  inductive reasoning 204
  mathematical proof 204
  necessity 205
  rule of inference 204
prospective cohort design 206–207
  cohort 206
  epidemiology, use in 207
  feature of 206
  population 206
  posttest-only research 206
  potential 207

Q
quadrat sampling 209
  ambulatory organisms 210
  ecosystem, sectioning off 209–210
  need for 209
  nonambulatory organisms 210
  sampling area 209
  sampling point 209
qualitative data
  case study 20
  histogram 107
  interpretive methods 125
  quantitative research 213
  positivist methods 195
  objectivity 181
  positivist methods 195

311

Index

replication, manipulation and randomization 232
qualitative research 210, 211–212
  benefits and burdens 212
  bias 210
  case study 211
  credibility 211
  descriptive research 60–61
  methods 211–212
  need for 212
  quantitative data 211
  subjectivity 211
  transferability 211
quantitative data
  case study 20
  descriptive research 60
  descriptive statistics 63
  explanatory research 74
  objectivity 181
  qualitative research 211
  replication, manipulation and randomization 232
quantitative research 213, 215
  bias 213
  descriptive research 60–61, 213
  empirical investigation 213–214
  experimental research 213
  inference 213
  qualitative data 213
  strengths and drawbacks 214
  US census 214–215
quasars 15
quasi-experiments 87, 269
quota sampling 175, 176

R

radioactivity 53
radium 52–54
random assignment 219
random effect 170
random sampling 175, 245
  error 217
    danger of 218
    experimentation 217
    reducing sampling error 217–218
    sample problem 218
    type I 217, 271
    type II 217, 271
  multistage sampling 164

probabilistic sampling 201–202
randomization 219
sampling design vs. experimental design 243
randomization 35–36, 72, 73, 219
  experiments 219
  need for 220
  random assignment 219
  random sampling 219
  replication, manipulation and 232
    control group 232
    experimental research 232
    generalization 232
    quantitative data 232
    sampling and experimental design 232
    strategies 234
    strengths and drawbacks 232–233
    survey sample example 234
  sample problem 221
  strategies 219–220
randomized block design (RBD) 135–136
  experimental group 258
  interaction 258
  representation 260
  sample problem 260
  strata 258
  testing 258–260
  uses 260
  variability 258
randomized complete block (RCB) design
  medical studies and 222–223
  replicate 221
  sample problem 223
  spatial effects 221
  subject 221
  treatment 221
  uses 222, 223
  variations, accounting for 221–222
randomness
  animal behavior experiments 245
  cluster sampling 244
  design strategy 245
  distribution 244
  stratified random sampling 244
  strengths and difficulties 245
  testing 244–245
realist ethnography 110
recursive function theory 274
redshift surveys 96

regression 3
regression discontinuity (RD) design 224
  analysis 225
  basics of 224
  before-and-after graph 224
  cutoff point 224
  practice 225
  sample problem 226
  sample size 224
  treatment 224
  uncommon design 224–225
regression modelling 226
  applications 228
  best-fitting model, stepwise regression for 228
  least squares 226
  limitations 228
  measurement 227
  multiple linear regression model 228
  relationship 226
  R-squared 226
  sample problem 229
  type and complexity 227
  variable 227
regression point displacement (RPD) 229
  action 230–231
  benefits and drawbacks 230
  community-based research 230
  importance 231
  sample problem 231
  statistical analysis 229
  treatment 229
  variable 229
regression point displacement designs (RPPDs) 229–231
relativity theory 70–72
repeated measures 47
replication
  positivist methods 196
  switching replications design 263
  manipulation and randomization 232
    control group 232
    experimental research 232
    generalization 232
    quantitative data 232
    sampling and experimental design 232
    strategies 234
    strengths and drawbacks 232–233
    survey sample example 234

reproduction 57
retrospective cohort design 234
  applications 236
  association, measures of 235–236
  cohorts and cohort studies 234, 235
  limitations 236
  population 235
  posttest-only research 235
  sample problem 237
robustness 162
R-squared value 226
rule of inference 204

S
sample distribution 183
sample group
  experimental research 72
  sample frame error 239
sampling
  area 209
  vs. census 248
    advantages and disadvantages 249–250
    convenience sampling 248
    data collection 249
    methods 249
    sample problem 250
    simple random sampling 248
    surveying 250
    utilization 250
    voluntary response sampling 249
  error 202
  frame 202
  framework 246
    attribute 246
    considerations and applications 247
    parameter 246
    sample frame error 239, 246
    sample problem 248
    sampling framework 246
    testing 246–247
    variable 246
  point 209
  systematic 266
    cluster sampling 266
    context and considerations 267
    sample problem 268
    simple random sampling 266
    techniques 266

uniform sampling 266
using 266–267
sampling design
  vs. experimental design 241
    cluster sampling 241–242
    control group 242
    design considerations 242–243
    experimental group 242
    gender bias 243
    importance 243
    random sampling 242
    variable 242
  randomness and interspersion
    animal behavior experiments 245
    cluster sampling 244
    design strategy 245
    distribution 244
    interspersion 244
    stratified random sampling 244
    strengths and difficulties 245
    testing 244–245
sample frame error 239
  dangers 240
  inference 239
  occurrence 240
  population 239
  problems with 239–240
  sample group 239
  sample problem 241
  sampling framework 239
selection 124
selection-maturation interaction 124
semiotic 191
significance levels
  action 252
  alpha 251, 252
  analysis of covariance 4
  analysis of variance 7
  confidence interval 251
  critical region 251
  error 251
  fractional factorial designs 91
  hypothesis-based study 118
  P value 251
  statistical significance 252–253
  testing 251–252
simple random sampling 30
  systematic sampling 266
  sampling *vs.* census 248

simple randomization 220
single-blind studies 182
single case study 269
skewed distribution 65, 66
Solomon four-group design 253
  applications 255
  causation 253
  external validity 253
  increased validity, increased costs 254
  internal validity 253
  minimizing confounding variables 253–254
  practice 254–255
  pretest-posttest design 253
  sample problem 255
split-plot factor 256, 257
split-plot type designs 255
  advantages and disadvantages 257
  applications 257
  block 255
  completely randomized design 255, 256
  factor 256
  improving research design 257
  sample problem 258
  split-plot factor 256, 257
  whole-plot factor 256
standard deviation
  descriptive statistics 63
  one-tailed and two-tailed *t*-tests 183
  Student's *t*-test 261
Standard Model of particle physics 105
statistical regression 124
steady-state theory 15
stepwise regression 228
strata 258
stratified random sampling 30, 202
  experimental group 258
  interaction 258
  randomness and interspersion 244
  representation 260
  sample problem 260
  strata 258
  testing 258–260
  uses 260
  variability 258
stratified randomization 220
Student's *t*-test 261
  comparing data, method of 261–262
  importance 263
  independent samples 261

nonrandom sampling 261
paired *t*-test 261–263
sample problem 262
sample size 261
standard deviation 261
types of 262–263
subjectivity 181
  positivist methods 196
  qualitative research 211
  interpretive methods 125
super-cooled helium 85
switching replications design 263
  action 264–265
  control group 263
  experimental design 263–264
  experimental group 263
  pretest-posttest design 263
  priming effect 263
  replication 263
  sample problem 265
  uses 264
syllogisms 59
synechism 192
systematic bias 181
systematic random sampling 30, 202
systematic sampling 245, 266
  cluster sampling 266
  context and considerations 267
  sample problem 268
  simple random sampling 266
  techniques 266
  uniform sampling 266
  using 266–267

T

tetrachloroethene 82
time-series design 143
  data points 269–270
  health and medicine, time series 270
  investigating efficacy 270
  longitudinal sampling 269
  quasi-experiment 269
  single case study 269
  testing 269
transferability 211
transmutation 56
trend line 40, 41
t-test 6–7
  one-tailed and two-tailed 183

action 185
alternative hypothesis 183
confidence interval 183–184
data sets 183
importance 185
limitations and considerations 184
null hypothesis 183
sample distribution 183
sample problem 185
standard deviation 183
Student's *t*-test 261
  comparing data, method of 261–262
  importance 263
  independent samples 261
  nonrandom sampling 261
  paired *t*-test 261–263
  sample problem 262
  sample size 261
  standard deviation 261
  types of 262–263
Turing, Alan
  early life 273
  history 273
  impact 274–275
  life's work 273–274
  Turing machine 273, 275
Turing machine 273
Turing test 274, 275
two-tailed *t*-tests 183
  action 185
  alternative hypothesis 183
  confidence interval 183–184
  data sets 183
  importance 185
  limitations and considerations 184
  null hypothesis 183
  sample distribution 183
  sample problem 185
  standard deviation 183
two-way interaction 193
tychism 192
type I and type II errors 217, 271
  critical region 271
  false positives and false negatives 272
  null hypothesis 271
  P value 271, 272
  random sampling 271
  sample size, importance 272
  testing 271–272

U
unequal allocation randomization 220
unified field theory 105
uniform sampling 266

V
variable 40
   correlation modelling 40
   experimental research 72
   hypothesis testing 115
   hypothesis-based study 117
   linear relationships 140
   nonlinear relationships 140
   regression modelling 227
   regression point displacement 229
   sampling framework 246
   sampling *vs.* experimental design, 242
voluntary response sampling 249

W
washout period 47
whole-plot factor 256
Wu, Chien-Shiung
   beta decay transition 278, 279
   early life 277
   history 277
   impact 279
   life's work 277–279

X
X-rays 52–53

Z
Zinjanthropus 139